Ian T.

Speech Processing

ESSEX SERIES IN TELECOMMUNICATION AND INFORMATION SYSTEMS

Series editors

Andy Downton
Ed Jones

Forthcoming Titles

Engineering the Human–Computer Interface
Computer Communication Networks
Image Processing
Satellite and Mobile Radio Systems

SPEECH PROCESSING

Edited by
Chris Rowden
Department of Electronic Systems Engineering
University of Essex

McGRAW-HILL BOOK COMPANY

London · New York · St Louis · San Francisco · Auckland
Bogotá · Caracas · Hamburg · Lisbon · Madrid · Mexico · Milan
Montreal · New Delhi · Panama · Paris · San Juan · São Paulo
Singapore · Sydney · Tokyo · Toronto

Published by
McGRAW-HILL Book Company Europe
SHOPPENHANGERS ROAD · MAIDENHEAD · BERKSHIRE · SL6 2QL · ENGLAND
TEL: 0628 23432
FAX: 0628 770224

British Library Cataloguing in Publication Data

Speech processing. – (Essex series in
telecommunications and information systems)
  I. Rowden, Chris   II. Series
006.54

ISBN 0-07-707324-X

Library of Congress Cataloging-in-Publication Data

Speech processing / editor, Chris Rowden.
    p. cm. – (Essex series in telecommunication and information systems)
Includes bibliographical references and index.
ISBN 0-07-707324-X
1. Speech processing systems.
I. Rowden, Chris   II. Series.
TK7882.S65S675   1991
006.4'54–dc20   91–29881 CIP

Copyright © 1992 McGraw-Hill International (UK) Limited. All rights reserved. No part of this publication may be reproduced, stored in a retrieval system, or transmitted, in any form or by any means, electronic, mechanical, photocopying, recording, or otherwise, without the prior permission of McGraw-Hill International (UK) Limited.

1 2 3 4 5 CUP 9 5 4 3 2

Typeset by Vision Typesetting, Manchester
and printed and bound in Great Britain at the University Press, Cambridge

Contents

Notes on contributors ix
Series preface xiii
Preface xv

1 Speech production and perception 1
Gill Waters

1.1 Introduction 1
1.2 The human vocal apparatus 1
1.3 Speech sounds 5
1.4 Acoustics of speech production 7
1.5 Phoneme production 10
1.6 Sound variations and combinations 20
1.7 Prosody and speaker variation 21
1.8 Feedback mechanisms 22
1.9 Speech perception 23
1.10 The physiology of the ear 23
1.11 The auditory nerve 26
1.12 Loudness perception 28
1.13 Pitch perception 29
1.14 Frequency selectivity and masking 30
1.15 The current view of hearing and speech perception 31
References and further reading 33

2 Analysis 35
Chris Rowden

2.1 Levels of speech analysis 35
2.2 Speech waveforms 36
2.3 Analysis at the signal level—stream processing 39
2.4 Analysis at the segment level—block processing 46
2.5 Separation of vocal tract resonances from speech excitation 60
2.6 Analysing speech at the utterance level 67
References 72

3		**Waveform coding for digital telephony** **Edwin Jones**	**74**
3.1		Current practice	74
3.2		Design objectives	77
3.3		Waveform sampling	80
3.4		Waveform quantizing	82
3.5		Differential coding	91
3.6		Comparisons and trends	95
		References	96
4		**Waveform coding—digital audio** **Malcolm Hawksford**	**97**
4.1		Introduction	97
4.2		The case for oversampling and noise shaping	98
4.3		Decimation and interpolation filter techniques	101
4.4		Noise shaping	114
4.5		Application of mild oversampling in ADC and DAC	129
4.6		Applications of high oversampling in ADC and DAC	135
4.7		Conclusion	150
		References and further reading	152
		Acknowledgements	157
5		**Parametric coding of speech** **Chris Rowden and Stephen Hall**	**158**
5.1		Parametric coding	158
5.2		An overview of linear predictive coding	159
5.3		Linear predictive analysis	162
5.4		Synthesis method for LPC speech	169
5.5		Coding the predictor parameters	171
5.6		Coding the residual sequence	174
5.7		Interaction between coding of filter parameters and coding of the residual	177
5.8		Sub-band coding: a hybrid between parametric and waveform coding	178
		References	182
6		**Synthesis** **Chris Rowden**	**184**
6.1		Introduction	184
6.2		An overview of text-to-speech synthesis	186

6.3	Textual input-to-speech synthesis systems	189
6.4	Converting words to phonemes (phonetic transcription)	192
6.5	Concatenation of stored speech units	201
6.6	Prosody	206
6.7	Electronic analogues of the vocal tract	213
6.8	Summary	219
	References	221

7 Recognition—the stochastic modelling approach — 223
Roger Moore
Copyright © Controller, HMSO, London, 1990

7.1	Introduction	223
7.2	Early approaches to automatic speech recognition	227
7.3	Towards a methodology	234
7.4	Speech pattern modelling	236
7.5	Future directions	248
	References and further reading	252

8 Pattern recognition and its application to speech — 256
Graham Leedham

8.1	Introduction	256
8.2	Pattern recognition techniques	257
8.3	Approaches to speech recognition	264
8.4	Isolated word recognition	266
8.5	Keyword spotting in connected speech	278
8.6	Connected word recognition	279
8.7	Speaker identification and verification	282
8.8	Summary	283
	References and further reading	284

9 Speech on packet networks — 287
Gill Waters

9.1	Introduction	287
9.2	An overview of existing packet networks	289
9.3	Local area networks	293
9.4	Approaches to integrated networks	299
9.5	Characteristics of packetized speech	304
9.6	Speech coding and protocol implications	308
9.7	Integrated packet networks	310
9.8	Magnet	313
9.9	Orwell	315

9.10	Metropolitan area networks	317
9.11	Conclusion	319
	References	319

10 Integrated services on local networks — 322
Stephen Ades

10.1	The office of the future	322
10.2	The ISLAND project	325
10.3	Design of a network	328
10.4	Providing office voice facilities	336
10.5	Control and reliability	344
10.6	High-level integrated applications	348
10.7	Where next?	356
	References	357
	Acknowldgements	359

11 Human factors — 360
Bob Damper and Graham Leedham

11.1	Introduction	360
11.2	Human factors methodology	361
11.3	Human capabilities	364
11.4	Technical capabilities in synthesis and recognition	369
11.5	The role of speech in the user interface	371
11.6	Design issues	376
11.7	The simulation of future speech systems	386
11.8	Conclusions	389
	References	391

Index — 394

Notes on the contributors

Stephen Ades BSc (Eng), PhD, CEng, MIEE, studied music at the Guildhall School, read electronic engineering at University College London and then carried out research for a PhD at the University of Cambridge Computer Laboratory. His research topic was integrated services local area networks and he also led the ISLAND Project.

After a period as a visiting scholar at XEROX Palo Alto Research Center (PARC) in California, he joined the PA Consulting Group to carry out product development and market assessment for new computer and communications products. He is currently sales and marketing manager for Chernikeeff Telecommunications, responsible primarily for cisco LAN-to-LAN routing products.

Bob Damper Bsc, MSc, PhD, DIC, CEng, MIEE, SenMIEE, FIOA is currently a Guest Researcher in the Department of Speech Communication and Music Acoustics, KTH, Stockholm, on leave from the University of Southampton where he is a Senior Lecturer in Electronics and Computer Science. He has wide research interests in speech and language processing including: speech aids for the handicapped, human factors of speech I/O, auditory modelling, text-phonemics correspondence, and applications of artificial neural nets in speech technology.

Stephen Hall BSc, PhD has studied in South Africa and Australia. He works for the Dindima Group, Australia, which designs and manufactures specialized video equipment. From 1988 to 1989 he was a lecturer in the Department of Electronic Systems Engineering, University of Essex, and from 1989 to 1990 he was a research contractor with OTC Ltd, Australia. His research interests include broadband integrated services digital networks, ATM switch design and protocols and video coding for ATM networks.

Malcolm Hawksford BSc, PhD, CEng, MIEE, FAES, FIOA is a reader in the Department of Electronic Systems Engineering at the University of Essex, where his principal interests are in the fields of circuit design, signal processing and audio engineering. He studied electrical engineering at the University of Aston in Birmingham, where he gained both his BSc and PhD degrees. Since his employment at Essex University, he has established the Audio Research Group, where research has included amplifier studies, digital signal processing and loudspeaker systems. Since 1982 research into digital crossover systems has also been undertaken together with more recent investigations into analogue-to-digital/digital-to-analogue conversion using oversampling and noise shaping. He is a technical adviser for *Hi-Fi News* and *Record Review*.

Edwin Jones BSc, MSc, PhD, CEng, MIEE, MIEEE is a senior lecturer in telecommunications systems at the University of Essex, having formerly been with GEC Research Limited. He has particular interests in digital transmission and is currently researching into efficient speech and data coding for local/wide area networks, signal processing for high capacity transmission systems and synchronization of digital multiplexes. Much of this is collaborative work with the UK telecommunications industry.

Graham Leedham BSc, MSc, PhD, CEng, MIEE obtained his BSc in electronic engineering from Leeds University in 1979. After a period in industry he returned to university, being awarded an MSc in 1981 and a PhD (for work on automatic transcription of Pitman's shorthand) in 1985, both from Southampton University. Since 1984 he has been a lecturer in the Department of Electronic Systems Engineering of the University of Essex. His research interests are in the man–machine interface and the automatic recognition of handwriting.

Roger Moore BA, MSc, PhD, AMIEE, FIOA studied computer and communications engineering at the University of Essex and was awarded the BA(Hons) degree in 1973. He subsequently received the MSc and PhD degrees from the same university in 1975 and 1977 respectively. Since 1980 he has been with the Royal Signals and Radar Establishment (now Defence Research Agency, Electronics Division) at Malvern, where he is currently a senior principal scientific officer and head of the Speech Research Unit. He has authored and co-authored over fifty scientific publications in the general area of speech technology applications, algorithms and assessment procedures. Roger Moore is a member of Council of the Institute of Acoustics and Visiting Professor in the Department of Phonetics and Linguistics, University College London. He is also a member of the editorial board for the journal

Computer Speech and Language and a founding member of the European Speech Communication Association (ESCA).

Chris Rowden BSc, CDipAF, CEng, MIEE, MIEEE graduated in electrical engineering from the University of Nottingham. He began his career as a research engineeer for Standard Telecommunications Laboratories, where his work on FDM telephone equipment led to the award of two patents relating to polyphase modulation and in-band signalling. In 1968 he joined the Department of Electrical Engineering Science at the University of Essex, holding a succession of posts, eventually becoming a lecturer in 1978, and senior lecturer in 1990. Research in the fields of speech processing and real-time applications of computers has produced a variety of publications, and attracted research grants. He works with students in his research group on various topics in speech recognition, speech coding for packet networks, and translation of sign language to spoken language. Chris has organized the Speech Processing short course at Essex since its inception in 1987.

Gill Waters BSc, MIEEE has been a lecturer in computer networks and real-time systems in the Department of Electronic Systems Engineering at the University of Essex since 1984. Formerly, she worked at the Rutherford Appleton Laboratory, Oxfordshire on a wide variety of computer communications projects. She has organized the short course in computer networks at Essex since its inception in 1987. Her research interests include multiparty protocols, broadcast and multicast applications, and the application of computer-supported collaborative work to teaching. She is also interested in integrated multiservice networks, particularly in their access protocols and the scope for novel applications. She is a member of the Association for Computing Machinery and the UK Women's Engineering Society.

Series preface

This book is part of a series, the *Essex Series in Telecommunication and Information Systems*, which has developed from a set of short courses run by the Department of Electronic Systems Engineering at the University of Essex since 1987. The courses are presented as one-week modules on the Department's MSc in Telecommunication and Information Systems, and are offered simultaneously as industry short courses. To date, a total of over 600 industrial personnel have attended the courses, in addition to the 70 or so postgraduate students registered each year for the MSc. The flexibility of the short course format means that the contents both of individual courses and of the courses offered from year to year have been able to develop to reflect current industrial and academic demand.

The aim of the book series is to provide readable yet authoritative coverage of key topics within the field of telecommunication and information systems. Being derived from a highly regarded university postgraduate course, the books are well suited to use in advanced taught courses at universities and polytechnics, and as a starting point and background reference for researchers. Equally, the industrial orientation of the courses ensures that both the content and the presentation style are suited to the needs of the professional engineer in mid-career.

The books in the series are based largely on the course notes circulated to students, and so have been 'class-tested' several times before publication. Though primarily authored and edited by academic staff at Essex, where appropriate each book includes chapters contributed by acknowledged experts from other universities, research establishments and industry (originally presented as seminars on the courses). Our colleagues at British Telecom Research Laboratories, Martlesham, have also provided advice and assistance in developing course syllabuses and ensuring that the material included correctly reflects industry practice as well as academic principles.

As series editors we would like to acknowledge the tremendous support we have had in developing the concept of the series from the original idea through to the publication of the first group of books. The successful completion of this project would not have been possible without the substantial commitment shown not only by individual authors but by the Department of Electronic Systems Engineering as a whole to this project. Particular thanks go to the editors of the individual books, each of whom, in addition to authorising several chapters, was responsible for integrating the various contributors' chapters of his or her book into a coherent whole.

July 1990

Andy Downton
Ed Jones

Preface

Speech Processing brings together in one volume the fundamentals of speech science and an introduction to areas of actual and anticipated development in speech technology. The book is written for those who need to assimilate quickly the achievements of speech processing; they may be working as design and development engineers, computer scientists or linguists, or studying those disciplines. This rather wide audience is a description of those who have attended the Speech Processing short course at Essex University over the last five years. To form the chapters of this book the lecture notes for that course have been considerably expanded, and great care has been taken to provide comprehensive and up-to-date references for further study. To cover a large field of study in one volume, the authors have been selective. Most chapters include necessary foundation material, which is described with the insight made possible as the subject matures. Among the more advanced topics there is a wealth of new material that does not appear in other books.

The first two chapters give a flavour of the multi-disciplinary nature of the subject, from human physiology, through signal processing to the elements of linguistics. An appreciation of the characteristics of the speech signal is made easier by understanding how it is produced. Similarly the science of speech perception can give clues as to the strengths and weaknesses of speech processing systems. A phonetic classification of speech sounds is described, which is used in later chapters. Chapter 2 on the analysis of speech extends the topic from signal analysis to higher levels, at which the syntax of spoken utterances can be analysed.

The next three chapters describe methods for coding speech signals. In Chapter 3, waveform coding techniques for digital telephony are described, leading into a discussion of adaptive coding. The next chapter introduces the techniques of oversampling and noise shaping, which contribute to achieving high-quality signal conversion in digital audio applications. This is becoming an important contribution to speech processing with the trend towards wideband speech systems. Chapter 5 covers parametric coding techniques, which represent the speech signal by modelling the vocal tract resonances and the excitation. These techniques produce very compact coding, and are used for transmission or storage wherever bandwidth is at a premium.

An appraisal of the synthesis and recognition of speech, areas which govern the speech interface between people and machines, appear in the next three chapters. Chapter 6 makes clear the range of problems involved in translating from written text to synthesized speech. Particular attention is paid to the generation of appropriate patterns of rhythm and intonation; an area where present performance is most limited, and where there is a pressing need for further research. The next chapter outlines the problems of speech recognition, and briefly records the development of the subject. It comes right up to date with a description of the stochastic modelling approach, using the Hidden Markov Model to deal most effectively with the variabilities that are always present in speech. Many commercially available speech recognizers are based on pattern recognition techniques. Some general principles of pattern recognition and examples of their application to speech appear in Chapter 8.

The telecommunication infrastructure has changed to become a mixture of circuit switched systems and packet switched systems, and this change has an important bearing on the way that speech signals are carried in present and future systems. Chapters 9 and 10 introduce the principles of packet networks and the provision of integrated services. Examples are given of network protocols and techniques for the transmission of speech on packet networks. The ISLAND project is discussed, which demonstrated integrated services, including voice, on a local area network. Guidance is offered for the design of new networks, with a discussion of their role in the office of the future.

The final chapter studies the human factors involved in using speech systems, and determines what is needed for ease of use and efficiency in operation. In considering the human factors we face the question of how speech technology, which has benefited so much from using models of human communication, can best be used to enhance communication with people and machines.

It is a pleasure to acknowledge with thanks the many contributions which led to the existence of this book. Don Pearson remoulded our MSc option courses into the one-week short course format, which opened the *Speech Processing* short course to a wide industrial audience as well as to our full-time and part-time students. Seminar speakers on the course have supported the venture; Stephen Ades, Bob Damper and Roger Moore have written chapters; Graham Tattersall reviewed an early draft and provided most useful comments. My colleagues at Essex University, Gill Waters, Stephen Hall, Malcolm Hawksford, Ed Jones and Graham Leedham, have expanded their lecture notes into chapters, in many cases after similar efforts for other books in the series. Pat Baker, Hilda Breakspeare, Georgina Swinbourne and Michael Sansom have helped in many ways with the preparation of text and diagrams. Andy Downton and Ed Jones as series editors have brought their experience to bear on the many organizational problems, while David Crowther and Camilla Myers of McGraw-Hill have provided continuing encouragement, practical help, advice and much more patience than we deserved.

Chris Rowden

1 Speech production and perception

GILL WATERS

1.1 Introduction

In this chapter, we take a look at the physiological and acoustic aspects of speech production and of speech perception, which will help to prepare the ground for later chapters on the electronic processing of speech signals.

The human apparatus concerned with speech production and perception is complex and uses many important organs—the lungs, mouth, nose, ears and their controlling muscles and the brain. When we consider that most of these organs serve other purposes such as breathing or eating it is remarkable that this apparatus has developed to enable us to make such a wide variety of easily distinguishable speech utterances. A good deal is known about the anatomy and physiology of speech production and perception but less is known about the interaction of the brain with the vocal and auditory apparatus, although there are theories that attempt to explain the complexity of these interactions. See, for example, Fry (1977) or Lieberman and Blumstein (1988).

1.2 The human vocal apparatus

Speech sounds are produced when breath is exhaled from the lungs and causes either a vibration of the vocal cords (for vowels) or turbulence at some point of constriction in the vocal tract (for consonants). The sounds are affected by the shape of the vocal tract which influences the harmonics produced. The way in which the vocal cords are vibrated, the shape of the vocal tract or the site of constriction can all be varied in order to produce the range of speech sounds with which we are familiar. Figure 1.1 shows the main human articulatory apparatus. The topics to be discussed in the following sections are treated in more

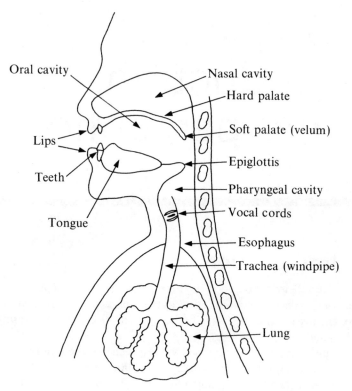

Figure 1.1 The human articulatory apparatus

depth in a number of texts, including Borden and Harris (1984) and Flanagan (1972).

1.2.1 Breathing

The use of exhaled breath is essential to the production of speech. In quiet breathing, of which we are not normally aware, inhalation is achieved by increasing the volume of the lungs by lowering the diaphragm and expanding the rib-cage. This reduces the air pressure in the lungs which causes air from outside at higher pressure to enter the lungs. Expiration is achieved by relaxing the muscles used in inspiration so that the volume of the lungs is reduced due to the elastic recoil of the muscles, the reverse movement of the rib-cage and gravity, thus increasing air pressure in the lungs and forcing air out.

The form of expiration achieved by relaxing the inspiratory muscles cannot be controlled sufficiently to achieve speech or singing. For these activities, the inspiratory muscles are used during exhalation to control lung pressure and prevent the lungs from collapsing suddenly; when the volume is reduced below that obtained by elastic recoil, expiratory

SPEECH PRODUCTION AND PERCEPTION

muscles are used. Variations in speech intensity needed, for example, to stress certain words are achieved by varying the pressure in the lungs; in this respect speech differs from the production of a note sung at constant intensity.

About 60 per cent of the breathing cycle for quiet breathing is expended on exhalation and 40 per cent on inhalation. During speech, the proportion of the breathing cycle used for expiration is increased to about 90 per cent, a fresh breath being taken at a convenient point in the speech stream. The vital capacity of the lungs is the volume of air used when taking a maximum inspiration and then producing a maximum expiration. Quiet breathing uses only about 10 per cent of the lungs' vital capacity; during conversational speech this increases to about 25 per cent and in loud speech to about 40 per cent. (Note that in addition to the lungs' vital capacity of about 5 litres there is a residue volume of about 2 litres which we are unable to expel.)

1.2.2 The larynx

There are two main methods by which speech sounds are produced. In the first, called *voicing*, the vocal cords located in the larynx are vibrated at a constant frequency by the air pressure from the lungs. The second gives rise to unvoiced sounds produced by turbulent flow of air at a constriction at one of a number of possible sites in the vocal tract. A schematic view of the larynx is shown in Fig. 1.2.

The vocal cords are at rest when open. Their tension and elasticity can be varied; they can be made thicker or thinner, shorter or longer and then can be either closed, open wide or held in some position

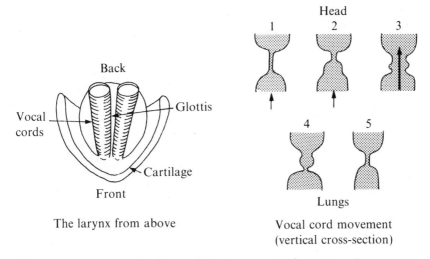

Figure 1.2 Schematic views of the larynx and of vocal cord movement during speech

between. The accepted theory of phonation (the production of voiced sounds) is called the myoelastic aerodynamic theory of phonation (where 'myo' refers to muscles). Early work on these mechanisms was carried out by Muller (1848), and more recently qualitative results have been produced by Van den Berg (1958). When the vocal cords are held together for voicing they are pushed open for each glottal pulse by the air pressure from the lungs; closing is due to the cords' natural elasticity and to a sudden drop in pressure between the cords (the Bernoulli principle). Considered in vertical cross-section the cords do not open and close uniformly, but open and close in a rippling movement from bottom to top as shown in Fig. 1.2.

The frequency of vibration is determined by the tension exerted by the muscles, the mass and the length of the cords. Men have cords between 17 and 24 mm in length; those of women are between 13 and 17 mm. The average fundamental or voicing frequency (the frequency of the glottal pulses) for men is about 125 Hz, for women about 200 Hz and for children more than 300 Hz. Figure 1.3 shows the range of

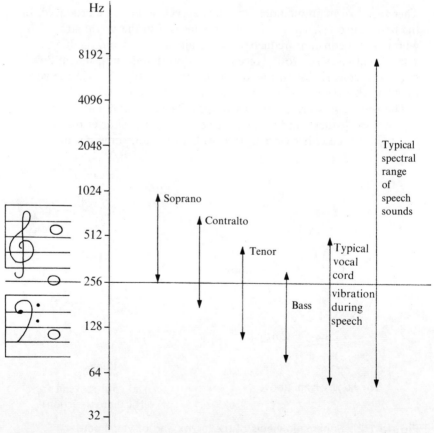

Figure 1.3 Frequency range of the human voice

fundamental frequencies produced by the various singing voices compared to musical notes, and the frequency range of speech sounds. When the vocal cords vibrate harmonics are produced at multiples of the fundamental frequency; the amplitude of the harmonics decreases with increasing frequency.

The quality of the voice is due partly to the way in which the vocal cords vibrate in addition to the effect of physical differences in the vocal tract. Heavy use of the voice at high intensity may cause irregularities or nodules on the vocal cords which affect voice quality.

1.2.3 The vocal tract

For both voiced and unvoiced speech sound that is radiated from the speaker's face is a modification of the original vibration caused by the resonances of the vocal tract. The oral tract is highly mobile and the position of the tongue, pharynx, palate, lips and jaw will all affect the speech sounds made which we hear as radiation from the lips or nostrils. The nasal tract is immobile, but can be coupled in to form part of the vocal tract depending on the position of the velum. Combined voiced and unvoiced sounds can also be produced as voiced consonants.

The major speech articulators are shown in Fig. 1.1. When the velum is closed the oral and pharyngeal cavities combine to form the voice resonator. The tongue can move both up and down and forward and back, thus altering the shape of the vocal tract; it can also be used to constrict the tract for the production of consonants. By moving the lips outward the length of the vocal tract can be increased. The nasal cavity is coupled in when the velum is opened for sounds such as /m/, in 'hum'; here the vocal tract is closed at the lips and acts as a side branch resonator.

1.3 Speech sounds

The smallest element of speech sound which indicates a difference in meaning is called a phoneme, and is written between slashes as, for example, /p/ in 'pan'. In fact, the sound produced for /p/ does vary depending where it appears in a word and is different in the words 'pan' and 'span'. About 40 phonemes are sufficient to discriminate between all the sounds made in British English; other languages may use different phoneme sets.

1.3.1 Phonemic representation

The representation of phonemes is shown in Table 1.1. For the purposes of this chapter we have chosen to use a representation easily

Table 1.1 Phonetic representation

Our symbol	IPA symbol	Key word	Our symbol	IPA symbol	Key word
Vowels			*Fricatives*		
/ee/	i	each	/f/	f	free
/i/	I	it	/th/	θ	thin
/e/	ɛ	end	/s/	s	see
/ar/	ɑ	hard	/sh/	ʃ	shall
/u/	ʊ	good	/v/	v	vine
/uu/	u	ooze	/dh/	ð	then
/er/	ɜ	bird (neutral)	/z/	z	zoo
/aa/	æ	had	/xh/	ʒ	azure
/a/	ʌ	bud	/h/	h	he
/aw/	ɔ	hoard	*Affricates*		
/@/	ə	allow (schwa)	/tsh/	tʃ	chair
/Q/	ɒ	hot	/dzh/	dʒ	jar
Diphthongs			*Semi-vowels*		
/ei/	ei	aid	/w/	w	we
/u@/	ʊə	pure	/y/	j	you
/au/	au	cow	/r/	r	red
/ou/	əu	own	/l/	l	live
Plosives			*Nasals*		
/p/	p	pie	/m/	m	me
/t/	t	ten	/n/	n	no
/k/	k	key	/ng/	ŋ	sing
/b/	b	be			
/d/	d	den			
/g/	g	go			

recognizable to those not familiar with phonetic symbols. The table also shows the international phonetic alphabet (IPA) representation. One of the problems of reading the literature on this topic is the variety of ways of representing the phonemes. This becomes a more difficult problem when presenting phonetic data for machine readable purposes, and an attempt has been made to define a system using standard typewriter keys to assist this problem (Wells, 1986).

1.3.2 Voiced, unvoiced and plosive sounds

As we have seen, voiced sounds, for example the vowel sounds /a/, /e/ and /i/, are generated by vibration of the vocal cords which are stretched across the top of the trachea. The pressure of air flow from

SPEECH PRODUCTION AND PERCEPTION

the lungs causes the vocal cords to vibrate. The fundamental pitch of the voicing is determined by the air flow, but mainly by the tension exerted on the cords.

The schwa /@/ is not a phoneme in the formal sense that it represents a logical unit which can differentiate the sound of one word from another. It is the rather neutral sound that an unstressed vowel becomes in informal speech; for example the vowel in the word 'but' would be a schwa sound in these circumstances, the word being transcribed as /b @ t/ rather than /b a t/.

Unvoiced sounds are produced by frication caused by turbulence of air at a constriction in the vocal tract. The nature of the sound is determined by the site of the constriction and the position of the articulators (e.g. the tongue or the lips). Examples of unvoiced sounds are /f/, /s/ or /sh/. Mixed voiced and unvoiced sounds occur where frication and voicing are simultaneous. For example if voicing is added to the /f/ sound it becomes /v/; if added to /sh/ it becomes /xh/ as in 'azure'.

Another form of unvoiced sound called a *plosive* is produced when the air flow is stopped by a closure in the vocal tract and pressure is built up behind the point of constriction which is then released in an explosion (e.g. /p/ and /k/).

Silence occurs within speech, but in fluent speech it does not occur between words where one might expect it. It most commonly occurs just before the stop in a plosive sound. The duration of these silences is of the order of 30 to 50 ms.

1.4 Acoustics of speech production

The vibration of the vocal cords in voicing produces sound at a sequence of frequencies, the natural harmonics, each of which is a multiple of the fundamental frequency. Our ears will judge the pitch of the sound from the fundamental frequency. The remaining frequencies have reducing amplitude. However, we never hear this combination of frequencies because as the sound waves pass through the vocal tract it resonates well at some frequencies and not so well at others and the strength of the harmonics that we hear are the result of the change due to these resonances. This effect is shown in Fig. 1.5 which will be discussed later.

1.4.1 Formant frequencies

The resonances in the oral and nasal tract are not fixed, but change because of the movement of the speech articulators described above. For any position the tract responds to some of the basic and harmonic frequencies produced by the vocal cords better than to others. For a particular position of the speech articulators, the lowest resonance is called the first formant frequency (f_1), the next the second formant frequency (f_2) and so on.

The formant frequencies for each of the vowel sounds are quite distinct but for each vowel sound generally have similar values regardless of who is speaking. For example, for a fundamental frequency of 100 Hz, harmonics will be produced at 200, 300, 400, 500 Hz, etc. For the vowel /ee/ as in 'he' typical values for f_1 and f_2 are 300 and 2100 Hz respectively. The tongue is near the front of the mouth when making this sound and the high second formant results from the small size of the vocal tract cavity. For the vowel /ar/ as in 'hard' the typical values of the corresponding formant frequencies are about 700 and 900 Hz. The tongue is kept much flatter, and a much rounder sound is produced.

The fundamental frequency will vary depending on the person speaking, mood and emphasis, but it is the magnitude and relationship of the formant frequencies which make each voiced sound easily recognizable.

There may be any number of formants between two and six, depending on the individual speaker, and these are subject to considerable variation in magnitude and frequency. The variation is centred around the values 500, 1500 and 2500 Hz for the first three formant frequencies. These values are very close to the first three formant frequencies for the neutral vowel sound /er/, where the oral tract is at its least restricted and the speech articulators are most relaxed. See Perkell (1969). We shall discuss the relative positions of the formant frequencies in more detail in Sec. 1.5 on phoneme production.

1.4.2 Resonances in an open-ended tube

A simplified model of the vocal tract is a simple lossless tube open at one end, in which plane wave propagation takes place along the axis of the tube. In this case, the relationship between sound pressure and volume velocity variation has been shown (Portnoff, 1973) to be given by the following pair of equations:

$$-\frac{\delta p}{\delta x} = \rho \frac{\delta(u/A)}{\delta t}$$

$$-\frac{\delta u}{\delta x} = \frac{1}{\rho c^2} \frac{\delta(pA)}{\delta t} + \frac{\delta A}{\delta t}$$

where

$p(x,t)$ = variation in sound pressure in the tube at position x and time t
$u(x,t)$ = variation in volume velocity flow at position x and time t
ρ = density of air in the tube

SPEECH PRODUCTION AND PERCEPTION

c = velocity of sound
$A(x,t)$ = value of the cross-sectional area of the tube

As a further simplification, if the cross-sectional area of the tube is taken to be uniform, we can use the knowledge that resonances in such a tube occur where the length of the tube is one-quarter of a wavelength and odd multiples of this value.

The length of the vocal tract is about 17 cm in a typical male speaker, and the speed of sound, c, is about 340 m/s. Thus at the first resonant frequency, the wavelength $\lambda = 4 \times 17$ cm gives

$$f_1 = \frac{c}{\lambda} = \frac{340 \times 100}{4 \times 17} = 500 \text{ Hz}$$

the second resonance has a wavelength with $3\lambda = 4 \times 17$ cm, giving $f_2 = 1500$ Hz, and the third resonance has a wavelength with $5\lambda = 4 \times 17$ cm, giving $f_3 = 2500$ Hz. The values 500, 1500 and 2500 Hz correspond to the formant frequencies of the neutral vowel sound.

This model is in fact very much a simplification; the walls of the vocal tract are not hard but are of soft tissue and the tract is of variable cross-section.

1.4.3 The source filter model of speech

A common approach to understanding speech production and the processing of speech signals is to use a source filter model of the vocal tract. This was first proposed by Muller (1848) and studied more recently by Fant (1960) among others. The model is usually implemented in electronic form but has also been implemented mechanically. In the electronic form an input signal is produced either by a pulse generator offering a harmonic rich repetitive waveform or a broadband noise signal is generated digitally by means if a pseudorandom binary sequence generator. The input signal is passed through a filter which has the same characteristics as the vocal tract (see Fig. 1.4).

The parameters of the filter can clearly not be kept constant, but must be varied to correspond to the modification of the vocal tract made by movement of the speech articulators. The filter thus has time-variant parameters; in practice the rate of variation is slow, with parameters being updated at intervals of between 5 and 25 ms.

The pitch of the voiced excitation is subject to control as is the amplitude of the output, in order to provide a fairly close approximation to real speech. The pitch period may vary from 20 ms for a deep-voiced male to 2 ms for a high-pitched child or female.

Either of the signal sources used in the source filter model will produce a broadband spectrum of energy in the frequency domain, as

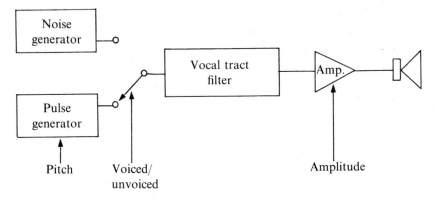

Figure 1.4 Source filter model of speech: block diagram

shown in Fig. 1.5. In the case of the pulse waveform it will consist of a regular pattern of lines which are spaced apart by the pitch frequency. For a noise waveform considered as the summation of a large number of randomly arriving impulses the distribution will approximate to a continuous function. In both cases the energy distribution decreases with an increase of frequency, but there are significant levels up to 15 to 20 kHz.

Frequency shaping is provided by the filter characteristic which is applied to the signal in the frequency domain. Typically the filter characteristic will consist of a curve where the various resonances of the vocal tract appear as peaks or poles of transmission. The frequency at which these poles occur represents the formant frequencies and will change for the various speech sounds. A typical vocal tract filter and the result of applying the filter to the two signal sources are also shown in Fig. 1.5.

The possibility of synthetic speech production has long held a fascination. Among early examples is the talking machine of Wolfgang von Kempelen which was built in about 1790 (see Linggard, 1985).

1.5 Phoneme production

The various categories of phonemes were shown in Table 1.1. These are the vowels, diphthongs, semi-vowels, stop consonants, fricatives and affricates. We will now look in more detail at the production of each of these categories.

SPEECH PRODUCTION AND PERCEPTION

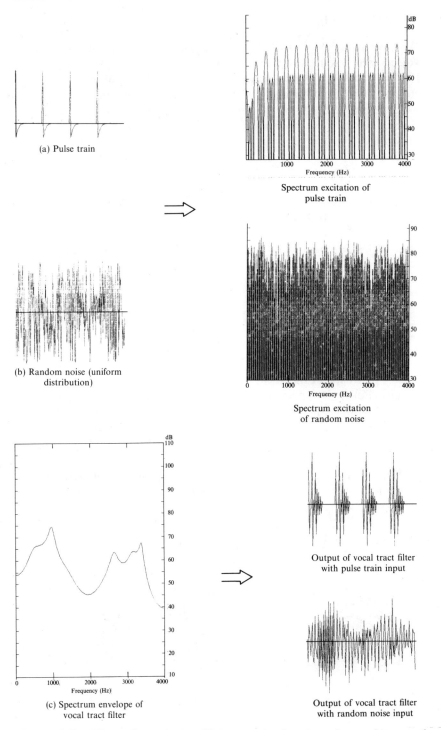

Figure 1.5 Effect of vocal tract filter on pulsed and random noise waveforms

1.5.1 Vowels

We have already discussed the production of vowels in general as voiced speech sounds. Table 1.2 shows the formant frequencies of English vowels for typical male speakers. The vocal tract reaches a steady configuration near the mid-point of producing most vowels, which is characterized by the formant frequencies listed in Table 1.2.

Table 1.2 Formant frequencies of some English vowels for typical male speakers

Vowel		f_1	f_2	f_3
/ee/	beat	280	2620	3380
/i/	bit	360	2220	2960
/e/	bet	600	2060	2840
/er/	bird	560	1480	2520
/ar/	father	740	1110	2640
/a/	hut	760	1370	2500
/u/	hood	480	740	2620
/uu/	loot	320	920	2200

Potter, Kopp and Green in their book on 'visible speech' show clearly the shape of the vocal tract in forming many of the speech sounds (Potter, Kopp and Green, 1947). Simplified schematic pictures for three vowels are shown in Fig. 1.6. As can be seen, the vowel quality comes mainly from the position of the tongue in the mouth. We will take as examples /ee/ as in 'key', /ar/ as in 'hard' and /uu/ as in 'food'.

When producing the /ee/ sound the tongue is moved forward and up to the roof of the mouth thus decreasing the size of the oral cavity.

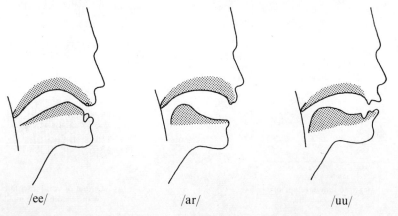

Figure 1.6 Typical vocal tract shapes for vowel production

SPEECH PRODUCTION AND PERCEPTION

This produces high second and third formant frequencies which give the sound its characteristic tightness and can become squeaky when stressed. In contrast, for /ar/ the tongue is moved back and is kept low, widening the oral cavity but narrowing the pharyngeal cavity. The mouth is opened wider which also increases oral cavity size. This configuration raises the lowest resonant frequency which is typically about 750 Hz, but the second formant is much nearer the first at about 1100 Hz.

For /uu/ both first and second formants are lowered due to a lengthened oral tract formed by protruding the lips and elevating the tongue towards the back of the palate. Typical values of the first three formants for /uu/ are 300, 900 and 2500 Hz.

The vowel triangle

Figure 1.7 plots the first two formant frequencies of the vowels, showing typical ranges produced by a number of different speakers. As can be seen, there is very little overlap which explains why vowels are so easy to distinguish from each other.

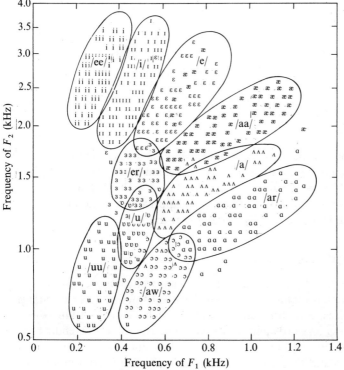

Figure 1.7 Formant positions for a range of speakers (from Lieberman and Blumstein, 1988, Figure 8.8; reprinted by permission of Cambridge University Press and the authors)

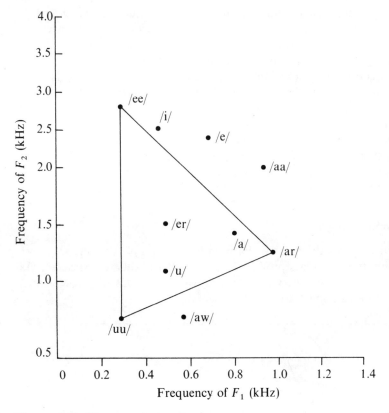

Figure 1.8 The vowel triangle

The graph can be further simplified to a triangle containing most of the vowel sounds with the vertices represented by the vowels /ee/, /ar/ and /uu/ which we have just discussed (Fig. 1.8). Note that the neutral vowel sound which most nearly approximates the uniform lossless tube is in the middle of the vowel triangle.

1.5.2 Diphthongs

Diphthongs are the concatenation of two vowel sounds and during their production require changes in the vocal tract in order to change the resonances. For those ending with /i/ such as /ei/ in 'train', the tongue moves forwards and upwards in the mouth. Those ending in /uu/ such as /au/ in 'cow' or /ou/ as in 'own' entail the tongue moving back and up and the lips moving forward. Voicing is continuous and the tract changes gradually from the first to the second resonating position. There are also a few tripthongs which consist of three concatenated vowel sounds.

1.5.3 Semi-vowels

The semi-vowels are so named because they are produced principally from resonances in the vocal tract as for vowels, but are in fact used in language as consonants. There are two classes—the glides /w/ and /y/ and the liquids /l/ and /r/.

During *glide* production, the movements in the vocal tract articulators cause the formant frequencies to glide up and down as shown in Fig. 1.9. For /y/, the tongue moves up and forward, a position nearer to /i/ than /a/ so that formant movement is more visible in /a y a/ than /i y i/. For /w/ the movement of the tongue is back and up and the lips are protruded.

The *liquids* are produced by raising the tongue to the ridge behind the top front teeth (called the alveolar ridge). In /l/ the top of the tongue rests lightly on the ridge, but in /r/ the tongue tip is grooved and does not quite touch the ridge. For /r/ the third formant is reduced considerably towards that of the second formant of the vowel, as shown in Fig. 1.10. For children, the liquids are more difficult to produce than the glides so that glides may be substituted as in 'I like to run', which might sound like 'I /y/ike to /w/un'.

1.5.4 Stops

Stop consonants or plosives are produced by forming a complete closure in the vocal tract, building up pressure from the lungs and suddenly releasing the pressure which is characterized by an explosion and aspiration of air. The point of constriction gives the specific speech sound. For /p/ the lips are held together, for /t/ the tongue is held against the alveolar ridge and for /k/ the back of the tongue is raised towards the palate, as shown in Fig. 1.11. Each of these stop consonants takes on the slightly different sounds, /b/, /d/ and /g/, with a change at the onset of voicing in relation to the stop.

The production of plosives may be modified by context; for example the /p/ in 'pot' emits much more air at the lips than in the modified version in 'spot'.

1.5.5 Nasals

The nasals /m/, /n/ and /ng/ are closely related to the stop consonants, as shown by the position of the vocal tract in Fig. 1.12. However, there are some major differences. The distinctive sounds are produced by lowering the velum and making a closure in the vocal tract, thus introducing resonances in the nasal cavity with the oral cavity acting as a side branch resonator, while sound radiation is produced from the nostrils. Because the nasal passage is open no pressure buildup occurs. Also because the vocal tract position must not be held so precisely as

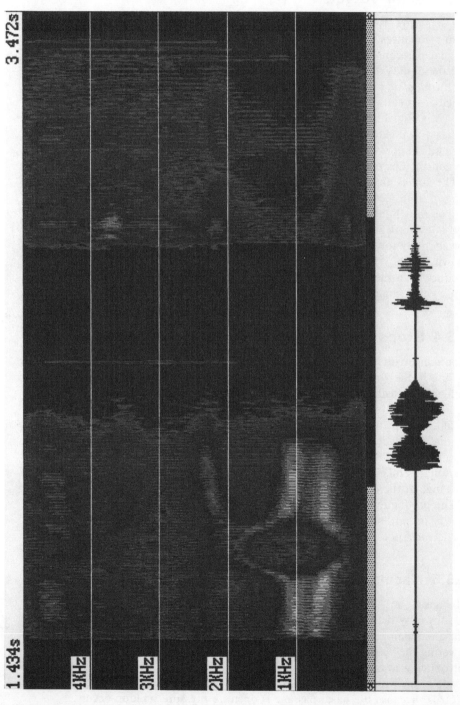

Figure 1.9 Spectrograms of /a y a/ and /i y i/ (produced by Professor Marcel Tatham of the Department of Language and Linguistics on a Loughborough Sound Images Workstation)

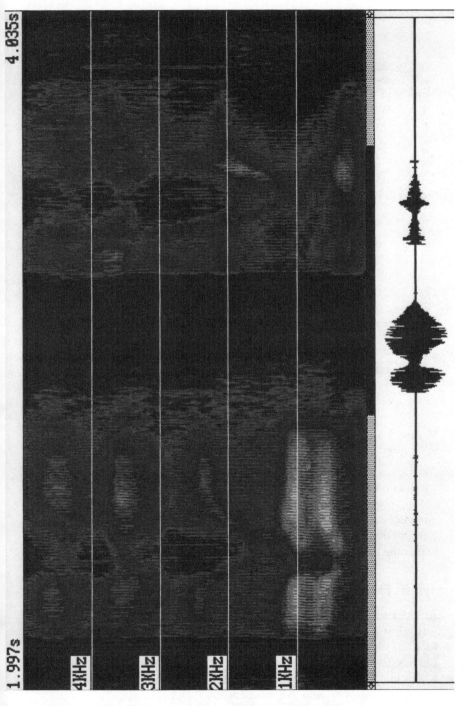

Figure 1.10 Spectrograms of /a r a/ and /i r i/ (produced by Professor Marcel Tatham of the Department of Language and Linguistics on a Loughborough Sound Images Workstation)

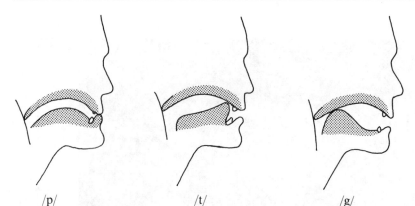

Figure 1.11 Typical vocal tract shapes for stop consonants

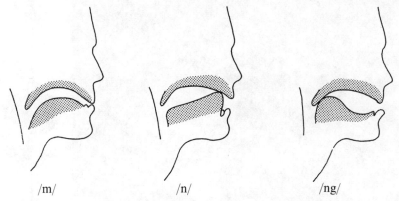

Figure 1.12 Typical vocal tract shapes for production of the nasals

for the stops it is possible to change the shape of the oral tract during production of the nasals.

The intensity of nasal sounds is lower than that of other speech sounds because the vocal tract is elongated and produces a broader band of frequency response, and also because the nasal cavity has very soft walls which absorb the sound.

1.5.6 Fricatives

In order to produce many of the familiar consonants a constriction is formed at a point in the vocal tract and air is forced past creating friction, which causes a noisy random vibration. Fricatives may be unvoiced as for /f/, /th/, /s/ and /sh/ or can be combined with voicing which in combination with the same constrictions produces the four sounds /v/, /dh/, /z/ and /xh/ respectively.

The constriction for /f/ and /v/ is formed by the lips, and for /th/ it is formed by the tongue pressing against the top of the incisors. For /s/ the tongue is pressed against the alveolar ridge and for /sh/ the tongue is held against the palate a bit further back than for /s/ in combination with a rounding of the lips. Thus the shape of the vocal tract for /f/ is similar to that for /p/, and the shape of the tract for /th/ is similar to that shown for /t/ in Fig. 1.11.

The /h/ sound as in 'house' is formed by a partial restriction in the larynx, typically at the vocal cords. It is normally unvoiced but can also be produced as a voiced sound when both preceded and followed by voiced sounds as in 'Ahoy'. During the production of the /h/ sound the vocal tract shape takes the form needed for the following voiced sound.

1.5.7 Affricates

There are two affricates in English. They are formed when the stops /t/ and /d/ are released more gradually, resulting in a longer period of frication. This gives the sound /tsh/ as in 'chair' and /dzh/ as in 'jar'.

1.5.8 Summary

Figure 1.13 summarizes the sites in the vocal tract important to the production of the various speech sounds. The vowels are formed by an

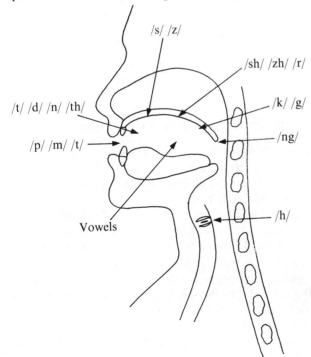

Figure 1.13 Location of various speech sounds

unconstricted vocal tract. A constriction at the lips is used for the production of phonemes such as /p/, /m/, /b/, /f/ and /v/. A constriction at the teeth produces phonemes such as /t/, /th/ and /n/. The phonemes /sh/, /s/ and /z/ are produced with a constriction at some point in the roof of the mouth. Phonemes /k/ and /g/ are produced when the back of the tongue is raised against the palate, and the restriction for /h/ is sited at the larynx.

1.6 Sound variations and combinations

1.6.1 Pitch and loudness

The pitch at which speech is produced depends on many factors such as the frequency of excitation of the vocal cords, the size of the voice box or larynx and the length of the vocal cords. Pitch also varies within words to give more emphasis to certain syllables.

The loudness of speech will generally depend on the circumstances, such as the emotions of the speaker. (We shout when we are angry.) Other factors are how far we need the speech to carry and the noise around us; we listen to this while speaking and adjust the loudness of our speech accordingly. Variations in loudness are produced by the muscles of the larynx which allow a greater flow of air, thus producing the 'sore throat' feeling when the voice has to be raised for a period to overcome noise. Loudness is also affected by the flow of air from the lungs, which is the principal means of control in singing.

1.6.2 Adaption, assimilation and co-articulation

In normal speech, articulations are made quickly and consecutively. Each articulation is therefore not discrete but is modified by neighbouring articulations. (This can be compared to the difference between typewritten text and handwriting.)

Adaptation of a phoneme results from the influence of neighbouring sounds which slightly change the vocal tract shape and articulatory positions. For example if the /k/ sound is followed by /ee/ as in 'key' the tongue may contact the palate further forward than if it is followed by /ar/ as in 'card' because the tongue is held further forward for the production of /ee/ than /ar/. When we speak quickly adaptation is more marked because the tongue does not quite reach the positions achieved in slower deliberate speech.

In extreme cases of adaptation a speech sound may change so that it actually takes on some of the characteristics of its neighbouring sound. This is called *assimilation*. An example occurs in the word 'ink' where the /n/ becomes more like /ng/ as the constriction in the vocal tract matches that of the following /k/ sound.

In *co-articulation* two speech articulators move at the same time in preparation for the production of two different phonemes. For example for the word 'two' the rounding of the lips for the /uu/ sound and the movement of the tongue to a position behind the alveolar ridge for the /t/ sound may occur at the same time.

The result of the combination of adaptation and co-articulation is that the sound produced is better (not so stilted). Also the speech produced is more impervious to noise because the information carried in a single articulation lasts for a longer time.

1.7 Prosody and speaker variation

We have been dealing with individual speech sounds as phonemes, each of which carries some meaning in language. However, if we consider speech in more general terms it is possible to view the combination of phonemes to produce prosody or meaningful spoken language consisting of words, phrases and sentences. Alternatively we can note the difference in sound production of the same word by different speakers using identical phonemes and look at phonetic characteristics.

1.7.1 Prosodic features

Important aspects of speech when considered as combined rather than individual segments are stress, intonation, duration and juncture. All of these are used to impart further meaning to spoken phrases or sentences.

Stressed speech segments result from an increase in pitch, intensity and duration and give emphasis to particular syllables. Even for a single word stress may vary; for example in the word 'permit' the first syllable is stressed when it is used as a noun and the second syllable when used as a verb. Stressed segments make fuller use of the speech articulators, whereas unstressed syllables are more noticeably affected by adaptation and co-articulation.

Intonation is the result of changing the fundamental frequency, causing the pitch of certain parts of a sentence to rise or fall. This gives additional meaning. For example declarations generally have a pitch that first rises and then falls as in 'I am cold today', whereas the pitch often rises at the end of questions as in 'Are you cold?'.

The *duration* of sounds assists stress, but variations in duration are also to be found because of neighbouring phonemes; for example the /ee/ in 'heave' is slightly longer than the /ee/ in 'heat' because of the necessity to build up air pressure for the plosive sound. Where duration and other changes are combined there is said to be a change in *juncture*. For example although 'an aim' and 'a name' consist of the same phonemes represented by /a/, /n/, /ei/, /m/, a speaker may stress the difference by elongating the /n/ in the first form; in normal speech the two may be indistinguishable.

All of these prosodic aspects are important for natural sounding speech, whether it is produced by humans or synthetically, and are referred to in Chapter 6 on speech synthesis.

1.7.2 Speaker variation

As the phonemic description of speech is a logical classification it does little to show the differences between different speakers. A phonemic description of a passage spoken by a Cornish lady would be identical to the same passage spoken by a Yorkshireman. It follows from this that there is a loss of information when a speech utterance is coded in terms of phonemes.

To describe the actual utterance made by such speakers it is necessary to transcribe it in terms of phonetic symbols. The set of phonetic symbols is much larger than the set of phonemes in order to accommodate the many variations. In this book we shall not venture further into this field as it has little relevance to current speech processing activity.

1.8 Feedback mechanisms

In order to control speech we are continuously monitoring its production. There are a number of ways in which this can be done—by hearing, touch, the sense of movement of the articulators and by central internal feedback in the brain.

Auditory feedback helps us to ensure that we are producing the correct speech sounds and that they are of the correct intensity for the environment. Speech can become quite difficult if this is interrupted as, for example, if echoes of a voice are introduced by the telephone system. It is still not clear how much tactile feedback is involved in speech but our experiences with local anaesthetics when visiting the dentist show that the lack of the sense of touch can make it appear to us as though our speech is slurred although it is generally still intelligible.

The effect of the sense of the movement of the articulators in feedback is also the subject of research but it is thought to operate both on reflex and voluntary levels. Within the brain, the ability to learn and use speech patterns (rather like learning a musical instrument) may make speech articulation easier. See, for example, Fry (1977) and Folkins (1985).

1.9 Speech perception

As we have seen, a speech signal is a complex combination of a variety of airborne pressure waveforms. This complex pattern must be detected by the human auditory system and decoded by the brain. The following sections look first at the anatomical and physiological construction of the ear and then discuss the mechanism whereby sounds are relayed to the brain. The perception of loudness and pitch and the phenomenon of masking (the concealment of some sounds by others) are then discussed. There are still many unanswered questions about the processes involved in speech perception owing to the difficulty of undertaking non-invasive experiments. However, an appreciable amount can be inferred from the results of psychoacoustic tests, and other results from studies with animals give a valuable insight into the human auditory system. The topics are necessarily discussed briefly here; the book by Pickles (1988) on the physiology of hearing and the complementary book by Moore (1989) on the psychology of hearing are recommended for those who wish to explore these topics further.

1.10 The physiology of the ear

The human ear can be considered to be divided into three parts—the outer ear with its protective visible covering, the middle ear which adjusts pressure levels between the outer and inner ear, and the inner ear or *cochlea* which contains the sensitive apparatus used to convert the sound energy into neural messages for the brain. These three parts can be clearly seen in Fig. 1.14. Sounds can be detected both by air conduction through the outer and middle ears and by bone conduction when vibrations travel through the bones of the head to the inner ear. Bone conduction is particularly important when a person is listening to his or her own voice.

1.10.1 The outer ear

The large visible outer part of the ear, called the pinna, offers protection (e.g. the hairs and wax in the ear prevent foreign objects from entering). It also helps to focus the sound energy slightly, and because it is a little more receptive to sounds from the front of the head than from those behind, the pinna can help with the localization of sound. It should, however, be noted that the primary means of localization is by the timing and intensity differences between pressure waves arriving at the two ears.

The external auditory canal progresses from the pinna to the eardrum (tympanic membrane) which separates the outer and middle ears. The canal is about 2.7 cm in length and causes a broad resonance effect which gives rise to an increase of sound pressure which is most effective

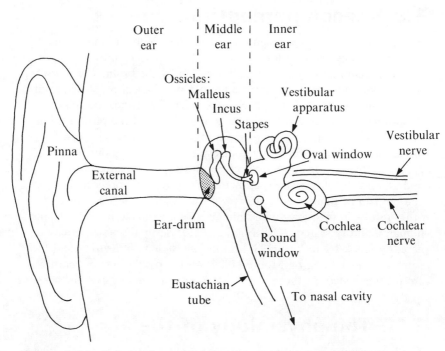

Figure 1.14 Schematic view of the human ear (not to scale)

between 2000 and 5500 Hz, rising to a peak of about 12 dB at around 4000 Hz (Weiner, 1947).

1.10.2 The middle ear

The middle ear overcomes air-to-liquid impedance to ensure that a detectable signal reaches the liquid-filled cochlea, to which it interfaces by two membranes called the oval window and the round window. Amplification is provided by a combination of two mechanisms. The first is by the action of the three small bones (or ossicles) which together act as a lever to amplify pressure from the tympanic membrane to the oval window. These three bones, the malleus, incus and stapes, are named because of their shape (hammer, anvil and stirrups respectively) and are the smallest bones in the human body. The stapes has a footplate of about $0.012\,\text{cm}^2$ which covers the oval window. The second amplification mechanism is provided by concentrating the pressure on to a smaller area, the tympanic membrane having an area about 18 times greater than the oval window. The overall effect is approximately a 30 dB pressure increase at the oval window over the pressure incident on the tympanic membrane.

There are also tiny muscles in the middle ear which act to reduce the effect of the ossicles if an intense sound is incident on the ear; this offers some protection against damage, but is not sufficiently quickly acting to guard against sudden noises as, for example, a pistol shot. The Eustachian tube (also shown in Fig. 1.14) equalizes the pressure on either side of the ear-drum and we are not normally aware of its action until there are rapid air pressure changes taking place as, for example, in an aeroplane.

1.10.3 The inner ear

The inner ear contains the cochlea which is of a coiled snail-like construction; when shown as though 'unrolled' the cochlea has a gradually tapering appearance from the base at the oval window to the apex (see Fig. 1.15). It consists of three principal fluid-containing enclosures which are separated by membranes. Pressure waves entering the cochlea at the oval window travel through the scala vestibuli along Reissner's membrane and through the narrow gap at the apex called the helicotrema to continue thought the scala tympani along the basilar membrane to the round window where the pressure is released. The basilar membrane has a large number of tiny hair cells to which nerve endings are attached. The mechanical energy involved in the shearing of these hair cells by the pressure wave is transduced into the form of energy interpreted by the brain. The scala vestibuli and scala tympani contain a fluid called perilymph which has similar properties to sea water, while the scala media contains a similar fluid called endolymph which has a greater viscosity. The inner ear also contains the vestibular apparatus consisting of three semicircular canals which help to maintain balance.

The pressure wave travels through the cochlear duct almost instantaneously so that the pressure difference occurs almost simultaneously at all places on the basilar membrane. The basilar membrane response is in the form of a travelling wave moving from base to apex whose amplitude increases gradually and then decreases rapidly. The peak of the pattern of vibration occurs at different places

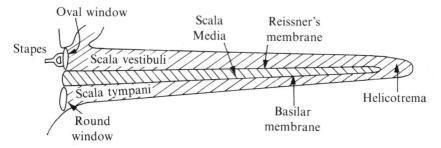

Figure 1.15 The 'unrolled' cochlea

along the basilar membrane for different frequencies, being nearer the base where the membrane is narrow and stiff for higher frequencies and nearer the apex for lower frequencies. Thus the basilar membrane performs a kind of Fourier analysis on the incoming signal. For a pure tone, all parts of the membrane will vibrate with the same frequency, but some vibrate with a greater amplitude. Early experiments by Von Bekesy (1942) indicated a rather broad response to a tone of a single frequency, but more recent experiments on animals (e.g. Khanna and Leonard, 1982) indicate that the response is quite sharply tuned. They also show that the response depends significantly on the physiological condition of the animal. Where tones consist of more than one frequency these can be detected unless they are very close together, in which case an increased amplitude is present for a longer distance along the basilar membrane.

1.11 The auditory nerve

A schematic view of a cross-section of the cochlea is shown in Fig. 1.16. On the basilar membrane is the *organ of Corti* which contains the hair cells. There is a total of about 15 000 hair cells consisting of one row of inner hair cells (near to the central cavity of the cochlea) and

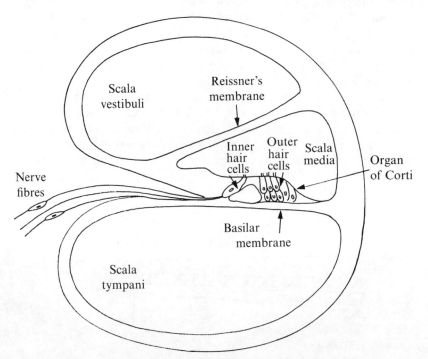

Figure 1.16 Schematic view of a cross-section of the cochlea

three rows of outer hair cells, the latter being more numerous. Each hair cell is about 35 μm long by 10 μm diameter. The nerve endings are situated near the base of the cells, with bundles of nerve fibres connecting the inner ear to the brain. Several nerve fibres are connected to each hair cell, and each hair cell supports three or more rows of hairs. The response of the hair cells is polarized; they produce a stimulus when the hairs are sheared in one direction, but none when sheared in the opposite direction.

The nerve fibres connected to the hair cells are both afferent (i.e. information is sent from the neurons to the brain) and efferent (information is sent from the brain to the neurons). There are more fibres attached to each inner hair cell than to each outer hair cell. There is evidence (which is not yet conclusive) that there is an active mechanical process through the efferent nerve fibres attached to the outer hair cells which influences the response of the inner hair cells and explains the sharpness of response to individual frequencies.

1.11.1 Auditory nerve responses

For incident tones of a single frequency, there are three important aspects concerned with how information is transmitted to the brain. Firstly, there is a background or spontaneous firing in the absence of incident noise. Secondly, an individual nerve fibre responds better to some frequencies than to others. Thirdly, the nerve firings are phase-locked to the basiliar membrane signal.

Where frequency is plotted against the SPL (sound pressure level) intensity required to stimulate a single neuron, the curve reaches a minimum at the 'characteristic frequency' with a steeper gradient for the rise in threshold for higher frequencies than for lower frequencies relative to the characteristic frequency. The discharge rate of a nerve fibre increases in response to an increased stimulus, but only up to a certain value. The range of intensity levels between threshold and saturation is typically between 20 and 50 dB.

The stimulus to the auditory nerve fibre occurs in a manner that is locked to the phase of the signal. It does not necessarily fire every cycle, but the time between firings depends closely on the frequency of the stimulating waveform. This effect occurs for incident frequencies up to about 4 to 5 kHz above which no phase locking takes place.

More complex sounds

Most studies of auditory nerve responses have been carried out using a single sinusoidal waveform. When a second tone of a similar frequency is presented to the ear, the response to the first tone may be reduced— an effect called *two-tone suppression* (Sachs and Kiang, 1968). Under these circumstances, if the second tone is very close to the first and within the normal range of the first's neuronal firing (i.e. within its 'tuning curve'), the tone is amplified. When it is just outside this range, suppression occurs. The effect is thought to be related to the non-linearity of the basilar membrane. A combination tone not physically present in the input, at a frequency of $2f_1 - f_2$ has been measured in several experiments (where f_1, f_2 are the two input frequencies and $f_1 < f_2$).

The response of the auditory nerve to *speech sounds* has been studied by a number of people (e.g. Young and Sachs, 1979) and can be understood by its response to simpler stimuli, with each fibre responding to the corresponding spectral components of the sound.

1.12 Loudness perception

The sensitivity of the human ear is not the same for tones of all frequencies. It is most sensitive to frequencies in the range 1000 to 4000 Hz. Low- and high-frequency sounds require a higher intensity sound to be just audible, and our concept of 'loudness' also varies with frequency. This is shown in Fig. 1.17. The scale of intensity used on the ordinate axis is defined relative to a zero at 10^{-12} W/m², all other intensities being related to this at a SPL in decibels. The lowest curve in Fig. 1.17 indicates the threshold below which sounds cannot be detected by a human ear. Note that this is not exactly the same for all people; variations in the threshold between individuals may be up to about 20 dB and their hearing would still be considered normal. Sensitivity to high frequencies decreases with age.

An indication of the way in which the perception of loudness varies with frequency is given by the two 'equal loudness contours' of Fig. 1.17 plotted at 40 and 60 phons. The phon is defined in units of decibels above 10^{-12} W/m² at 1000 Hz. Equal loudness contours are plotted by asking subjects to compare the perceived loudness of offered pure tones at a variety of frequencies. Figure 1.17 also shows the distribution of conversational speech sounds, most of which lie within the most sensitive range of the ear, as might be expected.

Small changes (0.5 to 2 dB) in SPL can be detected by the human ear and this degree of discrimination can be more than adequately explained by the capacity available in the firing rates of auditory neurons. Continued noise can affect the ear's responsiveness. One effect called fatigue produces an increase in the threshold detection level, a phenomenon that increases rapidly for intensities of about 90 to 100 dB SPL. After exposure to such sounds recovery may take several hours. Permanent hearing loss may result from exposure to intensities above about 110 to 120 dB SPL. This has recently been realized to be an important factor in the design of personal stereo systems, some of which can produce sounds at these high intensities of which the user may not be aware due to fatigue.

SPEECH PRODUCTION AND PERCEPTION

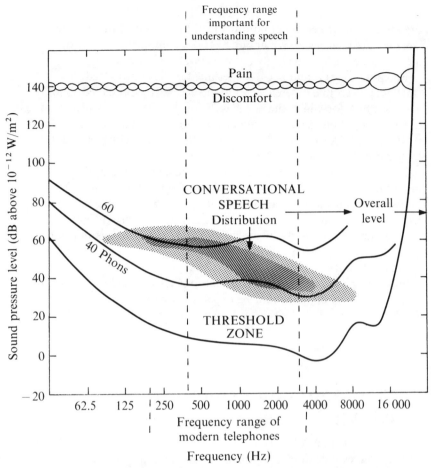

Figure 1.17 Threshold detection, equal loudness contours and the distribution of conversational speech sounds. (Figure 2-6 from *Hearing and Deafness*, Third Edition, by Hallowell Davis and S. Richard Silverman, copyright © 1970 by Holt, Rinehart and Winston, Inc., reprinted by permission of the publisher)

1.13 Pitch perception

Distinguishing audible tones by pitch enables us to order them on a musical scale. In general, for a pure tone the pitch corresponds to the frequency of the tone. For a periodic complex tone (containing several different frequencies) the pitch corresponds to the fundamental frequency. Pitch perception may be explained by two different theories. The place theory relates the pitch of a tone to the place of maximum excitation on the basilar membrane. The temporal theory relates pitch to the time patterns of neural impulses. Since both theories have drawbacks under certain conditions (e.g. the temporal theory does not work for frequencies above about 5 kHz) a combination of theories may be a better approach. (See, for example, Moore, 1989.)

For pure tones a number of experimenters have measured the smallest perceptible difference between successively presented tones. This is a task for which the ear is well equipped and a frequency difference of about 2 Hz can be detected for a 1000 Hz tone at about 60 to 70 dB SPL. Accuracy is better for middle range frequencies and for longer duration tones.

For complex tones, the ear is able to 'hear out' some of the individual partials (frequencies) within a complex tone. This ability is known as Ohm's acoustical law and it applies where the tones are not harmonically related or where the frequencies are very different.

A remarkable property of the human auditory system is that for some complex tones, even if the fundamental pitch is not present in the source signal, it may still be heard. A low pitch detected in this manner from a complex tone consisting of high pitches is called a *residue pitch* or periodicity pitch. It cannot be explained by the place theory alone. Explanations are based either on some form of pattern recognition or on the relative timings at the various sites of maximum excitation on the basilar membrane.

1.14 Frequency selectivity and masking

It is important that humans can detect a variety of information from sound sources including components at different frequencies. For example in listening to speech sounds it is necessary to hear the different combinations of formant frequencies in order to be able to identify vowel sounds. Fletcher (1940) first put forward the theory that the perception of frequency components in a sound is detected by a series of overlapping bandpass filters centred continuously at frequencies throughout the normal range of hearing. This theory has been strengthened by subsequent experimental work. The ability to discriminate between two simultaneously presented sounds which contain frequencies that are very close is limited to the width of one of these auditory bandpass filters—the *critical band*.

This concept of frequency selectivity explains a very commonly perceived effect—that of *masking*. Briefly, a sound is masked if it cannot be heard in the presence of another sound. We encounter masking, for example, when turning up the car radio so that it drowns out the noise of the engine. Masking is much more effective where the frequency components of the masker (the masking sound) are close to those of the maskee (the sound being masked). Masking has many useful applications. For example when clinically assessing the performance of one ear, the detection of sounds by the ear not under test can be suppressed by suitable masking sounds fed through headphones. A knowledge of masking can also help with the design of alarms for noisy environments, where the intensity of the alarm can be made sufficiently but not excessively high.

Masking is studied by finding the way in which the threshold of audibility of one sound is raised by the presence of a masking sound. Fletcher's experiments involved detecting the masked threshold of a sinusoidal signal (the level at which it just becomes audible) in the presence of a masker consisting of a band of noise. The width of the band was increased while keeping the noise power density constant. At first, as the noise bandwidth increases, the threshold increases, but then it flattens off, showing that frequencies outside this critical band contribute very little to the masking of the signal. These experiments have been repeated and a variety of others undertaken in an effort to find the shape of the auditory filters. In fact, critical bands are not the result of rectangular waveforms but of waveforms with rounded tops and sloping edges. Typically the critical band is within 10 to 20 per cent of the signal frequency.

It is thought that the physiological basis for the critical band comes from the frequency resolving power of the basilar membrane. Two mechanisms are put forward as reasons for the phenomenon of masking—swamping and suppression. In *swamping*, the masker produces significant activity in the neural fibres corresponding to its frequency, thus not leaving sufficient capability for the maskee to be detected. In *suppression*, the masker suppresses the normal neural activity caused by the masked signal alone, reducing the neural firing rate to a low level. The first of these seems to more easily explain the now widely accepted system of overlapping auditory filters, but the second may explain other related effects.

The masking described above is simultaneous masking. It is also possible for a sound to be masked by another sound which occurs just before it (forwards masking) or just after it (backwards masking). The mechanisms for non-simultaneous masking are not yet fully understood, but again such effects are achieved more easily at closely related frequencies.

1.15 The current view of hearing and speech perception

Speech is a multidimensional signal whose frequency, phase and amplitude components change rapidly in a complex manner. The patterns of sound contained within speech vary under the influence of neighbouring sounds. Perception of speech must therefore be based on a variety of cues. These include acoustic cues using the hearing mechanisms just discussed, but also include context-related cues such as a knowledge of the speaker and the expected linguistic content (both semantics and syntax). Examples of acoustic cues are the length of the silent interval before a plosive, the formant transitions between phonemes, the presence of a combination of formants, fluctuations in signal amplitude, etc.

There is increasing evidence (e.g. Liberman *et al.*, 1967) for 'categorical perception' of speech sounds; that is when presented with a range of sounds differing in acoustically equal steps, the listener tends to place these into categories such that small differences within a formant category are ignored. Categorical perception is one form of evidence that the perception of speech is special compared with the perception of other sounds (i.e. that speech is processed in a different way). The analysis of speech in the cerebral cortex is thought to take place mainly in Wernicke's area in the dominant cerebral hemisphere. This is evidenced by perception difficulties in people with localized brain lesions, and has recently been confirmed by the measurement of increased blood flow when neural activity is increased in response to a speech stimulus (e.g. Nishizawa *et al.*, 1982).

A number of models of speech perception have been proposed. The 'motor theory' (Liberman *et al.*, 1967) states that in perceiving speech sounds the brain follows the intended articulatory gestures of the speaker. At the other end of the spectrum of models, perception is seen as a sequence of processing stages. Firstly, the essential elements are extracted from the information presented by the auditory apparatus. Then the signals are compared with a range of property detectors (e.g. to detect formant frequencies). The next stage extracts the relevant phonetic features (e.g. voicing and nasality). Finally, segmental analysis and lexical searches take place. A review of the various models of speech perception is given in Klatt (1989).

Miller and Jusczyk (1989) discuss recent work in this area and suggest that there are already signs that a neurobiology of speech perception may eventually be forthcoming. They cite studies of categorical perception, studies which indicate that learning of speech perception takes place very early in life, and a variety of hints from animal studies. They also see formal modelling (e.g. of context-dependent processing) as another technique that should shed light on the complex processes involved.

References and further reading

References

Borden, G. J. and K. S. Harris (1984) *Speech Science Primer*, 2nd edn, Williams and Wilkins, Baltimore.

Davis, H. and S. R. Silverman (1970) *Hearing and Deafness*, 3rd edn, Holt Rinehart and Winston, New York.

Fant, G. (1960) *Acoustic Theory of Speech Production*, Mouton, The Hague.

Flanagan, J. L. (1972) *Speech Analysis Synthesis and Perception*, Springer-Verlag, Heidelberg.

Fletcher, H. (1940) 'Auditory patterns', *Review of Modern Physics*, vol. 12, 47–65.

Folkins, J. W. (1985) 'Issues in speech motor control and their relation to the speech of individuals with cleft palate', *Cleft Palate Journal*, vol. 22, 106–122.

Fry, D. (1977) *Homo Loquens: Man as a Talking Animal*, Cambridge University Press, Cambridge, UK.

Khanna, S. M. and D. G. B. Leonard (1982) 'Basilar membrane tuning in the cat cochlea', *Science*, vol. 215, 305–306.

Klatt, D. (1989) 'Review of selected models of speech perception' in *Lexical Representation and Process*, W. D. Marsley-Wilson (ed.), MIT Press, Cambridge, Mass.

Lieberman, P. and S. E. Blumstein (1988) *Speech Physiology, Speech Perception and Acoustic Phonetics*, Cambridge University Press, Cambridge, UK.

Liberman, A. M., F. S. Cooper, D. P. Shankweiler and M. Studdert-Kennedy (1967) 'Perception of the speech code', *Psychology Review*, vol. 74, 431–461.

Linggard, R. (1985) *Electronic Synthesis of Speech*, Cambridge University Press, Cambridge, UK.

Miller, J. L. and P. W. Jusczyk (1989) 'Seeking the neurobiological bases of speech perception', *Cognition: The International Journal of Cognitive Science*, vol. 33, nos 1–2, 111–137.

Moore, B. C. J. (1989) *An Introduction to the Psychology of Hearing*, 3rd edn, Academic Press, London.

Muller, J. (1848) *The Physiology of the Senses, Voice, and Muscular Motion with the Mental Faculties*, W. Baly (trans.), Walton and Maberly, London.

Nishizawa, Y., T. S. Olsen, B. Larsen and N. Larsen (1982) 'Left–right cortical asymmetries of regional cerebral blood flow during listening to words', *Journal of Neurophysiology*, vol. 48, 458–466.

Perkell, J. S. (1969) *Physiology of Speech Production: Results and Implications of a Quantitative Cineradiographic Study*, MIT Press, Cambridge, Mass.

Pickles, J. O. (1988) *An Introduction to the Physiology of Hearing*, 3rd edn, Academic Press, London.

Portnoff, M. R. (1973) 'A quasi-one-dimensional digital simulation for the time-varying vocal tract, M.Sc. Thesis, Department of Electrical Engineering, MIT, Cambridge, Mass.

Potter, R. K., G. A. Kopp and H. C. Green (1947) *Visible Speech*, Van Nostrand Co., New York.

Sachs, M. B. and N. Y-S. Kiang (1968) 'Two tone inhibition in auditory nerve fibers', *Journal of Acoustical Society of America*, vol. 43, 1100–1128.

Van den Berg, J. (1958) 'Myoelastic–aerodynamic theory of voice production', *Journal of Speech and Hearing Research*, vol. 1, 227–244.

Von Bekesy, G. (1942) Über die Schwingungen der Schneckentrennwand beim Präparat und Ohrenmodell', *Akust. Z.*, vol. 7, 173–186.

Weiner, F. M. (1947) 'On the diffraction of a progressive sound wave by the human head', *Journal of Acoustical Society of America*, vol. 19, 143–146.

Wells, J. C. (1986) 'A standardised machine-readable phonetic notation', Proceedings of IEE Conference on Speech Input/Output Techniques and Applications, March, Publication no. 258.

Young, E. D. and M. B. Sachs (1979) 'Representation of steady state vowels in the temporal aspects of the discharge patterns of populations of auditory nerve fibers', *Journal of Acoustical Society of America*, vol. 66, 1381–1403.

Further reading

Lieberman, P. (1984) *The Biology and Evolution of Language*, Harvard University Press, Cambridge, Mass.

Littler, T. S. (1965) *The Physics of the Ear*, Pergamon Press, Oxford, UK.

Matthei, E. and T. Roeper (1983) *Understanding and Producing Speech*, Fontana Paperbacks, London.

Rabiner, L. R. and R. W. Schafer (1978) *Digital Processing of Speech Signals*, Prentice-Hall, Englewood Cliffs, N.J.

2 Analysis

CHRIS ROWDEN

2.1 Levels of speech analysis

The analysis of speech has traditionally been divided into two topics, analysis in the time domain and analysis in the frequency domain. In this chapter we broaden the concept of speech analysis to cover other levels at which speech is processed, which form a hierarchy above the signal level. Classification of complex systems into organizational levels has been applied to computers, operating systems, networks of computers and recently to speech synthesis (Witten, 1986). The levels described by Witten are:

Level	Speech is ...
Application level	The interaction between speech systems and people
Utterance level	A message defined in spoken form
Segment level	Segments just large enough to discriminate one word from others
Signal level	An electrical analogue of pressure or velocity waves

The advantage of this scheme is very clear, and indeed it is the same advantage as is given by the application of levels in the other fields mentioned above. Processes that act at one level can be described and considered in terms of concepts defined at that level. The details of lower levels can be temporarily ignored. This abstraction is particularly useful in analysing a speech system and in the design of speech systems, where it turns out that most time domain analysis fits easily into the signal level, and most frequency domain analysis occurs at the segment level. Speech systems now have to incorporate information and knowledge at the level of words and sentences, and this promotes the need for analysis at the utterance level. Analysis at the application level is more properly dealt with in Chapter 11 on human factors. This chapter will give examples of speech waveforms, and then analysis of speech will be described in the context of the three lower levels.

2.2 Speech waveforms

The direct observation of a speech waveform on an oscilloscope is quite difficult because of the low frequencies involved and because there are often no significant edges in the signal which allow triggering of the timebase to provide a stable display. An oscilloscope with a slow timebase and a long-persistence phosphor screen, or a storage tube screen can be used to display some long-term characteristics of speech. For example the average magnitude can be assessed, as can an estimation of whether the speech waveform is random or periodic, and the location of silences within speech can be located. However, this is all rather limited.

With modern electronic technology, waveforms and spectrograms are more conveniently displayed on the graphics screen or printer of a small computer from a file of speech samples. Commonly used formats for speech files are either 12 or 16 bit/sample with a sampling rate between 8 and 14 kHz. It is usually not worth compressing the data to less than 16 bit/sample (2 byte/sample) because of the extra processing involved. On most small computers it is possible to digitize speech by sampling at 10 kHz with a 12 bit analogue-to-digital converter and storing the sample sequence directly to disk. In this way continuous speech files can be created up to the maximum capacity of one minute of speech for a 1.2 Mbyte floppy disk or 16 minutes of speech for a 20 Mbyte hard disk. For most practical applications such large speech files tend to be rather inconvenient, because the data held within files usually have to be read sequentially, and to access particular words or segments is easier and quicker in small files. Reading from and writing to files is usually much more efficient if done in a block of 512 or 1024 words depending on the particular computer, and so buffers of this length are commonly used. By coincidence a 512 word buffer corresponds to a sequence of speech samples of about 50 ms, which is a fairly representative duration of a phoneme.

Figure 2.1 shows a speech waveform consisting of 1024 samples which were assembled at the sampling rate of 8 kHz, so the duration of speech is 128 ms. Two clearly distinctive segments can be seen, which correspond to the fricative /xh/ and the schwa vowel /@/, as spoken in the word 'unusually'. The apparent diphthong implied by the letters 'ua', like so many other unstressed vowels in informal speech, is articulated very close to the neutral schwa.

The waveform shows examples of unvoiced speech and voiced speech. We can summarize the differences as low-amplitude, seemingly random values for unvoiced speech and high-amplitude values in a repeating pattern for voiced speech. Although the waveforms of other speech segments differ from the examples shown in Fig. 2.1, it is virtually impossible to identify which segment it is from the time waveform.

ANALYSIS

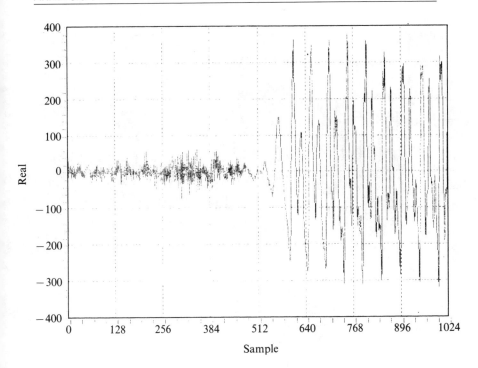

Figure 2.1 Waveform of two speech segments

The periodic segments of voiced speech are produced by the excitation of the vocal tract resonances by a sequence of glottal pulses. The frequency of the glottal pulse train, the pitch of the voice, is substantially independent of the vocal tract shape. However, the number of resonant waves produced by each pulse is strongly influenced by the vocal tract.

For most unvoiced speech sounds the signal resembles 'white' noise, with an approximately Gaussian distribution in the sample space. In the time domain the most significant features are the sudden release of a stop as in /t/, compared to the sustained frication of /s/. In the case of the stop, the release phase is preceded by 40 to 50 ms of silence, during which the tongue seals against the alveolar ridge.

In continuous speech segments are produced in quick succession, and the waveforms of the transition from one segment to the next often shows some surprising information. For instance in Fig. 2.2(a), which shows the sequence /ai m i/ from 'Prime Minister', it can be seen that although there are two letters 'm' in the text, in this example of continuous speech there is just a single segment /m/. Figure 2.2(b) shows the sequence /dh @ b @/ from 'the Bahamas' which illustrates just how little difference there is in the waveform of segments that are perceptually quite easy to distinguish from each other.

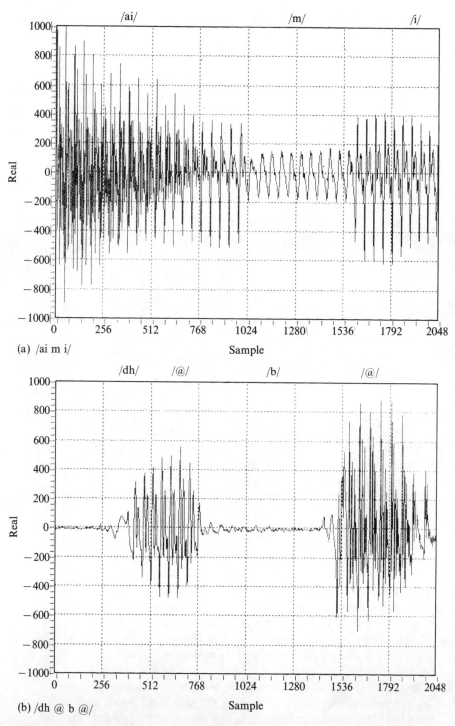

Figure 2.2 Waveforms of transitional sounds

2.3 Analysis at the signal level—stream processing

In many real-time applications of speech processing, sampled speech data will be presented to the processor as a continuous stream of data, and analysis of the data can be carried out on a sample-by-sample basis. The characteristics of speech are usually estimated as an expectation over a number of sample values using a moving average filter. These processes are invariably analysing signal level characteristics of speech, sometimes called short-time characteristics.

2.3.1 Average magnitude

The average magnitude of a sequence of speech data is perhaps the simplest way to determine whether speech is present and, if so, whether it is voiced or unvoiced. It can be computed by adding together the N most recent samples and dividing the result by N:

$$M(n) = \frac{1}{N} \sum_{m=n-N+1}^{n} |x(m)|$$

The effect of this calculation is to look at an infinitely long sequence of speech samples $x(m)$ through a window which is just N samples wide and to compute the average of the magnitude of the samples seen through the window.

An alternative formulation of the average magnitude makes this moving average calculation more explicit by convolving a window function $w(n-m)$ with the speech sequence $x(m)$:

$$M(n) = \frac{1}{N} \sum_{m=-\infty}^{\infty} |x(m)| w(n-m)$$

where $w(\ )$ is the window function, which selects a few values of the infinite sequence $x(m)$ for which $n-m$ is non-zero. It is well known that convolution in the time domain is the equivalent of multiplication in the frequency domain, so the short-time average magnitude $M(n)$ can also be estimated by filtering the magnitude of the signal with a linear filter of impulse response $w(\ )$, shown symbolically in Fig. 2.3(a).

The value of n is the index of the input signal stream, and also of the output stream of magnitude values. Using a rectangular window similar to that shown in Fig. 2.4, then each value of $M(n)$ would be the average of 11 samples. This would result in the sample sequence shown in Fig. 2.1 being smoothed as shown in Fig. 2.3(b). If $M(n)$ the original speech signal was sampled at f_S we can assume that there is an appropriate f_N at which the average magnitude can be sampled without loss of information, and that $f_N < f_S$. The process of resampling a signal at a lower rate is called down-sampling. As with the original sampling

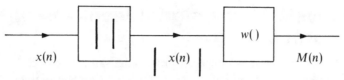

(a) Signal flow diagram

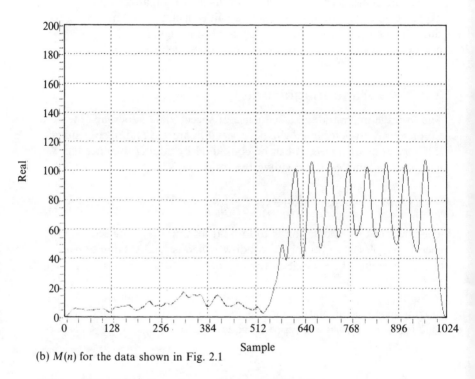

(b) $M(n)$ for the data shown in Fig. 2.1

Figure 2.3 Estimation of average magnitude $M(n)$

the Nyquist sampling theorem is important, and the signal must be band-limited to $f_N/2$ to avoid aliasing effects when the down-sampling is carried out. It is here that the filter analogy of the moving average process is useful because the filtering effect in the frequency domain is the Fourier transform of the convolution window in the time domain. It can be seen from Fig. 2.4 for a window of length 11 and a sampling rate of 10 kHz, in which it is clear that the output of the window filter is effectively low-pass filtered with a band-limit of 700 Hz. From the Nyquist sampling theorem the minimum sampling rate f_N for the average magnitude signal $M(n)$ is 1.4 kHz. In practice down-sampling by any ratio other than an integer division is difficult, so a compromise in this example would be to resample at 2 kHz, which corresponds to selecting every fifth sample value from the window filter output.

ANALYSIS

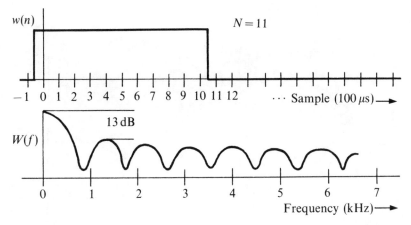

Figure 2.4 Rectangular window function and its Fourier transform

The rectangular window function is the simplest of window functions and is defined as

$$w(n) = \begin{cases} 1, & 0 \leq n \leq N-1 \\ 0, & \text{otherwise} \end{cases}$$

It is particularly easy to implement, as its application to a signal sequence requires multiplication by zero or unity, both of which can be achieved very quickly.

As seen in Fig. 2.4, the Fourier transform of the rectangular window consists of a main lobe and a number of subsidiary lobes which are not negligibly smaller than the main one. The Hamming window function will be described later which has subsidiary lobes that are substantially smaller than the main lobe. This is a very desirable feature as it produces less interaction, or 'leakage', between the wanted components in the frequency domain and unwanted components.

The average magnitude characteristic is useful as part of an automatic gain control system to ensure that speech signals are conditioned to provide the optimum level for further stages of processing. This is particularly important where speakers may move in relation to the microphone or where a number of different speakers may use a system.

2.3.2 Amplitude density function

The average magnitude measures a characteristic of speech over a short time. The characteristics over a long time are also of considerable interest, and the statistical distribution of sample amplitudes forms the amplitude density function. Knowledge of this function allows, for example, the correct choice of scaling of an analogue signal which is connected to an analogue-to-digital converter with linear quantization

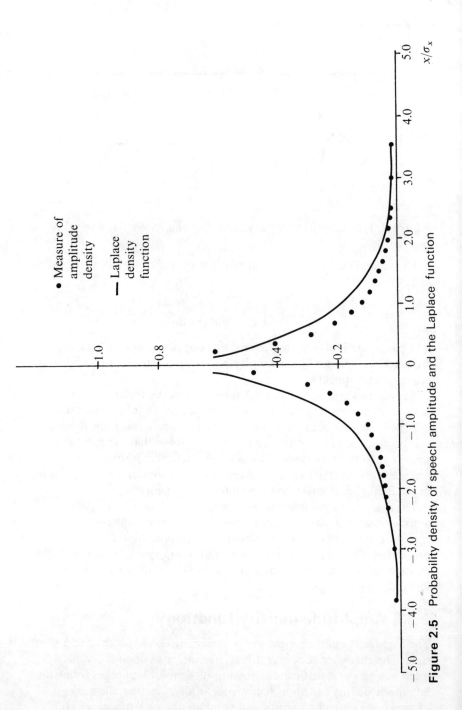

Figure 2.5. Probability density of speech amplitude and the Laplace function

intervals. The probability density function of a database of sampled speech signals can be estimated by determining the proportion $p(x)$ of speech samples x_i which lie in the interval $x < x_i \leq x + dx$. Clearly the value of x_i for samples of a real speech signal must all lie somewhere between $x = -\infty$ and $x = \infty$, so

$$\int_{-\infty}^{\infty} p(x)\,dx = 1$$

and the area under the probability density curve is equal to unity.

The empirical results of such a study for a speech file containing about 12 seconds of speech are shown in the graph in Fig. 2.5. In the file a male and a female speaker say each of the digits from 'zero' to 'nine' once each; the sampling rate is 16 kHz with linear quantization at 14 bit/sample. The values of x are normalized to make the r.m.s. value for the complete file equal to unity. Also shown on the graph is a commonly used approximation, the Laplace density function $p(x)$, which is also normalized to make the r.m.s. value, σ_x, equal to unity:

$$p(x) = \frac{1}{\sqrt{2}\sigma_x} \exp\left(-\frac{\sqrt{2}|x|}{\sigma_x}\right) \quad \text{Laplace density function}$$

The Laplace density function possesses the considerable advantage of being fairly amenable to analytical treatment, and is directly related to the r.m.s. value of the speech, σ_x.

A feature of the probability density computed from this experiment is the asymmetry about zero. It is quite common for the waveform of short segments of speech to display asymmetry about zero, but this varies with different speakers, the distance between the speaker and the microphone, and the acoustics of the recording studio. When the effects are averaged by computing the probability density over much greater lengths of speech with many speakers the results become symmetric (Davenport, 1952; Paez and Glisson, 1972).

2.3.3 Average energy—variance

The energy of a signal can be estimated from the variance, which is the square of the difference between the sample value and the mean value. So for a signal with a zero mean, the short-time energy can be defined as the average of the square of the sample values:

$$E(n) = \frac{1}{N} \sum_{m=-\infty}^{\infty} [x(m)w(n-m)]^2$$

where $w(\ )$ is a suitable window function, which selects only the values of $x(m)$, near $m = n$, for which $w(n-m)$ is non-zero. This may be rewritten as

$$E_n = \sum_{m=-\infty}^{+\infty} x^2(m)h(n-m)$$

44 SPEECH PROCESSING

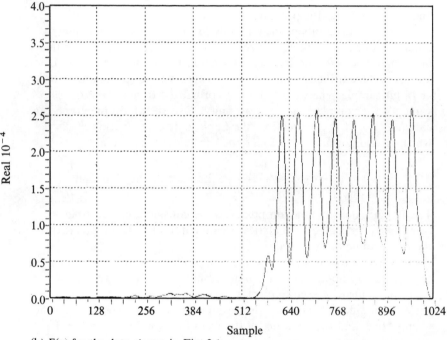

(a) Signal flow diagram

(b) $E(n)$ for the data shown in Fig. 2.1

Figure 2.6 Estimation of average energy $E(n)$

where

$$h(n) = \frac{1}{N} w^2(\ )$$

As shown in Fig. 2.6 the short-time energy $E(n)$ can be estimated by filtering the square of the signal with a linear filter of impulse response $h(\ )$, which theoretically is the square of an appropriate window function $w(\ )$. In practice it is usual just to use a suitable window function normalized to the window length N. The signal sequence in Fig. 2.6(b) was computed using a Hamming window function of length 32. A description of the Hamming window is deferred for the moment because the advantages become more evident in the context of the discrete Fourier transform discussed in Sec. 2.4.1.

ANALYSIS

Compared to the average magnitude shown in Fig. 2.3, the average energy of a stream of speech samples is a characteristic that shows a greater difference between voiced speech and unvoiced speech, and between unvoiced speech and silence. Therefore estimation of average energy is preferred in applications such as a speech detector, but usually in combination with a zero-crossing detector, described in the next section.

2.3.4 Zero-crossing rate

An important difference between the voiced and unvoiced segments of speech shown in Fig. 2.1 is the rate at which the signal changes sign,

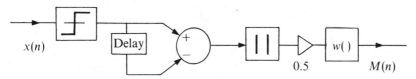

(a) Signal flow diagram

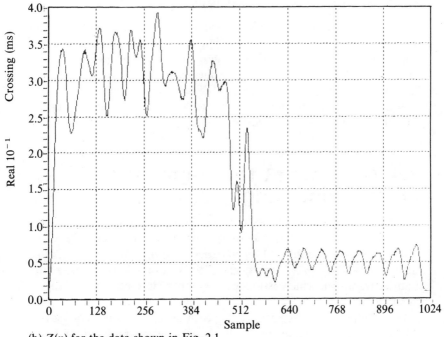

(b) $Z(n)$ for the data shown in Fig. 2.1.

Figure 2.7 Estimation of zero-crossing rate $Z(n)$

the zero-crossing rate. For the voiced segment the rate is about 0.5 crossings/ms and for the unvoiced segment around 3 crossing/ms. This characteristic can be estimated by a stream processing approach if the signal is processed by a single-step quantizer set at zero level; then the difference between each quantized sample and its predecessor is taken, and finally half the magnitude of that difference is presented to the window filter:

$$Z(n) = \frac{1}{2N} \sum_{m=-\infty}^{\infty} |\text{sign}(x(m)) - \text{sign}(x(m-1))| w(n-m)$$

where

$$\text{sign}(x(m)) = \begin{cases} 1, & \text{if } x(m) \geq 0 \\ -1, & \text{otherwise} \end{cases}$$

The estimation of $Z(n)$ for unvoiced speech can be significantly affected by mains ripple or any small d.c. offset to the speech samples, because of the low signal amplitude. Sensitivity to these effects can be significantly reduced by a high-pass filter stage, with a cut-off frequency around 70 Hz. The signal sequence in Fig. 2.7(b) was computed using a Hamming window function of length 32.

2.3.5 Summary of signal level analysis

This small selection of analysis techniques at signal level is sufficient to show that characteristics of speech signals can be estimated as a continuous process. This can be done by a moving average filter, which combines the selection of a recent sequence of data and low-pass filtering the characteristic values for the sequence. Because of the filtering action, the characteristic can be down-sampled to a lower rate than the original speech sequence.

2.4 Analysis at the segment level— block processing

There are some characteristics that relate to a segment of the speech signal and cannot be conveniently estimated as a continuous process. Analysis to estimate these characteristics must be done by processing a signal represented by a block of samples. An example is the discrete Fourier transform which estimates the amplitude spectrum of a segment of speech.

One of the earliest extensions of speech analysis to the frequency domain was the development of the speech spectrograph. In the first models the speech utterance was recorded onto a loop of magnetic tape, which was replayed many times as a tuned filter was gradually stepped through the range of frequencies. Synchronized to the tape was a drum carrying electrosensitive paper, onto which was permanently marked the filter output. The book *Visible Speech* (Potter, Kopp and Green, 1947) records a project in which a number of young women were trained to read speech from the spectrograph display. For this 'real-time' application of the technique, the spectrograph used a bank of fixed frequency filters in place of the single variable frequency filter. The output of each frequency band was displayed either on a cathode ray tube or by light patterns written onto a moving belt of phosphor by a set of miniature lamps.

With modern technology, the speech spectrum is most effectively estimated by one of the fast Fourier transform (FFT) methods using a block size that is an integer power of two (see for examples Chapter 1 of IEEE, 1979). Another form of block processing is linear predictive analysis, in which a sequence 100 to 200 speech samples can be processed to determine the optimal prediction filter for a segment of the speech signal (Atal and Hanauer, 1971). A brief overview of linear prediction as an analysis technique is given in Sec. 2.5 of this chapter; a fuller description appears in Chaper 5.

In block processing methods, an array of data values is assembled, the window function is then applied by multiplying each element in the array by the appropriate window weighting and then processing is carried out on the windowed array. A suitable window function is the Hamming window, named after R. W. Hamming, which is one of a family of raised cosine shaped window functions. The other commonly used member of the family is the Hann (sometimes called Hanning) window, named after J. von Hann. They both offer a wide main lobe and low magnitude of subsidiary lobes, when compared to a rectangular window of similar length. Discrete versions of raised cosine windows are defined by

$$w(n) = \begin{cases} a - (1-a)\cos\left(\dfrac{2\pi n}{N-1}\right), & 0 \leq n \leq N-1 \\ 0, & \text{otherwise} \end{cases}$$

For the Hamming window $a = 0.54$ and for the Hann window $a = 0.5$. The Hann window tapers smoothly down to zero values at each end of the range. The Hamming window has a more abrupt transition, as illustrated in Fig. 2.8; the 0th and the $(N-1)$th values are 0.08.

For any window function other than the rectangular window, the samples that lie near the ends of the window will be given a low weighting. If such samples describe a significant speech event of short duration, such as the sudden release of a stop, then the low weighting those samples receive means that the event will not feature largely in the block calculation. To ensure that all such speech events influence the block calculations, the blocks are usually overlapped so that any

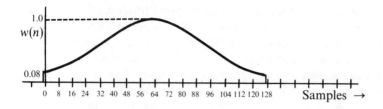

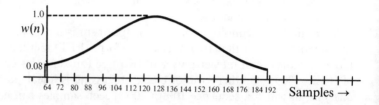

Figure 2.8 Overlapped Hamming window functions

event will be covered by at least two overlapping windows. An event that is at the edge of one window, and so receiving a low weighting, will be near the centre of a neighbouring window, within which it will receive the highest weighting.

2.4.1 Discrete Fourier transform of speech waveforms

The spectrum of a sampled periodic waveform can be estimated with the discrete Fourier transform (DFT), which is defined as

$$X(k) = \sum_{n=0}^{N-1} x(n) e^{-j2\pi nk/N}$$

The DFT (Fig. 2.9) generates a set of N values indexed by k in the frequency domain which is the transform of a periodic signal sampled in the time domain with index n. In the case of spectral analysis of speech waveforms we are dealing with $x(n)$ which is real, and for this case $X(k)$ and $X(N-k)$ are complex conjugates. Since the magnitude of complex conjugates is equal, then $|X(k)| = |X(N-k)|$, and so the magnitude of the upper half of the frequency range is the mirror image of the lower half. It is usual therefore to display only the lower $N/2$ values in the frequency domain of a signal transformed from an N point frame of speech. If the transform is used only for spectral display we are usually concerned only with the magnitude of the signal $|X(k)|$.

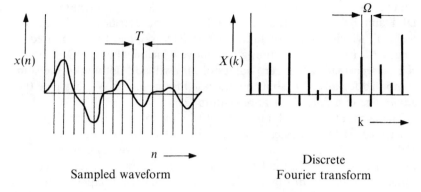

Figure 2.9 Discrete Fourier transform of a periodic signal

The inverse DFT is

$$x(n) = \frac{1}{N} \sum_{k=0}^{N-1} X(k) e^{j2\pi nk/N}$$

and again generates a set of N values indexed by n in the time domain. $X(k)$ is generally a sequence of complex values, and if the transform of a speech signal is later to be inverse-transformed back into the time domain then the complex value of the $X(k)$ must be preserved. All N values of $X(k)$ must be used because unless $X(k)$ displays the symmetry, the IDFT will not produce a sequence of real values.

The resolution of a line spectrum Ω is determined by the sampling interval $T(=1/f_s)$ and the number of values N:

$$\Omega = 1/NT \quad \text{Hz} \quad \text{or} \quad 2\pi/NT \quad \text{rad/s}$$

It is useful to be aware of the duality that exists between the time domain and the frequency domain. An impulse in the time domain transforms to a broad spectrum of frequencies, whereas a perfect sinusoid in the time domain transforms to a single line in the frequency domain. A difficulty that often arises with estimating the Fourier transform of a speech signal is that speech waveforms are not truly periodic, and it is only for periodic signals that the Fourier transform is defined. This, however, is a minor problem compared to the fact that the periodicity of speech is variable, which means that successive blocks of speech samples will not necessarily have the same number of periodic waves and will not have the same alignment to pitch events such as the largest cycle in the periodic wave. The art of obtaining meaningful frequency spectra of speech segments consists of overcoming these problems.

The usual approach to solving these problems lies in determining the optimum length of the sequence to be transformed and selecting a suitable window function to minimize the end effects produced by truncating the sequence of samples. To reconcile different lengths of a DFT and a signal sequence, the windowed final sequence can be 'zero-padded', which consists of adding zero-valued samples to any desired length N'. From the equation given above which defines the DFT, it can be seen that further zero values of $x(n)$ will add nothing to the sum computed for any $X(k)$, but the increased frame length N' does give the advantage of a finer spectral resolution Ω'.

The effect of the length and type of window and of zero-padding is illustrated by the sequence of waveforms and spectra shown in Fig. 2.10. For clarity the signal analysed is a single sinusoid, for which the Fourier transform of an indefinitely long (unwindowed) sequence is a single spectral line. In Fig. 2.10(a) is shown a 64 point sequence of a sampled sinusoidal signal, $x(n) = \sin(2\pi n/16)$ where $0 \leq n \leq 63$, windowed by the Hamming function $w(n) = 0.54 - 0.46 \cos(2\pi n/63)$ where $0 \leq n \leq 63$. The DFT of this sequence is shown in Fig. 2.10(b) with the 32 discrete points indicated by dots. As is usual, even with the *discrete* Fourier transform, the dots are connected together to form a continuous curve. The expected principal component at $f_S/16$ is clearly evident, but the expected single spectral line has been broadened to a main lobe, wide enough to give substantial components at $3f_S/64$ and $5f_S/64$. Most other components are quite small; ideally they should be zero.

The same windowed sequence is shown in Fig. 2.10(c) at the centre of a zero-padded sequence of 256 points, and the DFT of that sequence, shown in 2.10(d), now has 128 points in the frequency interval 0 to $f_S/2$. These points are so close together that the continuous curve is a satisfactory representation of the spectrum. Again it is clear that the principal component is centred on $f_S/16$, but although the spectral detail is finely resolved, the main lobe is quite broad, and most of the side lobes are more than 40 dB down on the main lobe. To confirm our preference for the Hamming window, compare Fig. 2.10(d) with 2.10(e) which shows the DFT for the same signal sequence but with a rectangular window function.

The width of the main lobe of the DFT is inversely proportional to the length of the windowed sample sequence. This is demonstrated by Fig. 2.10(f), which is a 128 point sinusoidal sequence, windowed and zero-padded, and Fig. 2.10(g), which is the corresponding DFT, with the main lobe and side lobes approximately half the width compared to the 64 point sequence shown in Fig. 2.10(d).

The foregoing description outlines the effect of the choice of window type and length, and of zero-padding on a DFT display of a frequency spectrum. A more mathematical treatment of this topic may be found in Chapter 3 of Candy (1988), but the examples here relate to a single spectral line at $f_S/16$, which corresponds to a 500 Hz sinusoid if the sampling rate is 8 kHz.

ANALYSIS

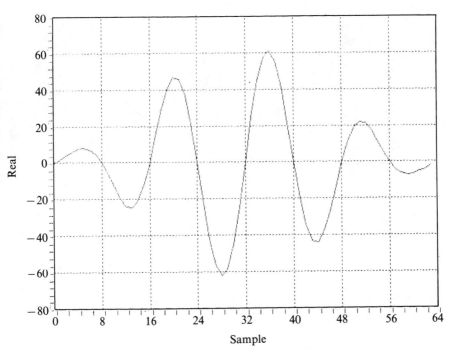

(a) A 64 point sequence of $\sin(2\pi n/16)$, Hamming window

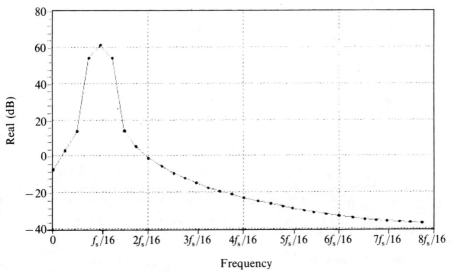

(b) DFT of sequence shown in (a)

Figure 2.10

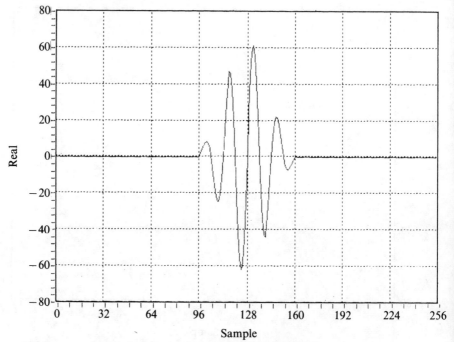

(c) Sequence shown in (a), zero padded for 256 points

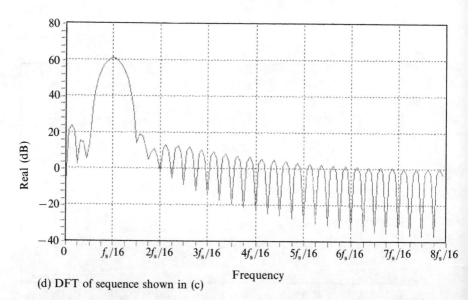

(d) DFT of sequence shown in (c)

Figure 2.10 (*contd*)

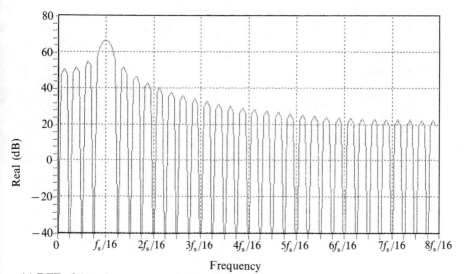

(e) DFT of 64 point sequence of sin $(2\pi n/16)$, rectangular window, zero padded to 256 points

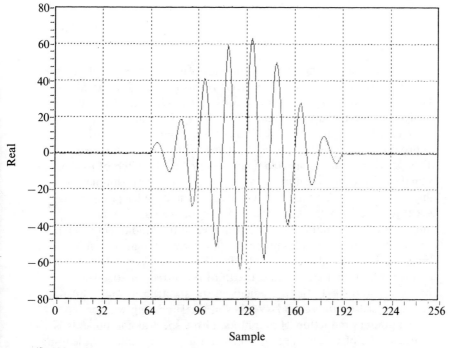

(f) 128 point sequence of sin$(2\pi n/16)$, Hamming window, zero-padded to 256 points

Figure 2.10 (*contd*)

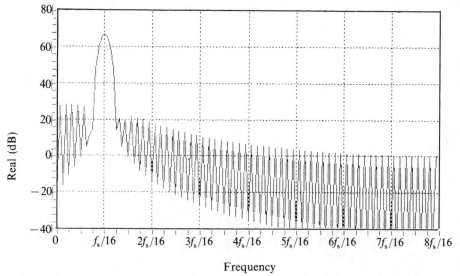

(g) DFT of sequence shown in (f)

Figure 2.10 (*contd*)

We now examine the spectrum of a sequence of voiced speech, shown in Fig. 2.11(a). The DFT in Fig. 2.11(b) was produced in exactly the same way used for Fig. 2.10(g), so we can transfer knowledge of that analysis to the real speech signal. The shape of the main lobe can be quite clearly seen at 1050 and 2030 Hz, and slightly less clearly at approximately 160 Hz intervals throughout most of the frequency band. This periodicity in the spectrum corresponds to the pitch of the speech signal, which can be confirmed as 160 Hz, or about 50 samples, from inspection of Fig. 2.11(a). Also evident are the broader peaks of the vocal tract resonances, but these can only be seen as they are 'illuminated' by suitably placed harmonics of the fundamental pitch. Three such harmonics illuminate the first formant, which is quite broad and is centred at about 320 Hz. The second formant is sharper and is centred on 1050 Hz, and the third is centred on 2050 Hz. Between 3 and 4 kHz the periodicity of the spectrum is quite different, and is caused by the secondary lobes of the Hamming window.

The gradual reduction of amplitude of the spectral components is characteristic of voiced speech, which can be a significant problem in speech analysis, as numerical calculations may be scaled to accommodate large low-frequency components, thus offering less accuracy to the high-frequency components. To produce a flatter spectrum the sequence can be preprocessed by a *preemphasis* filter

ANALYSIS

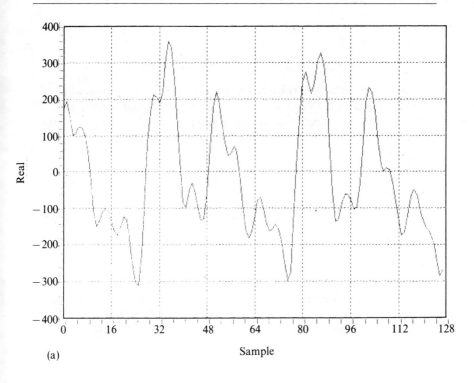
(a)

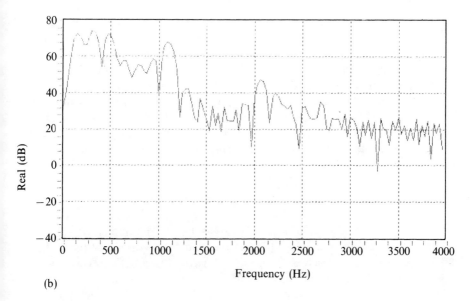
(b)

Figure 2.11 (a) 128 point sequence of voiced speech. (b) DFT of sequence shown in (a), Hamming window, zero-padded to 256 points

which has a transmission function, $H_{preemph}$, with one low-frequency zero:

$$H_{preemph}(z) = 1 - az^{-1}, \quad \text{where usually } 0.95 < a < 0.98$$

Where preemphasis has been applied, the spectrum is artificially boosted in the high frequency, and if a processed signal is converted back to a speech signal, then the opposite process of deemphasis filter is required to restore the original slope to the spectral components:

$$H_{deemph}(z) = \frac{1}{1 - az^{-1}}$$

However, for spectral display it is not necessary, nor indeed is it particularly convenient, to restore the original slope, and displays are often spectrally flattened by means of preemphasis.

There are applications for which a single DFT of a waveform is of interest, but it is also commonly required to present a sequence of DFT spectra displayed as a plot against the time axis. There are various forms of display; one is the speech spectrogram, in which amplitude is indicated by colour or grey level, as a function of both frequency and time. Examples of speech spectrogram output appear in Figs 1.9 and 1.10.

2.4.2 Auto-covariance and auto-correlation

For quasi-periodic sequences of voiced speech sounds, analysis in the time domain shows that there is some correlation between one sample and a few of its predecessors. The auto-covariance for a particular value of delay (k samples) is the average of the products of each sampled value $s(n)$ with the value which is distant by k samples, $s(n+k)$, and it is a function of the delay k. The summation is performed over a suitably long sequence of N samples, so that it is a measure of the signal over a segment of speech consisting of several periodic waves or an equivalent length of unvoiced speech. The auto-covariance function $C(k)$ is thus

$$C(k) = \frac{1}{N} \sum_{n=0}^{N-1} s(n)s(n+k)$$

for a sequence of samples $s(0)$ up to $s(N-1+k)$. The numerical value of $C(k)$ is relatively unlimited, as it is related to the square of the values of the samples themselves. To provide for easier comparison between one speech segment and another, the auto-covariance can be normalized to the largest possible value, which occurs for a delay $k=0$. This normalized form is called the auto-correlation function, $R(k)$:

$$R(k) = \frac{C(k)}{C(0)}$$

It should be noted that there are considerable differences in the way that the terms auto-covariance and auto-correlation are used in signal processing literature. The terms are often related to the theory of random processes, where an assumption of Gaussian distribution is commonly made. The definitions given above relate to those given by Blackman and Tukey (1958) and may appear in modified form due either to the assumptions of Gaussian distribution or due to the incorporation of a window function.

A speech waveform and its auto-correlation function are illustrated in Fig. 2.12 for a voiced speech segment /Q/. The fundamental pitch can be estimated at around 90 Hz from the separation between the similar waveshapes in Fig. 2.12(a), which has an 8 kHz sampling rate.

It is clear from the definition of $R(k)$ given above that $R(0)$ must always be unity. The auto-correlation function in Fig. 2.12(b) shows values close to $+1$ for a few sample delays, showing that a predictor for voiced speech could be quite accurate if it based the prediction on a small number of sample delays. From the figure it is also clear that there is another cluster of strongly positively correlated samples at a delay k_p that corresponds to the fundamental pitch of the voiced sounds. Therefore a predictor should also use samples delayed by a long value equivalent to the periodicity of the voiced speech. To use the long delays the system must be adaptive, as the pitch of the human voice can vary (for a range of speakers) from 50 to 500 Hz. For that range of frequencies, the positively correlated lobe could occur anywhere between 16 and 160 samples; in Fig. 2.12(b) it occurs at about 88 samples, corresponding to a pitch of 91 Hz, which agrees with the estimate made from Fig. 2.12(a).

The auto-correlation function of unvoiced speech, the segment /sh/, is shown in Fig. 2.13(b). This displays very different characteristics; the only point of similarity is that the value at zero delay is also $+1$, but this much is ensured by the definition of $R(0)$. There are, however, several non-zero values for small sample delays, so a short-term predictor can still give some usable prediction for this unvoiced speech. Following the duality between time and frequency domains, we may conclude that a short feature in the time domain corresponds to a large component in the frequency domain, and this is indeed the case, as the segment shown in Fig. 2.13(a) has a dominant resonance at about 3 kHz.

2.4.3 Power spectral density

The operation of the Fourier transform on the auto-correlation function $R(k)$ has been implied in the previous section, and indeed if the Fourier transform is applied to the auto-covariance function $C(k)$, the non-normalized form of $R(k)$, the result is an estimate of the power

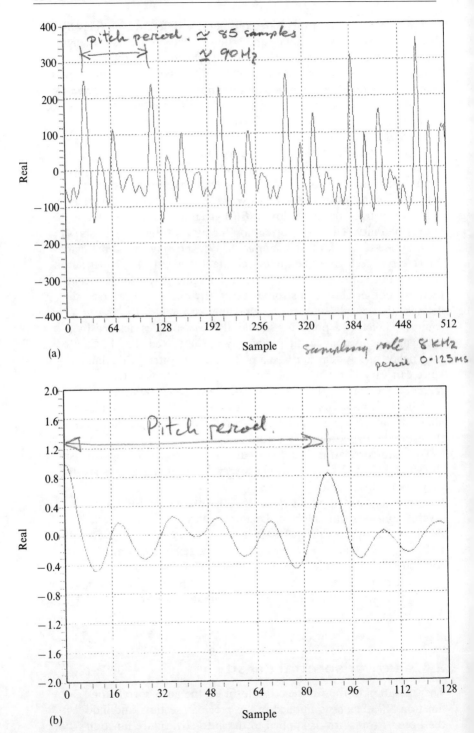

Figure 2.12 Waveform and auto-covariance of voiced speech

ANALYSIS

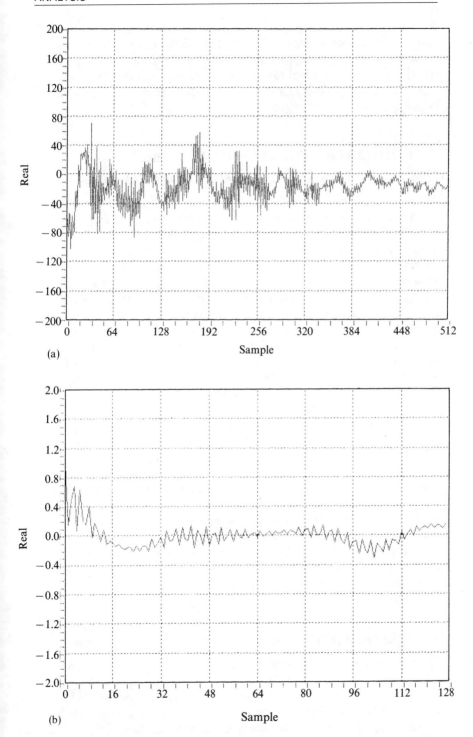

Figure 2.13 Waveform and auto-covariance of unvoiced speech

spectrum of the speech segment. This result is known as the Wiener–Khintchine theorem, which is described by most books on signal processing (see, for example, Candy, 1988). The advantage of a power spectrum is that it is a real function of frequency, and the power of signal components is sometimes a closer indication of the perceptual importance than the amplitude.

2.5 Separation of vocal tract resonances from speech excitation

The periodic segments of speech are produced by the excitation of the vocal tract resonances by a sequence of glottal pulses. The frequency of the glottal pulse train, the pitch of the voice, is substantially independent of the shape of the vocal tract. The analysis of voiced speech is often directed to the estimation of the pitch, F_0, and the formants, or resonant frequencies of the equivalent filter to the vocal tract, $F_1, F_2, F_3, \ldots$ for as many as desired. For some purposes the amplitude of each formant as well as the centre frequency is required; for others it is ignored. The characterization of the unvoiced segments is sometimes simpler because of the random nature of the signal. For some unvoiced speech sounds the signal resembles 'white' noise, with a uniform distribution through the spectrum. This is particularly true of the fricatives produced at the front of the mouth, /th/, /s/, /f/, for which the turbulence of the air flow is minimally affected by the resonances of the mouth. Fricatives produced further back, /h/, /sh/, show a very non-uniform distribution through the spectrum, and formants are more evident.

The relative independence between the shape, and therefore the resonances, of the vocal tract and the excitation means that these two components of the speech signal can be analysed separately from one another. The representation of these separate components by parameters that change at a slow rate is the basis for the coding advantage achieved by parametric coding techniques, which are described in Chapter 5. The two common methods of analysing speech to achieve this separation, linear prediction and cepstral filtering, are described in this section.

2.5.1 Linear prediction

From the examples of the auto-correlation function of voiced speech and unvoiced speech, it is clear that the value of a sample of a speech signal can be predicted from previous sample values. This prediction can be based upon a linear combination of the most recent sample values, so the process is called linear prediction.

Let us assume that a sequence of speech samples s_n is applied to the

ANALYSIS

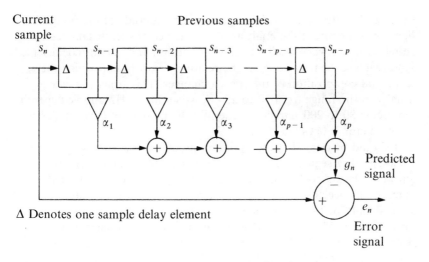

Figure 2.14 Linear prediction of a speech signal

system shown in Fig. 2.14. The system has p delay elements, each of which is clocked at the same rate as the speech was sampled. At each clock instant the sample value present at the input to the delay element is placed in the memory of the element, and its value appears at the output of the element. The action of the delay element is rather like that of a D-type flip-flop in a shift register; the difference is that the data is a speech sample consisting of 12 or 16 bits rather than the single bit that can be stored by a single flip-flop.

The delay elements effectively remember the p most recent values of the sample sequence, and these values are then linearly combined according to the weightings α_1 to α_p to produce a prediction of the value of the next sample g_n. This describes the simplest form of linear predictor in which the predicted value g_n is a linear combination of past speech samples, s_{n-k}:

$$g_n = \sum_{k=1}^{p} \alpha_k s_{n-k} \qquad (2.1)$$

This linear predictor has the same structure as a finite impulse response (FIR) digital filter, or transversal filter. In starting the summation from $k=1$, we have deliberately not used the current value of the speech sample s_n, so that if we subtract the predicted signal g_n from that current value s_n we can obtain an error signal e_n, which represents the difference between the actual signal from that which is predicted. This error signal is also called the *residual* signal:

$$e_n = s_n - \sum_{k=1}^{p} \alpha_k s_{n-k} \qquad (2.2)$$

By careful choice of the weightings α_k, the magnitude of the residual signal resulting from the analysis of a sequence of speech samples can be reduced to a very small value. These optimum predictor coefficients α_k for each sequence change at the relatively slow rate at which the segmental composition of the speech changes. This *segmental rate* is usually in the range from 10 to 25 ms, which for 8 kHz sampled speech represents 80 to 200 samples. Generally the number of delay stages p in the predictor is between 10 and 20.

It should be clear that the optimum predictor coefficients α_k, where $1 < k < p$, for a segment of speech will be related in some way to the auto-correlation coefficients $R(k)$, where $1 < k < p$, calculated for that segment of speech. In Chapter 5 that relationship will be examined in some detail; for the purposes of this chapter it is sufficient to say that for a segment of a speech waveform in which the characteristics are relatively stationary, the sequence of speech samples s_n, where $1 < n < N$, can be represented by p coefficients of a predictor filter α_k, where $1 < k < p$, and a sequence of N error samples e_n, where $1 < n < N$, which is the residual signal. The residual signal sequence for voiced speech segments contains generally quite small values except for a large discontinuity which coincides with the start of each glottal pulse. The residual sequence can be approximated by a sequence of impulses. For unvoiced speech the residual sequence resembles random noise at a fairly low level.

2.5.2 Distance (distortion) measures of speech segments

A *distortion measure* is an objective measurement that relates to the quality of processed speech. One area of application of distortion measures is the assessment of speech coders, in which it is required to measure the difference between a 10 to 20 ms segment of coded speech and the equivalent segment of the original speech. Another use is for speech recognition which requires a measure of the similarity between an incoming frame of speech and one of a number of reference frames. In speech recognition applications the measure is usually called a *distance measure*.

Desirable properties of a distance measure $d(S_a, S_b)$ are:

1. Symmetry, $\qquad d(S_a, S_b) = d(S_b, S_a)$
2. Triangle inequality, $\qquad d(S_a, S_c) \geq d(S_a, S_c) + d(S_b, S_c)$
3. Positive definiteness, $\qquad d(S_a, S_b) > 0 \quad \text{for } S_a \neq S_b$
   $\qquad\qquad\qquad\qquad\qquad\quad = 0 \quad \text{for } S_a = S_b$
4. Computationally efficient, evaluation is not too costly
5. Physically meaningful, related to some tangible quality of the speech segment.

A measure that possesses all of the first three properties is called a *metric*, but there are many quite usable distance measures that are not metrics.

Taking a vector of 10 features, $S_i(s_{i1}, s_{i2}, \ldots, s_{i10})$, which might, for example, be the energy at the outputs of a filter bank, each vector can be represented by a point in a ten-dimensional space. The *Euclidean distance* between that point and another, which, for example, may be a reference vector, $S_c(s_{c1}, s_{c2}, \ldots, s_{c10})$, is the square root of the sum of the squares of the feature differences:

$$d_{\text{euclid}}(S_i, S_c) = \sqrt{\sum_{n=1}^{10} (s_{in} - s_{cn})^2}$$

The Euclidean distance is a metric if the features used for S_i and S_c are orthogonal features; that is to say, there is no interaction between any pair of features, a condition which is not satisfied for the filter bank example. This measure is relatively easy to compute, but it suffers from the disadvantage that the features do not all have the same perceptual significance.

The use of Euclidean distance can lead to instances where the closest match as measured by $d_{\text{euclid}}(\ )$ is not the best match in a perceptual sense. A solution to this problem is to apply a different weighting to each of the features, such as computed by the *Mahalanobis distance*:

$$d_{\text{mahalanobis}}(S_i, S_c) = (S_i - S_c)^T W^{-1} (S_i - S_c)$$

The matrix W^{-1} is the inverse of the auto-covariance matrix of the reference vector S_c, and it provides the means for giving a separate weighting for the interactions between features. In general use of the method, for each reference vector S_c a separate matrix W is required, but in speech recognition applications it is more practical to use one compromise matrix W computed from the auto-covariance of all reference vectors. The Mahalanobis distance is costly to compute, because $O(N^2)$ multiplications must be performed for an N-dimensional vector, compared to $O(N)$ multiplications for the Euclidean distance. An apparent advantage of the Mahalanobis distance compared to Euclidean distance is the absence of the square root operation, but in practice this is neglected in most applications of the Euclidean distance measure, because magnitude comparisons on the squared Euclidean distance, $d^2_{\text{euclid}}(\ )$, will be the same as magnitude comparisons on $d_{\text{euclid}}(\ )$. Indeed, it has been shown (Moore, 1986) that under certain assumptions, a maximum likelihood classifier can be replaced by a nearest neighbour classifier which uses the squared Euclidean distance. It can readily be seen that for the special case where W^{-1} is the identity matrix, $d^2_{\text{euclid}}(\ )$ and $d_{\text{mahalanobis}}(\ )$ are the same.

Either of the distance measures mentioned above could be applied to features derived from linear predictive analysis of speech, but there are alternatives that apply specially to such parametric representation of speech. The *Itakura–Saito distance* measure is one that appears in a number of different formulations, which are mathematically related under certain sets of assumptions, but not in a simple way. It is more costly to compute, but it is related to the spectrum of the speech signal:

$$d_{IS}(X, H) = \int_{-\pi}^{\pi} \left[\frac{X(e^{-j\phi})}{H(e^{-j\phi})}\right]^2 - \ln\left[\frac{X(e^{-j\phi})}{H(e^{-j\phi})}\right]^2 - \ln\frac{d\phi}{2\pi}$$

where X is the spectrum of the speech signal and H the spectrum of the filter.

2.5.3 Cepstral analysis

Cepstral analysis is another way of separating the vocal tract filter characteristics from the excitation sequence (Noll, 1967), and it can be explained by assuming that the speech signal is composed of an excitation $e(t)$ applied to the vocal tract filter, which has an impulse response $v(t)$. This is a time domain view of the source filter model of speech, shown in Fig. 2.15.

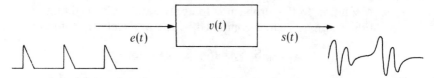

Figure 2.15 Source filter model of speech in the time domain

The speech signal in this model is $s(t)$ which is the convolution of $e(t)$ with $v(t)$:

$$s(t) = e(t) * v(t)$$

In the frequency domain this is simply

$$S(\omega) = E(\omega)V(\omega)$$

where $S(\omega)$, $E(\omega)$, $V(\omega)$ are Fourier transforms of continuous time functions $s(t)$, $e(t)$, $v(t)$ or the discrete Fourier transforms of sampled time sequences $s(n)$, $e(n)$, $v(n)$.

The excitation $E(\omega)$ and the vocal tract filter $V(\omega)$ are combined multiplicatively, which makes it difficult to separate them. However, by taking the logarithm of $S(\omega)$ the excitation and vocal tract functions become additive:

$$\log[S(\omega)] = \log[E(\omega)] + \log[V(\omega)]$$

Figure 2.16 shows the fundamental pitch-related component added to the vocal tract response. The additive property of the log spectrum still

ANALYSIS

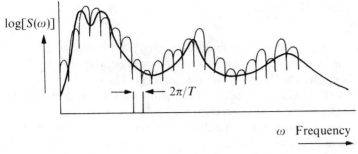

Figure 2.16 Log spectrum of voiced speech

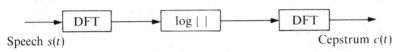

Figure 2.17 Estimation of cepstrum

applies when a further Fourier transform is applied to it, and the result of that operation is called the cepstral function or *cepstrum*. Thus the process of estimating the cepstrum is summarized by the signal flow diagram shown in Fig. 2.17.

The strongly periodic component in the log spectrum in Fig. 2.16, at a frequency interval equivalent to the reciprocal of the pitch period T, becomes a sharp peak in the cepstrum. The independent variable (x axis) of the cepstral function has dimensions of time and is given the name *quefrency*. For most voiced speech a clear distinction can be made between that pitch-related excitation component and the vocal tract contribution which appears as a cluster of components at low values of quefrency. Quefrency is commonly, but not necessarily, plotted on the scale 0 to 15 ms for speech signals, as shown in the example of Fig. 2.18.

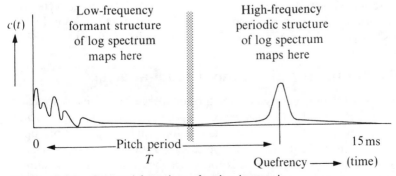

Figure 2.18 Cepstral function of voiced speech

2.5.4 Cepstral smoothing of vocal tract transfer function

As the vocal tract transfer function and the excitation function of speech appear in separate parts of the quefrency scale, they can be separated from one another by a filtering process. The cepstrum is made up of a set of discrete cepstral coefficients, which were the output set of the final DFT process. Therefore we could apply the filter by means of a rectangular window function. The Fourier transform of the rectangular window indicates that such a process would generate unwanted side lobes to the required components in the log frequency domain, so a more gradual windowing function would be preferred. If a cepstrum signal is filtered in this way and the inverse discrete Fourier transform taken of the resulting signal, then a smoothed version of the log spectrum of the vocal tract filter is produced, such as shown in Fig. 2.19. Cepstral smoothing can be used to estimate the spectral envelope curve shown in Fig. 2.11 as an alternative to the discrete Fourier transform.

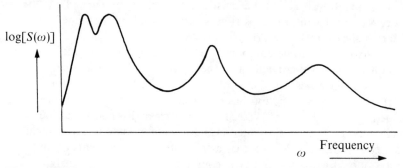

Figure 2.19 Cepstrally smoothed log spectrum of voiced speech

The frequency, bandwidth and amplitude of the formants can easily be determined from the spectral envelope curves. These parameters are of importance, for example, in providing the control information for the Klatt parallel formant synthesizer described in Chapter 6.

2.5.5 Summary of segment level analysis

Estimating the frequency characteristics of speech by means of applying the Fourier transform to a sequence of speech samples provides insights not available from time domain analysis, for example the relationship between formant frequencies and the categorization of speech sounds into phonemes. If the sequence is too short then insufficient resolution is given in the frequency domain; if the sequence is too long then the nature of the signal will not be stationary for the duration of the sequence. The window function used to select a sequence can have a significant effect on the results obtained; resolution of the components is inversely proportional to the length of the window, but finer resolution of the spectrum can be achieved quite independently by zero-padding the windowed speech samples. Zero-padding can also be used to reconcile the variable periodicity of the speech signal with the fixed length of the FFT for which the optimum length is an integer power of two.

The relatively slow changing characteristics of the vocal tract can be separated from the relatively faster changing excitation by two methods, linear predictive analysis or cepstral analysis. The vocal tract shape for speech sounds can be used for parametric coding or as a speech classifier for automatic speech recognition. Either method can be used, speech segments being represented by linear prediction coefficients, or more conveniently the reflection coefficients of a lattice filter, or by cepstral coefficients.

2.6 Analysing speech at the utterance level

At the utterance level ideas and information are represented by sentences constructed to conform to a *language*. In a formal sense a language is the collection of all of the possible sentences. Since this is likely to be an enormous collection for real languages, we need a concise mechanism for determining whether a sentence is part of the language or not. A formal mechanism for doing this is a *grammar*, which provides a way of defining all of the sentences that are accepted as belonging to the language. A hierarchy of grammars in various levels of complication has been described by Chomsky (1957), and are named type 3, 2, 1 and 0.

Type 3 is a finite-state grammar, which defines a regular language. This is overrestrictive as a method of analysing natural language, as it uses a finite number of states to summarize the history of symbols (words) that have been encountered. The finite number of states are sufficient only to represent one structure for a sentence, which does not allow one structure to be embedded within another structure. This, the reader will be aware, is a common occurrence in natural language.

Type 2 is a context-free grammar, with rules that apply regardless of the context in which words or phrases may occur. Rules may be applied recursively, which allows the embedding of one structure within another, to any depth required. Computation with grammars of this type is quite achievable with stack organized data structures. The toy grammar described in the next section is of this type.

Type 1 is a context-sensitive grammar, which is rather like a context-free grammar except that some of the rules are only applicable within a defined context. The practical problems associated with type 1 languages relate to the difficulty of assembling large context-dependent sets of rules, but as a theoretical model it is used extensively in the science of linguistics.

A type 0 grammar is a recursively enumerable set, which is a very general description of the enormous number of ways that the finite number of words in a language can be combined into a valid string, but again at present it appears to be a theoretical rather than a practical concept.

2.6.1 Context-free grammar

A context-free grammar can be described (after De Roeck, 1983) rather mathematically as a quadruple $\langle V_n, V_t, P, S \rangle$ in which:

- V_n is a finite set of non-terminal symbols, usually parts of speech (e.g. noun, adjective, etc.) or larger constructs (e.g. noun phrase, verb phrase, sentence).
- V_t is a finite set of terminal symbols, e.g. the dictionary of available words.
- P is a finite set of rules called productions, which have the general form $\alpha \rightarrow \beta$ (rewrite α as β), where both α and β represent a string of elements of V_n or V_t.
- S is the starting symbol. S must be a member of V_n and must occur at least once on the left side of the rewrite symbol in the set of productions. In the context of natural languages S is usually the sentence.

A grammar G is a very compact way of defining a language $L(G)$ to be the set of all possible strings of terminal symbols. Now if we consider the task of constructing a grammar that defines the English language, it is unclear whether it is possible or not. However, we can examine some of the characteristics of grammar by way of an example context-free grammar, which generates the sentence:

> People who like this sort of thing will find this the sort of thing they like.[†]

The toy grammar, shown in Fig. 2.20, describes the set of productions in two parts: the lexicon rules, which rewrite the terminal symbols as non-terminal symbols, and the syntax rules, which rewrite

---

[†]The sentence was written well over a hundred years ago by Abraham Lincoln about a book *Collections and Recollections*, by G. W. E. Russell. Cited in the *Oxford Dictionary of Quotations*, Oxford University Press, Oxford, 1979.

ANALYSIS

Lexicon rules

V_t terminal symbols		V_n non-terminal symbols
find	→	NOUN|VERB
like	→	NOUN|VERB
of	→	PREP
people	→	NOUN|VERB
sort	→	NOUN|VERB
the	→	ARTIC
they	→	PRON
thing	→	NOUN
this	→	PRON|ADJEC
who	→	PRON
will	→	NOUN|VERB|MODAL

Syntax rules

V_n non-terminal symbols

1 PREP+NOUN	→	PREP PHRASE
2 NOUN+PREP PHRASE	→	NOUN PHRASE
3 ADJEC+NOUN PHRASE	→	NOUN PHRASE
4 ARTIC+NOUN PHRASE	→	NOUN PHRASE
5 NOUN	→	NOUN PHRASE
6 MODAL+VERB	→	VERB
7 VERB+NOUN PHRASE	→	VERB PHRASE
8 VERB	→	VERB PHRASE
9 PRON+VERB PHRASE	→	QUAL PHRASE
10 PRON	→	NOUN PHRASE
11 NOUN PHRASE+QUAL PHRASE	→	NOUN PHRASE
12 NOUN PHRASE+NOUN PHRASE	→	NOUN PHRASE
13 NOUN PHRASE+VERB PHRASE	→	SENTENCE

Figure 2.20 A toy grammar that generates the example sentence

concatenations of non-terminal symbols as non-terminal symbols. Dealing firstly with the capacity of the toy grammar to generate sentences in a language, it is clear that as well as generating the example sentence it is capable of generating many other sentences. A few examples follow:

The people sort like this.

Who will like the people they find?

They like the will.

These sentences obey the rules of the grammar, so they are well formed, and they are also meaningful. Meaningful sentences are not,

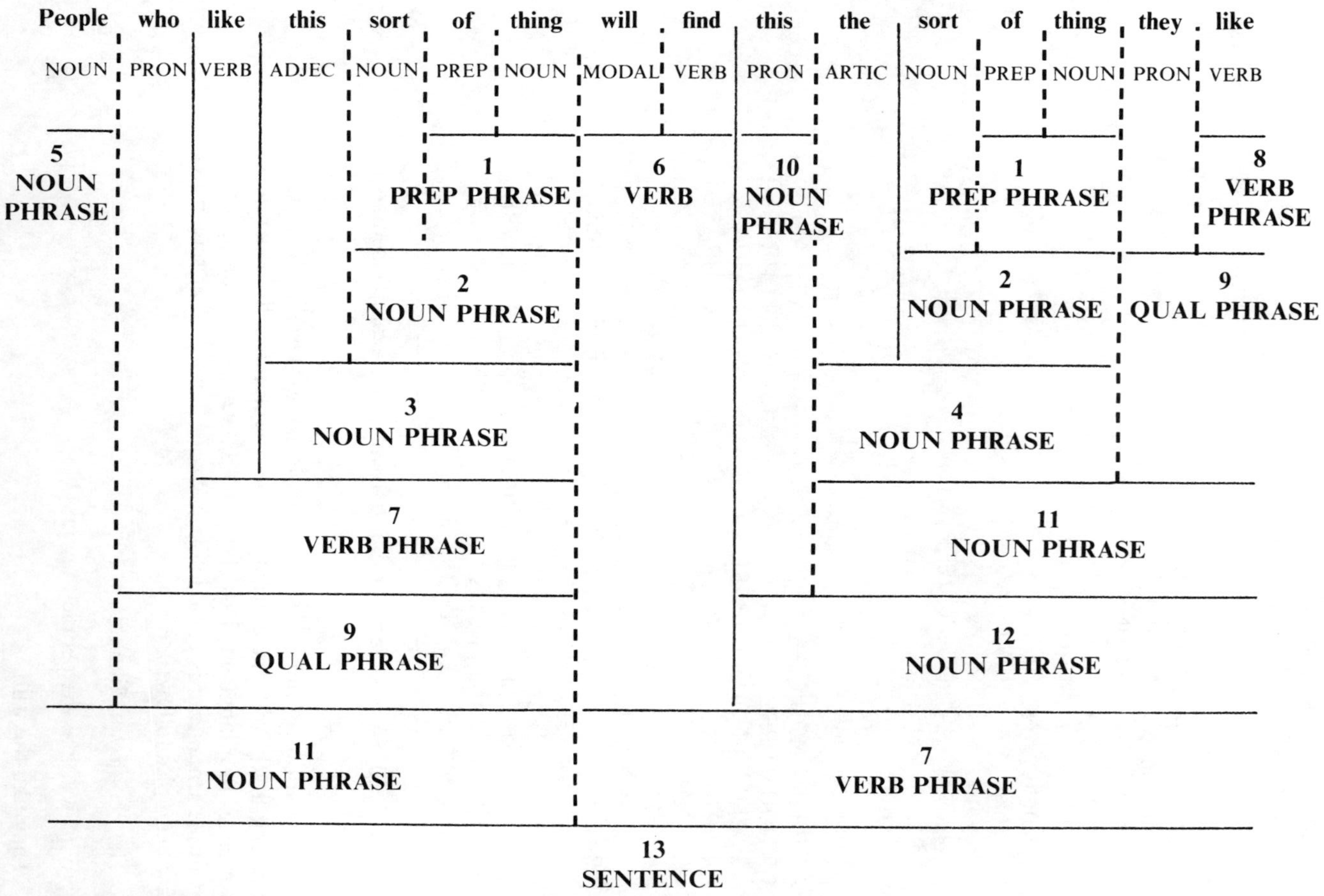

Figure 2.21 Analysis of the example sentence using the toy grammar

however, guaranteed by the grammar, which defines only a surface structure; the meaning of sentences is determined by a deep structure. The further examples:

This like will sort find.

People the will of the like.

obey the syntax, and in some sense appear to be 'correct' sentences, but are meaningless.

In the context of this chapter it is the use of the grammar to analyse sentences that should receive greatest attention. The detail of an analysis of the example sentence is shown in Fig. 2.21. Each operation of a rule is shown by bracketing the non-terminal symbols involved, and the index number of the rule is shown in a circle. Detailed study of Fig. 2.21 will show that the order in which the rules are applied is important. For instance if rule 10 was used to rewrite (who: PRON) to (who: NOUN PHRASE) and then rule 12 used to rewrite (people NOUN PHRASE)+(who: NOUN PHRASE) to (people who: NOUN PHRASE), we would find that rule 13 could rewrite (people who: NOUN PHRASE)+(like this sort of thing: VERB PHRASE) to (people who like this sort of thing: SENTENCE), and the grammar then offers no way of incorporating the rest of the example sentence into the syntactical structure. If a parsing program ever reaches this state it is said to have halted, and this is one of the greatest problems that arise in parsing natural language. Clearly the parser must be flexible enough to investigate alternative ways of applying the rules. A data structure that allows for these alternatives to be represented and compared is a *chart* (Varile, 1983).

Another important problem can be seen in the entries for *find*, *like*, *people* and *sort*, which indicate that they could be classified as either a noun or a verb. This is very common in the English language and gives rise to one of the principal values that a parser can perform, namely to disambiguate such words. The rule:

find → NOUN|VERB

means that there are two possible rules:

find → NOUN
find → VERB

Which is the right one to use can only be determined by trying each in the sentence context. As such examples of alternative rules are not rare, over half of the words in the lexicon exhibit this ambiguity, it is clear that even in a moderate length sentence, the number of possible rule applications that must be tried in the search for the correct one multiplies to a very large number.

Why do we need to analyse sentence structures with a parser? In speech synthesized from text, the rhythm and intonation of the word generated from information about the type of sentence and the phrase boundaries within the sentence, the details appear as in Chapter 6. Such information is not explicit in written text, and so an important application of natural language parsing is to determine sentence type and phrase boundaries. In this particular application it is sufficient to use the parser only up to phrase level; the last stage of grouping phrases to form a sentence is not necessary. As this is the stage at which the halting problem usually appears it means that parsers that are a little less than perfect can still be usable.

In the field of speech processing a grammar may be used by a parser to check that a sequence of words received from a speech recognizer conforms to the subset of natural language required by the particular application. With the present state of speech recognition, a regular language with a finite state grammar can be sufficient to deal with finite vocabulary recognizers if recursive structures are not expected.

2.6.2 Summary of utterance level analysis

The analysis of speech in lexical or textual form is less well established than other forms of analysis described in this chapter. The development of parsers and grammars has been largely in the context of developing an understanding of natural language, or in the context of defining and using computer programming languages. The latter field has given a stimulus to the development of computer-based parsers, which perhaps would not have been so extensive if the interest was solely in natural language. In practical applications it is not possible to make any assumptions about the strings of words that are presented to a parser beyond the fact that if the word can be found in a dictionary then the set of possible parts of speech can be identified. The parser then has the double task of disambiguating the part of speech, for most words can serve as more than one, and building the syntactical structure necessary for the particular application. The two common application areas at present are the establishment of a framework for the generation of appropriate rhythm and intonation for synthesized speech, and as a stage of verification in automatic speech recognition, perhaps offering a predictive capability to large vocabulary systems.

References

Atal, B. S. and S. L. Hanauer (1971) 'Speech analysis and synthesis by linear prediction of the speech wave', *Journal of the Acoustical Society of America*, vol. 50, 637–655.
Blackman, R. B. and J. W. Tukey (1958) 'The measurement of power spectra from the point of view of communications engineering', *Bell System Technical Journal*, vol. 37, no. 1, 185–282 and no. 2, 485–569; reprinted with corrections (1959), Dover, New York.

Candy, J. V. (1988) *Signal Processing—The Modern Approach*, McGraw-Hill, London.
Chomsky, N. (1957) *Syntactic Structures*, Mouton, The Hague.
Davenport, W. B. (1952) 'An experimental study of speech-wave probability distributions', *Journal of the Acoustical Society of America*, vol. 24, 390–399.
De Roeck, A. (1983) 'An underview of parsing' in *Parsing Natural Language*, M. King (ed.), Academic Press, London.
IEEE (1979) *Programs for Digital Signal Processing*, edited by Digital Signal Processing Committee of the IEEE Acoustics, Speech, and Signal Processing Society, IEEE Press, New York.
Moore, R. K. (1986) 'Computational Techniques', in *Electronic Speech Recognition*, G. Bristow (ed.), Collins, London.
Noll, A. M. (1967) 'Cepstrum pitch determination', *Journal of the Acoustical Society of America*, vol. 41, 293–309.
Paez, M. D. and T. H. Glisson (1972) 'Minimum mean squared error quantization in speech,' *IEEE Transactions on Communications*, vol. COM-20, 225–230.
Potter, R. K., G. A. Kopp and H. C. Green (1947) *Visible Speech*, Van Nostrand, New York.
Varile, G. B. (1983) 'Charts: a data structure for parsing' in *Parsing Natural Languages*, M. King (ed.), Academic Press, London.
Witten, I. H. (1986) *Making Computers Talk: An Introduction to Speech Synthesis*, Prentice-Hall, Englewood Cliffs, N.J.

3 Waveform coding for digital telephony

EDWIN JONES

3.1 Current practice

The conversion of telephone quality speech into digital form for transmission in a pulse code modulation (PCM) format is a well-established practice. However, the benefits of digital transmission and processing can only be realized if cost-effective (and so reasonably simple) analogue-to-digital coding processes are used that are compatible with an acceptable end-to-end subjective quality. So far, telephone system administrations have favoured a regular sampling of the speech waveform and then direct conversion into digital form, that is *waveform coding*, as the most cost-effective solution.

For many years, coders have been used that digitize each sample separately. However, more recently with the advent of large-scale integration, more complex arrangements have become economically feasible in which the coding process simultaneously takes into account more than one sample of the speech waveform. Within this category, adaptive differential pulse code modulation (ADPCM) is an attractive waveform coding technique because it offers a good compromise between coding complexity and transmission bandwidth efficiency. The economic balance in modern telephone systems is further changing as the analogue-to-digital (A/D) conversion process moves nearer to the speech source. Thus, opportunities for sharing the relatively expensive encoding equipment between several talkers are being removed. As digital transmission permeates the whole network, an encoding equipment will be needed in each telephone terminal. When this occurs, bandwidth efficient cost-effective A/D conversion will become particularly crucial.

A functional diagram for a PCM system employing waveform coding

WAVEFORM CODING FOR DIGITAL TELEPHONY

is shown in Fig. 3.1. Also shown are representative signals at the interface points within the waveform coding system. The *analogue input* signal is first passed through a *band-limiting filter* which ensures that the frequency content is limited to a known range. This enables the *sampler*, sampling at clock frequency f_s in accordance with the sampling theorem, to produce a satisfactory pulse amplitude modulated version of the analogue signal. A *sample stretcher* then produces flat-topped versions of the amplitude modulated samples ready for the A/D conversion process. In practice the sampling and stretching functions may be realized in a single *sample and hold* circuit. The *A/D converter* or *quantizer* will then convert the amplitude modulated samples to digital form, usually expressing the result as an n-bit binary word. The signal is then ready for any further processing (not shown in Fig. 3.1) which may be required in connection with its passage over the digital transmission channel. At the receiving terminal, a *digital-to-analogue* (D/A) converter will transform the n-bit binary words back into a pulse amplitude modulated format. A low-pass *reconstruction filter* then removes unwanted high-frequency sampling sidebands to produce an *analogue output* signal which is an approximation to the original analogue input signal.

The word 'approximation' in the last sentence is rather significant. The PCM system has taken an analogue input signal which is continuously variable in both time and amplitude. The sampling process has produced a discrete time representation of this waveform while the quantizing process has converted the continuous range of waveform amplitudes into a finite set of digital codewords. We see later in this chapter that, at least in theory, regular time sampling of a continuous waveform can be arranged to avoid any information loss. However, amplitude quantizing is an approximating process; each analogue sample is labelled with the nearest quantized value. This approximation can be made arbitrarily small by using a large number of codewords; an n-bit binary codeword will permit 2^n quantized values. In practice, a compromise must be sought between the subjective effects of quantizing distortions and codeword length with its resulting transmission bandwidth requirements. Thus, PCM offers a trade-off between the advantages of digital transmission (principally cost-effective processing and reliable transmission) and the terminal distortion associated with the quantizing process, a trade-off that increasingly leans in favour of PCM.

Waveform coding of telephone-quality speech is a commercially important topic. It is not surprising, therefore, that many books and technical articles deal with this subject. For detailed treatment beyond the scope of this introductory chapter the reader is referred to books such as: Cattermole (1969), Schwartz (1980), Owen (1982) and Jayant and Noll (1984), and also, more recently, the review article by Jayant (1990) and its references. Detailed specifications of international

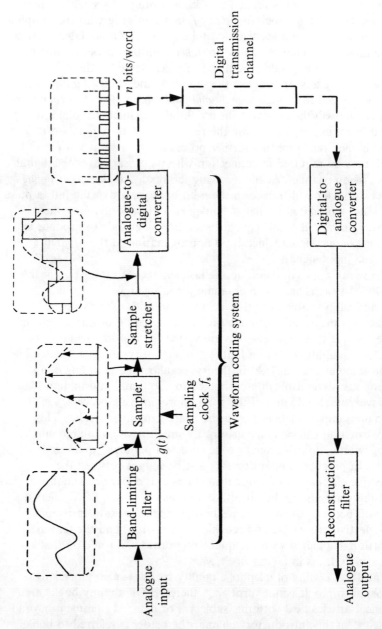

Figure 3.1 Functional diagram for a PCM system

standards are to be found in the 'Recommendations' of the International Telegraph and Telephone Consultative Committee (CCITT, 1988).

3.2 Design objectives

The provision of an acceptable *telephone-quality* speech communication system requires the careful selection of a number of design parameters. Here a balance must be struck between the need to provide a satisfactory true-to-life speech channel and the economic considerations of transmission bandwidth efficiency and terminal equipment complexity. Much research has been put into these considerations, initially for analogue and later for digital telephone systems; see, for example, Richards (1973 and 1968). Here we consider the outcome.

3.2.1 Speech bandwidth

The long-established *analogue telephone bandwidth* of 300 to 3400 Hz was also adopted, in the late sixties, for PCM telephone systems. The main contribution to the intelligibility of speech resides in spectral components around 1 kHz. The telephone bandwidth provides for good intelligibility and also, by including some higher frequency components, it gives a fair rendering of the characteristics of a speaker's voice. Furthermore, it is a bandwidth constraint that has become familiar to telephone users, and so perhaps we remain willing to accept it.

For applications such as audio-visual conferencing, however, where one would like to approach the quality of face-to-face spoken communication, a 50 to 7000 Hz bandwidth is regarded as more appropriate. Known as *wideband audio*, this relatively new standard is described in Recommendation G.722 (CCITT, 1988) and in Mermelstein (1988). Considerably enhanced quality is achieved, but of course at the expense of extra transmission bandwidth and extra terminal processing.

3.2.2 Dynamic range

The variation in amplitudes, or dynamic range, of a speech waveform can be quite large. For realistic speech a good proportion of this range must be reproduced. In addition, allowance must be made for the variation in loss which can be incurred in the path between the speaker and the analogue-to-digital encoding equipment. This can arise through the variability of microphone sensitivities and through losses in the local transmission path. In practice, we must cater for a dynamic range of at least 30 dB.

3.2.3 Signal-to-distortion ratio

To ensure satisfactory user opinion, a digital telephony system also needs to yield a signal-to-distortion ratio of some 30 dB or better. Furthermore, this must be maintained over the range of input signal levels mentioned above; thus the permitted level of distortion will be a function of the signal level. That is to say, the human ear can accept some distortion provided that it is always *masked* to an adequate extent by the wanted speech signal.

Now in a PCM system, where the bit error ratio arising from transmission is typically 10^{-5} or better, this will contribute very little to the overall distortion. Thus, virtually all the distortion is quantizing distortion; that is it arises from the approximation involved in the analogue-to-digital process. CCITT specify an acceptance bound as shown in Fig. 3.2; all PCM systems in a network are required to conform to this specification. Thus, over an adequate range of input signal levels a minimum performance is assured. We consider this characteristic in more detail later. For now we simply note that over part of the input signal range the bound requires a performance in excess of 30 dB. This is to cater for tandem connections; that is an allowance must be made for the possibility that a number of PCM systems (each one analogue–digital–analogue) will be connected in succession.

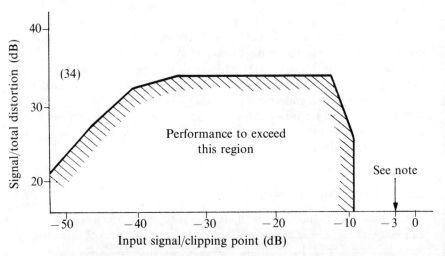

(*Note*: a sine wave of this level would just avoid peak clipping. This point corresponds to the +3.1 dBm0 power level of CCITT Recommendation G.712.)

Figure 3.2 CCITT performance bound for digital telephony (for random noise input)

WAVEFORM CODING FOR DIGITAL TELEPHONY   79

3.2.4 Subjective assessment

The user's opinion of the quality of a speech communication system is a crucial parameter which a designer must take into account. For this purpose a subjective testing scheme has been established; see, for example, Kitawaki and Nagabuchi (1988) and their references. Listeners conduct a series of tests over the telephone system and award a score from 0 to 4 (or 1 to 5 in the case of some authors) in accordance with the verbal descriptions of quality and impairment given in Table 3.1.

Table 3.1 A subjective measure for quality and impairment

Score	Quality	Impairment
4	Excellent	Imperceptible
3	Good	Just perceptible but not annoying
2	Fair	Slightly annoying
1	Poor	Annoying but not too objectionable
0	Bad	Very annoying

To allow for the considerable variability between user opinions, results are combined to yield a *mean opinion score* for the system under test. A high-quality digital telephone system which is indistinguishable

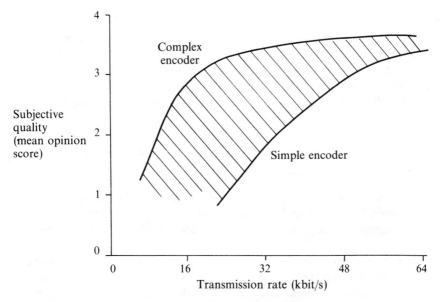

Figure 3.3 Range of performance for waveform coding

from band-limited speech (300 to 3400 Hz) will give a mean opinion score of about 3.5. In general, digitized systems yielding a score in excess of 3.0 can be regarded as worthy of consideration for commercial telephony. Subjective performance figures achieved in practice will depend upon the sophistication of the analogue-to-digital coding equipment. Figure 3.3 shows what is currently achievable as a function of transmission rate.

It should be noted that while this subjective scoring scheme gives a useful measure for commercial quality telephony it may not be suitable for other systems. For quality audio, for example, the score region 3 to 4 will be too coarse. It also has limitations when it comes to distinguishing between poor quality systems, where intelligibility tests may also be required.

3.3 Waveform sampling

3.3.1 Theorem

The sampling theorem states that *a waveform containing frequencies no higher than f_m can be completely recovered from amplitude samples taken at rate f_s, provided that f_s is greater or equal to $2f_m$*. This important fundamental theorem is usually associated with Nyquist and Shannon. Its various extensions and applications are reviewed in a tutorial paper by Jerri (1977). In its basic form as stated above it can be demonstrated as follows.

Referring to Fig. 3.1, let $g(t)$ be the band-limited analogue input waveform to the sampler, where f_m is the highest frequency present in its spectrum $G(f)$, as seen in Fig. 3.4(a). Let the sampling signal be a sequence of impulses of repetition rate f_s (Fig. 3.4(b)); this can be written in Woodward's notation as $\text{rep}\,\delta(t)$. Sampling is a multiplication process; thus the output waveform of the sampler is

$$g(t)\,\text{rep}_{1/f_s}\delta(t) \tag{3.1}$$

and its Fourier transform gives the corresponding frequency spectrum

$$\rightarrow G(f)\,{}^*f_s\,\text{rep}_{f_s}\delta(f) = f_s\,\text{rep}_{f_s}G(f) \tag{3.2}$$

as shown in Fig. 3.4(c). Thus, by comparing the frequency domain representations of Fig. 3.4(a) and (c) we see that the original signal of (a) (shown shaded) can be completely recovered from the sampled output of (c) if $f_s \geq 2f_m$.

The recovery process will take place at the distant terminal where the *reconstruction filter* of Fig. 3.1 must transmit the wanted signal and stop the unwanted sidebands centred on the sampling frequency f_s and on multiples of it.

WAVEFORM CODING FOR DIGITAL TELEPHONY

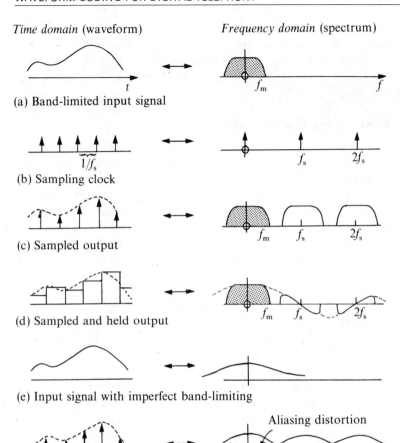

Figure 3.4 Sampling—waveforms and spectra (amplitude scaling factors not shown)

3.3.2 Sample and hold

The system shown in Fig. 3.1 incorporates a sample stretcher or hold circuit which provides flat topped pulses for the A/D converter. The output of this circuit can be found by convolving its input waveform with a single pulse of width $1/f_s$ (denoted $\text{rect} f_s t$); this leads to the Fourier transform pair

$$g(t) \text{rep}_{1/f_s} \delta(t) * \text{rect} f_s t \leftrightarrow \text{rep}_{f_s} G(f) \operatorname{sinc} \frac{f}{f_s} \qquad (3.3)$$

(defining the sinc function as $\operatorname{sinc} x = \sin \pi x / \pi x$). This is illustrated in Fig. 3.4(d) where it is seen that the output spectrum has simply been multiplied by a sinc function. Since no frequency scaling is involved the requirements of the sampling theorem remain unchanged. Strictly, there will be a small spectral loss in the wanted signal band which increases as the frequency approaches f_m. For speech systems this loss has a negligible effect and can be ignored.

3.3.3 Aliasing distortion

Practical filters have a finite rate of cut-off. Thus, in reality, the input filter in Fig. 3.1 will be unable to guarantee perfect band-limiting to f_m. Hence, the input signal spectrum of Fig. 3.4(a) will, in practice, look more like Fig. 3.4(e). This will lead to the sampled output spectrum of Fig. 3.4(f); that is, some of the unwanted sideband spectral components will interfere with the wanted signal spectrum. This overlapping of spectral components is known as *aliasing distortion*; once present it cannot easily be removed. A compromise is therefore necessary between the sampling rate and the band-limiting filter cut-off rate to ensure that aliasing distortion is kept below an acceptable level.

For the conventional 300 to 3400 Hz bandwidth telephone signal a sampling frequency of 8 kHz has been adopted as a realistic compromise requiring a modest transmission bandwidth and yet reasonable filter performance. Details are given in Recommendation G.712 (CCITT, 1988) which specifies a wanted signal-to-aliasing distortion ratio of at least 25 dB in the band up to 3400 Hz. For the 50 to 7000 Hz wideband telephone standard, referred to earlier, a sampling rate of 16 kHz has been agreed.

3.4 Waveform quantizing

Amplitude quantization of the continuously variable time samples takes place in the analogue-to-digital converter of Fig. 3.1. The range of input signals which can be coded (the *working range* of Fig. 3.5) is considered to be divided into N *quantizing intervals*. All sample amplitudes within a given quantizing interval (δ_r) are represented by the single *quantized value* (x_r). The discrepancy between the true and quantized samples is termed the *quantizing error* or *quantizing distortion*. This is the major impairment in a PCM system and is controlled by selecting an appropriate value for N. In addition, the quantizing intervals may be uniformly or non-uniformly distributed over the working range depending upon the application. Signals that fall outside this working range exceed the *clipping point* and so cannot be encoded correctly; this gives rise to *clipping distortion*.

3.4.1 Quantizing distortion

Figure 3.5 defines our terms. In addition, let the input signal to the quantizer have a probability density function $p(x)$ and assume that δ_r is

WAVEFORM CODING FOR DIGITAL TELEPHONY

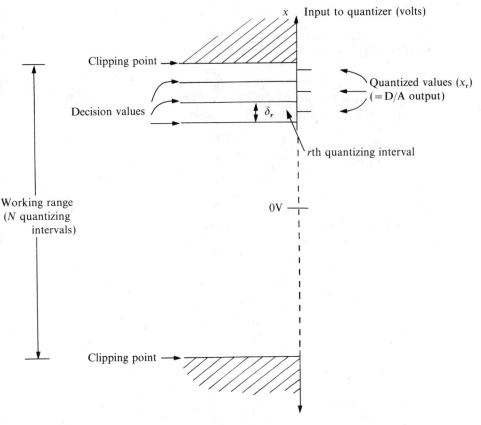

Figure 3.5 Amplitude quantizing—definition of terms

small so that $p(x) = p_r$; that is it is constant over the rth quantizing interval.

Consider an input signal in the range

$$x_r - \frac{\delta_r}{2} \leq x \leq x_r + \frac{\delta_r}{2}$$

Its instantaneous squared error is $(x - x_r)^2$. Thus, the mean square error for signals in the rth quantizing interval:

$$E_r = \frac{1}{\delta_r} \int_{x_r - \delta_r/2}^{x_r + \delta_r/2} (x - x_r)^2 p_r \, dx = \frac{1}{12} p_r \delta_r^2 \quad (3.4)$$

Therefore, the total mean square quantizing error over the whole coding range is

$$E_q = \sum_r E_r = \frac{1}{12} \sum_r p_r \delta_r^2 \quad (3.5)$$

Using this general result, we consider two cases of practical interest.

Uniform quantizing

Here the quantizing intervals are uniformly distributed over the working range so that for each interval

$$\delta_r = \delta$$

We assume that the probability of input signals falling outside the working range is negligibly small such that

$$\sum p_r = 1$$

Thus, from Eq. (3.5),

$$E_q = \frac{\delta^2}{12} \tag{3.6}$$

that is the quantizing error is independent of the input signal level.

Now, let S be the mean square input signal level; the signal-to-quantizing distortion (power) ratio is then

$$\frac{S}{E_q} = S \frac{12}{\delta^2} \tag{3.7}$$

Any signal of amplitude exceeding the working range will suffer peak clipping. We define a *mean square clipping level* S_p, such that

$$\text{the zero-to-peak working range} = \sqrt{S_p} = \frac{N\delta}{2} \tag{3.8}$$

Substituting this in Eq. (3.7) and noting that it is usual to encode signal samples into n-bit binary words such that $N = 2^n$ gives

$$\frac{S}{E_q} = \frac{S}{S_p} \times 3.2^{2n}$$

or in decibels

$$\left[\frac{S}{E_q}\right]_{dB} = \left[\frac{S}{S_p}\right]_{dB} + 4.8 + 6n \tag{3.9}$$

This is plotted for a range of values of n in Fig. 3.6. The $-3\,\text{dB}$ input signal level corresponds to a sine wave which just fills the working range. Input signals such as speech or band-limited noise (for test purposes) will have a larger peak-to-mean ratio. Thus, as their input level is increased they will exhibit clipping before the $-3\,\text{dB}$ sine wave clipping point. The consequences of this are discussed later.

Logarithmic quantizing

We see from the above that with uniform quantizing, the resulting quantizing distortion is constant so that the signal-to-distortion ratio is directly *proportional* to the input signal level. As has already been

WAVEFORM CODING FOR DIGITAL TELEPHONY

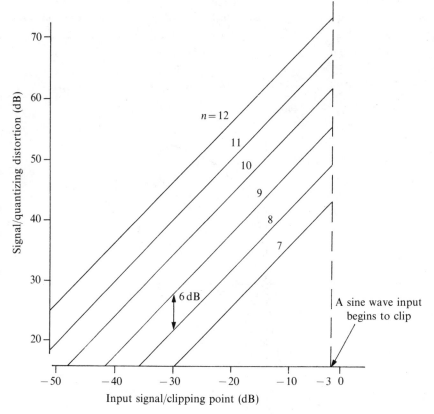

Figure 3.6 Performance of a uniform quantizer

mentioned, to take advantage of the distortion masking ability of the human ear, we would like to have a *constant* signal-to-distortion ratio over a range of input signal levels. To achieve this, we need to change the size of the quantizing intervals δ_r (and so the quantizing distortion) in sympathy with the input voltage x; that is small input voltages must be finely quantized while large input voltages can be coarsely quantized.

This can be achieved by using either a non-uniform A/D converter or by inserting a *compressor* before a uniform A/D converter as shown in Fig. 3.7(a). In both cases the complementary non-uniform operation must be performed at the receive terminal in order to provide the requisite overall linear transfer characteristic. (In PCM systems the term *companding* is often used to describe this compression and subsequent expansion of a signal.) For the purposes of analysis, we now consider the model given by Fig. 3.7(a).

In accordance with the above requirements for a constant signal-to-distortion ratio, δ_r needs to increase with x. This is achieved with a compressor of the form shown in Fig. 3.7(b), where it is seen that an

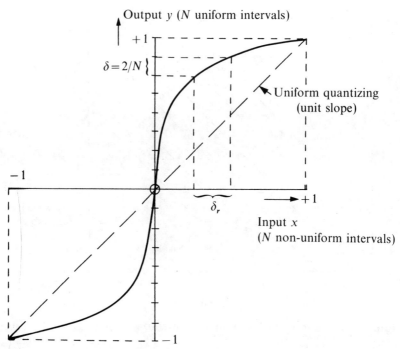

(a) System configuration

(b) Required compressor transfer characteristic

Figure 3.7 Logarithmic quantizing

input quantizing interval δ_r is transformed into a quantizing interval δ before being presented to the A/D converter. The effect of such a precoding amplitude compressor can be analysed as follows.

Assume that the number of quantizing intervals (N) is large and let the compressor have a normalized input/output working range of $x = y = \pm 1$. Thus

$$\frac{dy}{dx} \approx \frac{\delta}{\delta_r} = \frac{2}{N\delta_r} \tag{3.10}$$

that is

$$\delta_r = \frac{2}{N}\left(\frac{dx}{dy}\right)_{x=x_r} \tag{3.11}$$

Substituting this in Eq. (3.5) gives

$$E_q = \frac{1}{3N^2} \sum_r P_r \left(\frac{dx}{dy}\right)^2_{x=x_r}$$

and again, for fine quantization, this can be approximated by the integral

$$E_q = \frac{1}{3N^2} \int p(x) \left(\frac{dx}{dy}\right)^2 dx \qquad (3.12)$$

Now, the mean square signal power

$$S = \int p(x) x^2 \, dx \qquad (3.13)$$

Thus, from Eqs (3.12) and (3.13) the signal-to-quantizing distortion ratio

$$\frac{S}{E_q} = \frac{3N^2 \int p(x) x^2 \, dx}{\int p(x)(dx/dy)^2 \, dx} \qquad (3.14)$$

This can be made independent of x as required if the slope of Fig. 3.7(b) is made proportional to $1/x$, that is for the positive quadrant, when

$$\frac{dy}{dx} = \frac{1}{kx} \qquad (3.15)$$

is substituted in Eq. (3.14); in other words,

$$\frac{S}{E_q} = \frac{3N^2}{k^2} = \text{constant} \qquad (3.16)$$

The companding law required to achieve this is found by integrating Eq. (3.15) and applying the boundary condition $x=y=1$, which gives

$$y = 1 + \frac{1}{k} \ln x \qquad (3.17)$$

Similarly, consideration of the negative quadrant of Fig. 3.7(b) leads to the 'mirror image' requirement.

Equation (3.17) cannot be realized in practice because when $x=0$, a companding law with infinite slope is implied, that is with infinitely small quantizing intervals. Two solutions which yield realizable approximations, known as the *A-law* and the *μ-law*, have been accepted for the purposes of international standardization; these are defined in Recommendation G.711 (CCITT, 1988).

The A-law specifies a constant finite slope and so uniform quantizing in the vicinity of $x=0$. This then merges into the logarithmic law requirement of Fig. 3.7(b) at a point $x=1/A$; that is for the positive quadrant:

$$y = \frac{Ax}{1+\ln A} \qquad 0 \leq x \leq 1/A \qquad (3.18)$$

$$= \frac{1+\ln Ax}{1+\ln A} \qquad 1/A \leq x \leq 1 \qquad (3.19)$$

where $A = 87.56$.

The μ-law obtains a finite slope in the region of $x=0$ by specifying a quasi-logarithmic characteristic of

$$y = \frac{\ln(1+\mu x)}{\ln(1+\mu)} \qquad 0 \leq x \leq 1 \qquad (3.20)$$

where $\mu = 255$ (note that ln means $\log_e$).

The performance of the two is very similar. The reader is again referred to Cattermole (1969) and Jayant and Noll (1984) for comprehensive analyses and comparisons between the two approaches. By way of example, we consider the A-law in a little more detail.

Equation (3.18) corresponds to the small signal uniform portion of the compressor characteristic. On differentiation,

$$\frac{dy}{dx} = \frac{A}{1+\ln A} \qquad 0 \leq x \leq 1/A$$

with $A = 87.56 \qquad = 16$

that is this portion has a slope of 16 times that of the characteristic which would give uniform quantizing intervals over the whole signal range (the unit slope line on Fig. 3.7(b)). This, if we specify 8 bits per speech sample, we can have 256 quantizing intervals on the y axis. With a slope of 16 this means that we will achieve the equivalent of 12-bit coding for signal amplitudes $x \leq 1/A$.

For the logarithmic law portion given by Eq. (3.19), differentiation gives

$$\frac{dy}{dx} = \frac{1}{(1+\ln A)x} \qquad 1/A \leq x \leq 1$$

Relating this to Eq. (3.15) gives $k = 1 + \ln A$ and on substituting in Eq. (3.16):

$$\frac{S}{E_q} = \frac{3N^2}{(1+\ln A)^2} \qquad (3.21)$$

For $N = 2^8$ and $A = 87.56$,

$$\left[\frac{S}{E_q}\right]_{dB} = 38.2 \, dB$$

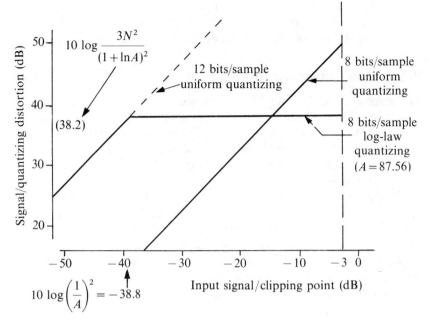

Figure 3.8 Comparison of uniform and logarithmic law quantizing

This constant signal-to-quantizing distortion ratio will apply for signal amplitudes $x \geq 1/A$, as shown by the horizontal portion of the log-law quantizing characteristic in Fig. 3.8. This figure compares the performance of an A-law quantizer with that of a uniform quantizer. We see from the two solid lines that a trade-off has been effected. Both characteristics require the transmission of an 8-bit sample word and so occupy the same transmission bandwidth. However, the logarithmic law exhibits the desirable constant signal-to-distortion ratio over a range of input levels. In the lower portion of this range a better distortion ratio is obtained relative to uniform quantizing, but this is achieved at the expense of an apparently poorer performance for higher input signal amplitudes.

In practice the compressor at the input to the A/D converter and the complementary expander of the D/A output (Fig. 3.7(a)) have to be matched very precisely. This is not practicable with continuously variable log-law characteristics. The practical solution adopted for both A-law and μ-law companded systems is to use a *piecewise linear* approximation to the transfer characteristic of Fig. 3.7(b). For illustrative purposes the vertical axis has been divided into 16 uniform intervals. In practice the logarithmic curve within each of these intervals is replaced by a straight line segment, with each segment, on the vertical axis, also being divided into 16 uniformly distributed quantizing intervals; that is a total of 256 quantizing intervals requiring an 8-bit transmitted word, as already mentioned. This segmented approximation to the required logarithmic law causes a ripple of up to 2 dB deviation from the flat portion of the plot of Fig. 3.8.

3.4.2 Peak clipping distortion

The graphs of Figs 3.6 and 3.8 plot quantizing distortion for input signals up to −3 dB relative to the clipping point. Thus, strictly, they only apply to sine waves, for it is at this point that a sine wave would just fill the working range. Any attempt to increase the amplitude of the sine wave beyond this point would attract the rapid onset of *peak clipping distortion*.

It has already been noted that speech and random noise test signals exhibit a much larger peak-to-mean ratio. Thus, clipping will occur at lower signal levels and become progressively more severe as the input signal level is increased. This can be analysed as follows.

Let the input signal have a symmetrical probability density function $p(x)$ centred on the working range. From Eq. (3.13), the mean square error due to clipping

$$E_c = 2 \int_1^\infty p(x)(x-1)^2 \, dx$$

Now, for a random noise test signal (assuming Gaussian amplitude distribution) of mean square signal power S, we have

$$p(x) = \frac{1}{\sqrt{2\pi S}} e^{-x^2/2S}$$

that is

$$E_c = \frac{2}{\sqrt{2\pi S}} \int_1^\infty e^{-x^2/2S}(x-1)^2 \, dx \qquad (3.22)$$

Figure 3.9 combines this result for peak clipping with the *A*-law quantizing distortion graph of Fig. 3.8 to show the three distinct regions that make up the overall performance characteristic for a logarithmically companded system. For comparison the CCITT performance bound of Fig. 3.2 is also shown. The difference between the two curves allows for quantizer imperfections and for the use of a piecewise linear rather than the theoretically ideal logarithmic law compressor characteristic.

However, we must note that speech signals do not have a Gaussian amplitude distribution; their larger peak-to-mean ratio causes clipping to become apparent at even lower input levels. The precise speech amplitude distribution depends upon microphone details. Subjective tests suggest that, typically, clipping distortion becomes 'just objectionable' at a relative input level of about −9 dB (Richards, 1968).

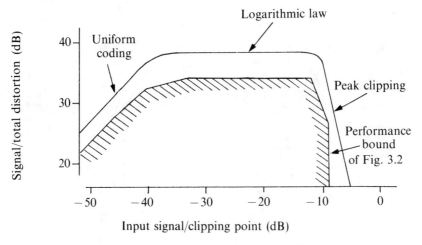

Figure 3.9 Performance of an *A*-law compander (for random noise input)

3.5 Differential coding

So far, this chapter has considered analogue-to-digital encoding methods which produce a direct digitized version of each time-sampled value. Such memoryless arrangements are relatively simple to realize but they do not take advantage of the considerable amount of correlation and thus predictability that can exist between the neighbouring time samples of a speech waveform. The correlation between *adjacent* samples is especially strong and this has led to the development of a range of *differential* coding methods in which the difference between adjacent samples is encoded. The dynamic range and variance of this difference will usually be significantly smaller than that of the original signal. Thus, for a given subjective quality of coding, fewer signal quantizing intervals will often suffice; that is relative to memoryless PCM, a shorter codeword per sample can be transmitted, resulting in a reduced channel bandwidth requirement.

Figure 3.10 gives a block diagram for a differential PCM system. It is seen that compared with the memoryless system of Fig. 3.1, the encoder incorporates a feedback loop. This loop reconstructs the *previous sample* value which is subtracted from the current input to provide the requisite *difference signal*. An A/D converter then produces a *coded difference* for transmission to the distant terminal. This coded difference is also used to reconstruct the previous sample by integrating past differences.

We see that the decoder at the distant terminal is similar to the feedback loop of the encoder. Thus a digital error during transmission will affect the received difference signal and hence successive integrated

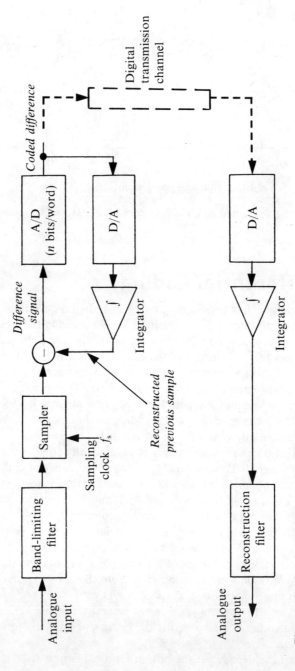

Figure 3.10 A differential PCM system

WAVEFORM CODING FOR DIGITAL TELEPHONY

differences. Leaky integrators, with carefully chosen time constants, are used to minimize the cumulative effect of such errors by allowing them to dissipate as rapidly as possible.

Some compromises must be made when selecting the sampling rate and the quantizing precision of the A/D converter. To minimize transmission bandwidth we may wish to minimize the sampling rate, but this will mean that the difference between adjacent samples may not always be small; this in turn implies a larger number of quantizing intervals in the A/D process. Figure 3.11 illustrates the dilemma (for $n=3$; that is eight quantizing steps giving a maximum coding range of $\pm 4\delta$). Over the *waveform tracking* portion of the figure, the input signal is changing slowly so the difference signal will be small; thus δ should also be small to ensure low quantizing distortion. By contrast, when the signal is changing more rapidly the difference signal will be large and the quantizer may only be able to follow it if δ is large. If the quantizer fails to follow the signal we get *slope overload* distortion. This can be reduced either by raising the sampling rate so that the difference signal is smaller or by specifying more bits from the quantizer; both will lead to an increased bandwidth penalty.

Various bandwidth efficient differential speech coding systems are in use, but they tend to yield rather poor speech quality. This arises from the difficulty in meeting the opposing demands of accurate tracking while still maintaining an adequate protection against excessive slope overload; that is it is difficult to obtain a satisfactory compromise between sampling rate and quantizing accuracy. However, more recently, adaptive differential PCM (ADPCM) systems have been developed that adapt the quantizing process in accordance with the changing needs of the speech waveform. We now consider the use of this technique for coding telephone-quality speech.

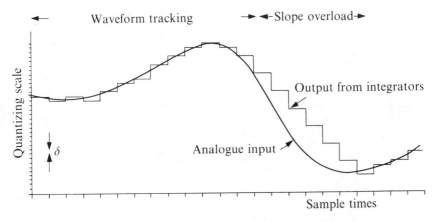

Figure 3.11 Waveforms for differential PCM (for $n=3$)

3.5.1 Adaptive differential pulse code modulation (ADPCM)

A 32 kbit/s ADPCM telephone standard is currently available. This gives a similar subjective performance and yet allows a factor of two saving on the transmission rate relative to the older A-law and μ-law PCM standards described earlier. Here we provide an outline description of the system; further details are to be found in Recommendation G.721 (CCITT, 1988) and, for example, in Papamichalis (1987) which describes how it can be realized using digital signal processing integrated circuits.

As with the conventional PCM standard the waveform sampling rate remains at 8 kHz; the transmission bandwidth saving is achieved by coding to only 4 bits (rather than 8 bits) per sample. A simplified block diagram of the encoder is given in Fig. 3.12. This diagram shows that the system is considerably more sophisticated than the basic differential PCM method of Fig. 3.10, in that the ADPCM standard requires an *adaptive predictor* and an *adaptive quantizer*. The predictor provides a *signal estimate* which is subtracted from the *input* to provide a *difference signal*. The predictor ensures that the dynamic range and variance of this difference is small so that a satisfactory approximation is obtained with only 16 non-uniform quantizing levels in the adaptive quantizer. Thus, a 4-bit word is produced for transmission to the distant decoder. This output signal also stimulates the feedback loops within the encoder, initially to control the adaptive quantizing and then to provide an input for the adaptive predictor.

The *adaptation control* circuit influences the adaptive quantizer in two ways. It provides a quantizing step-size adaptation which is logarithmically based to optimize the signal-to-quantizing distortion

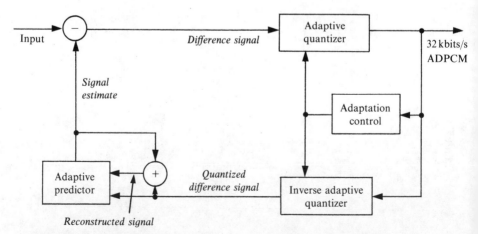

Figure 3.12 A simplified block diagram of an ADPCM encoder

ratio. It also incorporates a speed control mechanism which uses short- and long-term averages to determine how rapidly the input signal is changing and then adjust the rate of adaptation accordingly. Special care has to be taken to ensure that non-voice inputs (such as carrier-based voiceband data signals), which produce difference signals with rather different characteristics, are dealt with appropriately.

An *inverse adaptive quantizer* provides a *quantized difference signal* from the 4-bit transmitted word. The adaptive predictor and its associated local feedback loop provide a *reconstructed signal* which, with the aid of a gradient algorithm, is used to compute a *signal estimate* of the current input value. Thus, the adaptive encoding loop is completed.

This backward adaptive arrangement means that the transmitted signal is the sole stimulus for the various adaptation algorithms. It follows that no adaptation control data has to be sent to the distant decoder. The decoder is thus similar to the feedback loop of the encoder with the reconstructed signal available from one of the inputs to the adaptive predictor, as shown in Fig. 3.12.

Finally, we must note that in practice the input signal to the encoder is already in digital form, it having been obtained from a conventional PCM encoder. This arrangement means that all the complex adaptive real-time processing required within the ADPCM encoding loop can be realized in the digital domain.

3.6 Comparisons and trends

Referring to Fig. 3.3, the performance of the above ADPCM standard corresponds approximately to the 32 kbit/s point on the *complex encoder* curve. The A-law and μ-law PCM standards achieve performance scores that lie between the two bounds at the 64 kbit/s point. Under error-free conditions the perceived subjective quality of speech over a 32-kbit/s ADPCM link is slightly inferior to that over a 64 kbit/s PCM link. However, for transmission bit error ratios higher than about 10^{-4} the ADPCM system is found to be marginally better (Yamamoto and Wright, 1990). Furthermore, for up to four digital speech systems connected in tandem, the ADPCM standard maintains a performance that is only slightly inferior to a single link. This is made possible by a *coding adjustment* within the ADPCM decoding algorithm which reduces the cumulative effect of distortion arising from tandem codings. In contrast, successive connections of PCM systems result in an approximately proportionate increase in quantizing power for each additional system.

Following the *complex encoder* curve of Fig. 3.3 down to 16 kbit/s suggests an achievable mean opinion score of just below 3.0. This is hardly adequate for a telephone-quality speech link. However, there is much pressure to conserve transmission bandwidth, especially from the designers of mobile and satellite telephone links, where transmission bit rates of 16 kbit/s and below are seen as desirable design objectives. It is expected that acceptable performance will only be achieved at such rates at the expense of yet more sophisticated terminal processing.

The technique of *sub-band coding*, which splits the incoming speech band into a number of parts and encodes each separately, is one possibility. This permits more flexibility in the encoding algorithms and also allows some distortion products to be removed at the decoder. (The *wideband audio* standard, G.722, mentioned at the beginning of this chapter uses this technique.) An alternative is to turn from waveform coding to the generally more process and delay intensive parametric, or model-based, coding methods discussed in later chapters. At the time of writing, hybrid methods appear particularly promising in that they can combine the high-quality potential of waveform coding with the transmission bandwidth efficiency of parametric coding techniques. It remains to be seen which speech coding configurations will emerge as capable of providing the requisite balance between subjective performance, transmission bandwidth, resilience to transmission errors and processing complexity.

References

Cattermole, K. W. (1969) *Principles of Pulse Code Modulation*, Iliffe, London.
CCITT (1988) *Blue Books*, Geneva.
Jayant, N. S. (1990) 'High quality coding of telephone speech and wideband audio', *IEEE Communications Magazine*, vol. 28, 10–20, January.
Jayant, N. S. and P. Noll (1984) *Digital Coding of Waveforms*, Prentice-Hall, Englewood Cliffs, N.J.
Jerri, A. B. (1977) 'The Shannon sampling theorem—its various extensions and applications: a tutorial review', *Proceedings IEEE*, vol. 65, 1565–1596, November.
Kitawaki, N. and H. Nagabuchi (1988) 'Quality assessment of speech coding and speech synthesis systems', *IEEE Communications Magazine*, vol. 26, 36–44, October.
Mermelstein, P. (1988) 'G.722, a new CCITT coding standard for digital transmission of wideband audio signals', *IEEE Communications Magazine*, vol. 26, 8–15, January.
Owen, F. E. (1982) *PCM and Digital Transmission Systems*, McGraw-Hill, New York.
Papamichalis, P. E. (1987) *Practical Approaches to Speech Coding*, Prentice-Hall, Englewood Cliffs, N.J.
Richards, D. L. (1973) *Telecommunication by Speech*, Butterworth, London.
Richards, D. L. (1968) 'Transmission performance of telephone networks containing PCM links', *Proceedings IEE*, vol. 115, 1245–1258.
Schwartz, M. (1980) *Information, Transmission, Modulation and Noise*, 3rd edn, McGraw-Hill, New York.
Yamamoto, Y. and T. Wright (1990) 'Error performance in evolving digital networks including ISDNs', *IEEE Communications Magazine*, vol. 27, 12–18, April.

4 Waveform coding— digital audio

MALCOLM HAWKSFORD

4.1 Introduction

A high-performance digital audio system is limited ultimately by the resolution and hardware accuracy of the interfaces between analogue and digital domains. Consequently it is paramount to address the inherent distortions in these processes and in particular to identify conversion techniques that have both the potential to meet the desired resolution and the means to minimize and decorrelate the errors incurred by hardware constraints. The task addressed in this chapter therefore is to investigate the digitization of audio signals with a specific emphasis on noise shaping and oversampling techniques (Adams, 1986; Matsuya, 1987; Tewksbury and Hallock, 1978; Ritchie, Pandy and Ninke, 1974; Hawksford, 1977, 1985, 1987).

The audio industry has at present adopted three basic signal formats that are classified in terms of the sampling rates, 48, 44.1 and 32 kHz (Watkinson, 1988). The highest sampling rate is generally used for master recording, 44.1 kHz is for consumer applications including compact disc (CD), while the lower rate of 32 kHz is used in digital sound for television and broadcast applications. Although these sampling rates are standardized the methods of analogue-to-digital conversion (ADC) and digital-to-analogue conversion (DAC) are not so restricted, where our discussion demonstrates the advantage in choosing higher rates of oversampling at the domain interface.

Rather than just present alternative conversion strategies, our approach attempts a unification of the constituent processes where the emphasis is to seek *more natural conversion topologies*. Of specific relevance is the technique of delta modulation (Deloraine, van Miero and Derjavitch, 1947–1948; Philips, 1949; de Jager, 1952) and the derivative system, delta–sigma modulation (Inose and Yasuda, 1963; Inose, Yasuda and Murakami, 1962). This process can be generalized to encompass oversampling together with high-order noise shaping and can form the basis of a powerful converter topology where systematic distortions are randomized and near-optimum antialiasing and reconstruction filters implemented.

4.2 The case for oversampling and noise shaping

ADCs and DACs have performance bounds determined by both amplitude resolution and maximum conversion rate, where the general trend sees a reduction in amplitude resolution at higher sampling rates due to finite system response and settling times. For example, a direct audio DAC may resolve 16 bits at a 176 kHz sampling rate while a video DAC reduces to 8 bits at 13 MHz, although improved technology is continuing to extend the performance boundary where flash converters can operate well in excess of a 100 MHz conversion rate (Peetz et al., 1986). In designing audio conversion systems, the aims are to maximize dynamic range, minimize and decorrelate hardware dependent distortion and to implement optimal antialiasing and recovery filters; this together with the desire to match the amplitude resolution and conversion speed of available ADC/DACs for audio band encryption is motivation for investigating oversampling and noise-shaping architectures.

The applications of oversampling and noise shaping to ADC/DAC represent a means by which the performance optimization as a function of conversion speed and amplitude resolution can be made. Thus, although the amplitude resolution may diminish as conversion rate is increased, the noise shaper can be designed to take advantage of this interchange and to code signals whose spectra are located well below the sampling frequency, by placing the quantization distortion into the redundant frequency space created by oversampling. In Sec. 4.4.2, a prediction is made of the signal-to-noise ratio (SNR) as a function of noise-shaper parameters; this in turn can be used to assess the conversion performance of an ADC as a function of amplitude resolution and conversion speed. As a result, the SNR attainable from an ADC/DAC is not a simple function of the number of bits as noise shaping can be used to exercise the converter such that the noise spectrum is redistributed non-uniformly and concentrated in a region predominantly outside the audio band. Also self-dither (Adams 1984) or intelligent dither (Hawksford 1985, 1987) can reduce the correlation between system-induced distortion and the input signal to form a noise-like residue, and the use of oversampling can enable antialiasing and reconstruction filters to be implemented digitally. There is also an opportunity for a psychoacoustic advantage since the noise spectrum can be shaped to match the low-level sensitivity of human hearing.

The demands upon analogue filters in realizing both antialiasing and reconstruction filters for near-Nyquist sampling are considerable, where both minimum guard bands and the desire for phase linearity to maintain a well-formed impulse response set difficult to achieve criteria. The problem is further compounded when multichannel systems require identical channel characteristics; where observed in the time domain, the objective is to achieve time synchronized impulse responses of identical form, while viewed in the frequency domain, the filters should exhibit high rates of out-of-band attenuation together with low in-band amplitude ripple.

To achieve a performance that is repeatable, exhibits long-term stability and is environmentally stable without the need for periodic tuning, requires the application of digital technology where the precision of signal processing can be predetermined and the use of FIR (finite impulse response) symmetrical filters guarantees precise phase linearity.

However, a digital antialiasing or reconstruction filter is inappropriate for Nyquist sampling and demands a higher sampling rate in order to relax the constraints placed upon analogue antialiasing and reconstruction filters. In a system where the sampling rate is greater than the Nyquist rate (e.g. Goodman, 1969b), the signal is said to be *oversampled*, where the oversampling factor R_N is defined as

$$R_N = \frac{\text{selected sampling rate}}{\text{Nyquist sampling rate}} \quad (4.1)$$

A basic structure for using mild oversampling is shown in Fig. 4.1, where, for illustration, oversampling is applied to both ADC and DAC. The analogue input signal is initially band-limited by a low-order analogue filter and then converted to digital at a sampling rate $R_{N1}f_s$ hertz where R_{N1} is the oversampling factor and f_s hertz is the Nyquist rate. The oversampled and digitized signal is next converted to the lower rate f_s hertz by a decimation filter, which in essence is a high-order, digital low-pass filter followed by subsampling. Consequently, the decimation filter and the input analogue filter form a composite antialiasing filter, though the majority of filtration is performed in the digital domain. Because the output of the decimation filter is band-limited, the digital signal can be resampled at the lower rate f_s hertz to produce the Nyquist format, digital audio signal.

A similar process is also feasible at the DAC stage to generate an oversampled signal where, following the insertion of $(R_{N2}-1)$ zero amplitude samples between the Nyquist samples to form an oversampled rate $R_{N2}f_s$ hertz, a low-pass filter or interpolation filter computes the intermediate samples. Decimation and interpolation are therefore complementary and, because they are implemented in the digital domain, can take advantage of the impulse response symmetry necessary for linear phase.

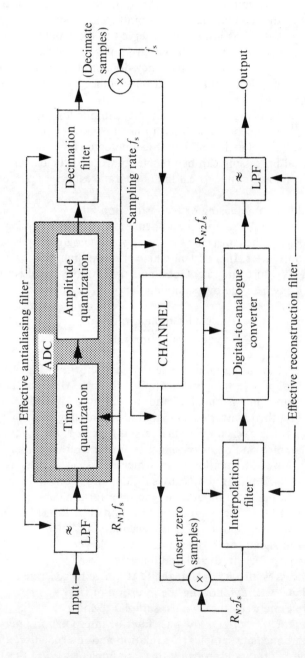

Figure 4.1 ADC/DAC system using mild oversampling and substantially digital antialiasing and reconstruction filters

So far the application of oversampling has been used to transfer signal processing from the analogue to the digital domain where generally mild oversampling ratios are used together with direct forms of ADC and DAC. However, as already briefly discussed, there is a potential interchange between amplitude resolution and conversion speed which can be used to optimize the performance of a given converter and thus have a radical impact upon converter architecture.

If Nyquist sampling is performed, then each data conversion carries unique signal information, where any conversion error or distortion has no means of recovery. However, an oversampled signal carries redundant signal information and allows a degree of intersample averaging, which can enhance signal resolution when observed within the audio band. For example, if the oversampling ratio is R_N, then during each Nyquist interval there are R_N samples which subsequently can be averaged by the antialiasing/decimation filter.

This averaging process can be implemented either as an open-loop system or as a closed-loop system where both techniques have application in ADC and DAC and are discussed in Secs. 4.5 and 4.6. The potential advantages of oversampling and noise shaping to digital audio systems are summarized as follows:

1 Relaxation of performance requirements/signal degradation imposed by excessive analogue circuitry in antialiasing and recovery filters.

2 Low in-band ripple, high stop-band attenuation digital filters outperform their analogue equivalents and guarantee stability, zero tuning, temperature independence, repeatability and definable dynamic range.

3 Noise shaping can optimize the balance between sampling rate and amplitude resolution for a given ADC/DAC to maximize dynamic range and distortion.

4 By producing ADC/DAC oversampled and noise-shaped converters with a performance in excess of 16 bits by 44.1/48 kHz, bandwidth and SNR in association with digital dither are determined in the digital domain to closely match the maximum performance of the channel (Vanderkooy and Lipshitz, 1984; Lipshitz and Vanderkooy, 1986).

4.3 Decimation and interpolation filter techniques

Sampling rate conversion requires the use of digital, low-pass filter techniques to enable the sampling rate to be either increased or decreased without distorting the audio band and by maintaining the minimum of out-of-band intermodulation products. Because the

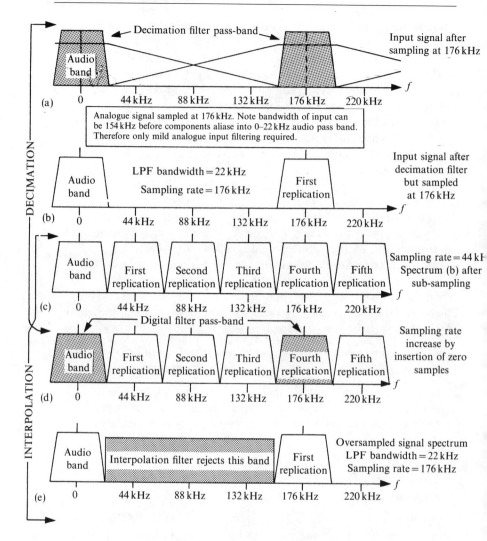

Figure 4.2 Times-four decimation and interpolation using a single-stage filter represented in the frequency domain

frequency spectrum is replicated when sampled, there are numerous methods that can be employed to reduce the computational overhead and some of these are discussed in this section. However, to illustrate the basic processes, Fig. 4.2 shows both a ×4 decimation and interpolation conversion presented in the frequency domain, where, observing the frequency replication inherent in sampled data systems, the frequency bands requiring rejection are identified.

In principle the same overall rejection filter response can be used for both interpolation and decimation, though for higher conversion ratios

WAVEFORM CODING—DIGITAL AUDIO

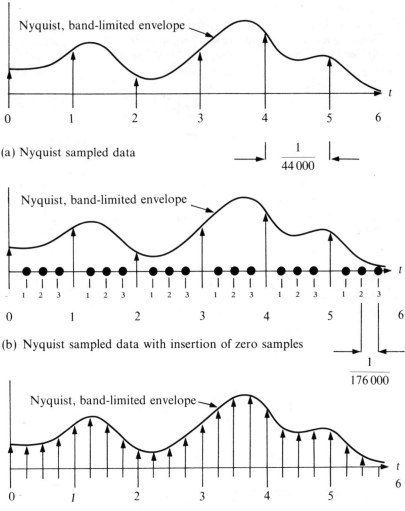

Figure 4.3 Time-domain representation of interpolation process showing zero pulse insertion

the decomposition of the filter process into a set of cascaded filters will yield differences of detail. For integer R_N, the interpolation conversion requires an additional $(R_N - 1)$ samples uniformly spaced between each input sample, where R_N is the sampling rate conversion ratio defined by Eq. (4.1). The process illustrated in Fig. 4.3 inserts zero samples between Nyquist samples and then uses a digital filter to suppress redundant out-of-band spuriae, thus effecting the calculations of intermediate samples. Zero intermediate samples are chosen to eliminate aperture distortion that would otherwise result if the additional input samples were held constant over a Nyquist period $1/f_s$ second, where the linear-distortion transfer function $A_a(f)$ would take the form

$$A_a(f) = \frac{1}{f_s} \text{sinc}\left(\frac{\pi f}{f_s}\right) \tag{4.2}$$

where $\text{sinc}(x) = \sin x / x$ and result in a linear phase attenuation of high audio-frequencies.

The low-pass filters (or equivalents) used in sampling rate conversion can be designed using a range of digital filter techniques. However, it is evident that as both the number of filter coefficients in a FIR filter and R_N increase, the number of multiplications and additions also increase. This high computational overhead can prove a formidable problem and does not represent an efficient use of DSP devices; consequently, a search for more efficient methods is demanded.

Several procedures for improving the computational efficiency of oversampling filters have been identified and are now discussed with reference to interpolation, although similar techniques are equally applicable to decimation.

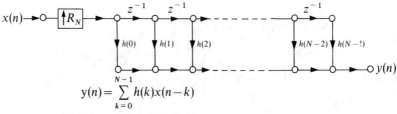

(a) Direct form of FIR filter

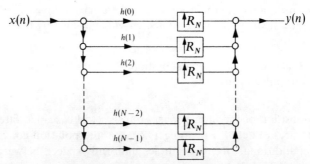

(b) More efficient FIR filter in an interpolation process

Figure 4.4 Signal flow graphs showing more efficient FIR filter and interpolation process that saves multiplications

4.3.1 Transposed FIR structure

The signal flow graphs of both a direct realization and a transposed FIR structure are shown in Fig. 4.4. In the direct structure, each zero sample undergoes a redundant multiplication by the filter coefficients, while after transposition only the Nyquist data samples are multiplied, which reduces the multiplication burden imposed by the oversampling ratio R_N; for example for $R_N = 4$, a 96 tap FIR interpolation filter requires only 24 multiplications per Nyquist period (Goedhart, van de Plassche and Stikvoort, 1982).

4.3.2 Multistage filters

In systems that require higher orders of oversampling ($R_N > 4$), it is more efficient to decompose the interpolation process into a number of cascaded stages (Crochiere and Rabiner, 1975; Hansen, 1987) and to allow the sampling rate to increment at each stage. In Fig. 4.5 this process is illustrated where $\times 4$ oversampling is performed using a two-stage interpolator, though the technique is readily extended to more stages. The diagram shows that after $\times 2$ interpolation the in-band distortion components associated with $2f_s$ only occupy one half of the band between 0 and $4f_s$ where frequency space emerges with virtually no signal components. These guard bands or 'don't care' regions relax the design requirement of the second-stage interpolation filter H_2. Either H_2 can be designed as a low-pass filter, with a more gentle attenuation region and hence a reduction in filter length, or it can be designed as a multiband comb filter with rejection regions coincident with the undesired spectral replication bands.

The interpolation process can be generalized by factoring the interpolation ratio R_N into the product

$$R_N = \prod_{i=1}^{I} R_i \tag{4.3}$$

where R_i is a factor, often chosen to be 2, and I is the number of factors. Such architecture can reduce both computation and storage requirements, reduce truncation errors due to finite word length in the processor and simplify the design problem.

4.3.3 Half-band filters

In data-transmission systems, there is a need to design channel filters with ideally zero intersymbol interference (ISI). Viewed in the time domain, this requires the target channel filter to exhibit an impulse response with uniformly spaced zero crossings, where, by forcing a subset of the interpolation filter impulse response samples to zero, multiplications can be saved.

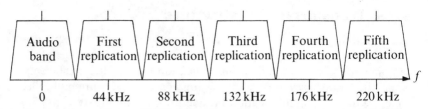
(a) Audio signal sampled at 44 kHz, showing spectrum replication to 220 kHz

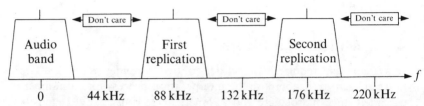
(b) Replicated spectrum after times-two interpolation showing 'Don't care' bands

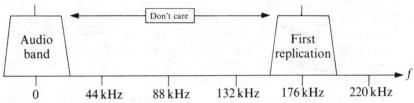
(c) Replicated spectrum after a further times-two interpolation

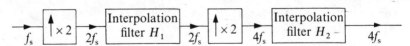
(d) Two-stage interpolation process

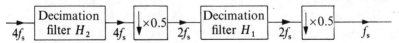
(e) Two-stage decimation system using symmetrical filter reversal scheme of sec. 4.3.4.

Figure 4.5 Two-stage interpolation with complementary decimation system using reverse filter sequence

The target interpolation filter can be expressed as the sum of two parallel filter structures as shown in Fig. 4.6(a), where $A_o(f)$ is an ideal 'brickwall' filter and $A_\varepsilon(f)$ a transfer error function modelling target filter non-ideality. Both $A_o(f)$ and $A_\varepsilon(f)$ are assumed to be linear phase and the corresponding impulse response can be expressed as

$$h(t) = \text{sinc}(\pi f_c t) + x(t)\sin(\pi f_c t) \tag{4.4}$$

The expression for $h(t)$ is divided into two components: a sinc (...) function representing the idealized filter impulse response and an error component $x(t)\sin(\pi f_c t)$ where the multiplying sine function constrains the error response to exhibit time-domain zeros, uniformly spaced at an interval of $1/f_c$ second. If the response $x(t)$ is the impulse response of a low-pass filter, then viewed in the frequency domain the Fourier transform $F[x(t)]$ shows a frequency translated spectrum centred on $\pm f_c/2$ hertz, where an example error spectrum is illustrated in Fig. 4.6(b) together with the composite spectrum in Fig. 4.6(c).

The important characteristic of this filter response is the odd symmetry about the points p, q resulting directly from the constraint for uniformly spaced time-domain transmission zeros. If a second constraint is imposed requiring $A_\varepsilon(f) = 0$ for $f \geq f_c$, then from Nyquist's sampling theorem the minimum sampling rate is $2f_c$, where setting $f_s = 2f_c$ and sampling symmetrically about $t = 0$ an impulse response with alternate sample values of zero is generated as shown in Fig. 4.6(d).

Hence, by arranging for the spectrum of the impulse response to have odd symmetry about $f_s/4$ hertz and to have a bandwidth constrained to $f_s/2$ hertz, alternate coefficients of the filter are zero and results in a 2:1 saving in multiplications (Ansari, 1988; Bellanger *et al.*, 1974; Goodman and Carey, 1977; Mintzer, 1982; Sun, 1986; Vaidyanathan and Nguyen, 1987).

4.3.4 Interpolation/decimation symmetry

The ×4 sampling rate converter illustrated in Fig. 4.2 indicated that the same (in this case single) filter could be used for both interpolation and decimation. This process can readily be extended to multistage systems (Crochiere and Rabiner, 1983), though the order of the filter cascade is reversed as shown in Fig. 4.5(e). Hence, having selected a coefficient set for an interpolation, the same filters can now be used for decimation, where, as an example, ×8 decimation is shown in Fig. 4.7, in which filters similar to those of Fig. 4.2 are identified. The key to the system is that, although aliasing distortion occurs, providing the folded components are excluded from the audio band there is no error due to aliasing. However, of greater significance is the symmetry between interpolation and decimation and their corresponding relationship to oversampling in DAC and ADC systems.

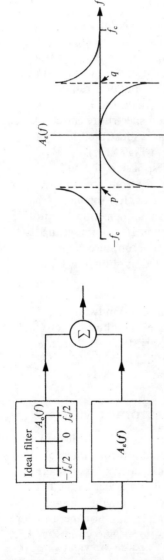

(a) Sum of ideal filter and error filter giving zero ISI.

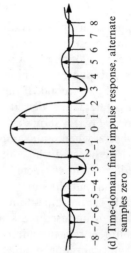

(b) Error spectrum after frequency translation

(c) Composite responses showing symmetry about $f_c/2$

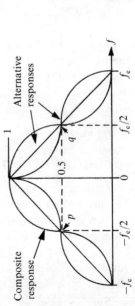

(d) Time-domain finite impulse response, alternate samples zero

Figure 4.6 Half-band filter showing filter symmetry

WAVEFORM CODING—DIGITAL AUDIO

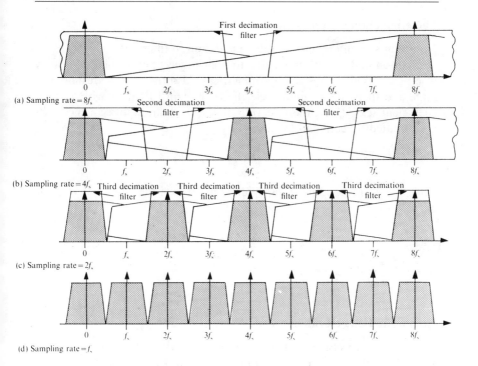

Figure 4.7 Three-stage decimation system represented in the frequency domain

These four subsections have illustrated techniques that can be used to reduce the design burden and computational requirements. However, to realize the actual filters it is necessary to calculate the coefficients for each filter stage and in order to perform that task a range of design methods are available. In this text, a number of these techniques are outlined and references suggested to enable a broader knowledge base to be developed.

4.3.5 Digital filter design

Digital filters can be classed into three groups.

Infinite impulse response (IIR)

Recursive filters are generally considered less attractive for decimation/interpolation as they require correction for the non-linear phase, can suffer word length truncation errors with associated limit cycles and offer no advantage through coefficient reduction as with half-band filters.

Finite impulse response (FIR)

Non-recursive filters offer a wide application in interpolation/decimation and are attractive due to their inherent exact phase linearity when the impulse response is symmetrical. Also the impulse response is finite and there are no limit cycle effects. For an FIR filter with a finite coefficient set $\{h(k)\}_N$, the time response $y(n)$ for an input sequence $x(k)$ is computed by convolution as

$$y(n) = \sum_{k=0}^{N-1} h(k)x(n-k) \qquad (4.5)$$

while the corresponding transfer function $H(f)$ of the FIR filter is given by

$$H(f) = \sum_{k=-N/2}^{N/2} h(k)e^{-j2\pi kf/f_s} \qquad (4.6)$$

where $1/f_s$ is the sampling period.

Composite IIR/FIR filters

Generalized structures that combine both an IIR and an FIR filter can be used to give a degree of group-delay equalization. However, this filter class is not considered in the present discussion as the design requirement of the linear phase is met precisely by the FIR structure.

The following discussion is a brief review of some of the tools available for designing FIR filters.

4.3.6 Window design

An idealized low-pass filter exhibits an impulse response of the form sinc($\pi f_c t$) where the bandwidth extends from $-f_c/2$ to $f_c/2$. Such a filter is unrealizable and also unsuitable for digital audio because of infinite time dispersion in the precursive response. By truncating the impulse response as shown in Fig. 4.8(a), a finite response filter results which can now be translated directly to an FIR structure, where the truncation is equivalent to applying a rectangular window. Unfortunately, such abrupt truncation leads to excessive amplitude ripple in the frequency domain as shown in Fig. 4.8(c), when the resulting spectrum is the convolution of the idealized rectangular response and the sinc(. . .) function derived from the Fourier transform of the rectangular window. To reduce the amplitude ripple, alternative windows have been proposed that exhibit a more gradual attenuation of the idealized impulse response, yet still allow a finite truncation.

Of the available windows, the Kaiser window offers an excellent

WAVEFORM CODING—DIGITAL AUDIO

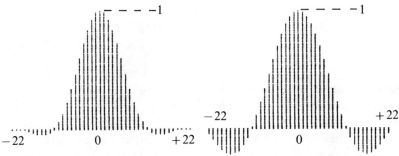

(a) Kaiser weighted, 43 coefficient FIR LPF filter

(b) Rectangular weighted, 43 coefficient FIR LPF filter

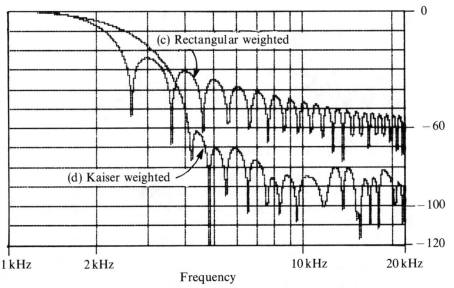

(c), (d) Rectangular and Kaiser weighted, low-pass filter frequency response (43 coefficients)

Figure 4.8 Time- and frequency-domain FIR filter with 2 kHz cut-off frequency and 44.1 kHz sampling frequency

compromise of filter complexity, attenuation shape and ripple. The sample sequence $W_K(n)$ that defines the Kaiser window is calculated by

$$w_K(n) = \frac{I_0\{\beta[1-(n/N)^2]^{0.5}\}}{I_0[\beta]}, \quad |n| \leq N$$
$$w_K(n) = 0, \quad |n| > N \qquad (4.7)$$

where I_0 is the first kind and zero-order modified Bessel function, β is a factor determining frequency response (normally $4 \leq \beta \leq 9$), $2N+1$ is the number of samples and the response is symmetrical where $W_K(-n) = W_K(n)$. The Bessel function $I_0(x)$ is evaluated as

$$I_0(x) = 1 + \sum_{P=1}^{M} \left[\frac{(0.5x)^P}{P!} \right]^2 \qquad (4.8)$$

where M is chosen such that the last term in the series $< 10^{-8}$.

A low-pass filter weighted by the Kaiser window where $\beta = 2\pi$ and $N = 43$ is shown in Fig. 4.8(a) and (d), which has the same number of samples as the rectangular window of Fig. 4.8(b) and (c); the improved performance is self-evident.

A second class of window is described by the generalized Hamming window and is given by

$$W_H(n) = \begin{bmatrix} \alpha + (1-\alpha)\cos(n\pi/N) & |n| \leq N \\ 0 & |n| > N \end{bmatrix} \qquad (4.9)$$

for $0 \leq \alpha \leq 1$. If $\alpha = 0.54$ the function is a Hamming window while for $\alpha = 0.50$ it is designated a Hanning window. However, filter designs (using identical N) reveal superior performance using the Kaiser window functions described by Eq. (4.7).

4.3.7 Equiripple FIR design

A popular FIR design procedure is the Remez exchange method which uses a Chebyshev approximation to the idealized low-pass filter response. The scheme employs a tolerance method as shown in Fig. 4.9 where pass-band and stop-band ripple δ_p, δ_s and the attenuation region ΔF are specified. The program (McClellan, Parks and Rabiner, 1973) can calculate, for a prescribed number of coefficients, a filter having the shortest transition region for a given filter length, where the error function $E(f)$ to be minimized is

$$|E(f)| = \max \tilde{W}(f) |\tilde{H}(f) - H(f)| \qquad (4.10)$$

where

$\tilde{W}(f)$ is a weighting factor influencing δ_s, δ_p and
$\tilde{H}(f)$ is the desired magnitude response.

The input parameters consist mainly of transition frequencies, filter length and weighting factors, where for an interpolation filter for digital audio typical parameters are

$F_p = 20 \text{ kHz} \qquad f_p = F_p/(R_N F_{sa})$
$F_s = 22 \text{ kHz} \qquad f_s = F_s/(R_N F_{sa})$
$F_{sa} = 44.1 \text{ kHz} \qquad f_{sa} = 1/R_N$
$R_N = 4 \qquad \Delta f = f_s - f_p$

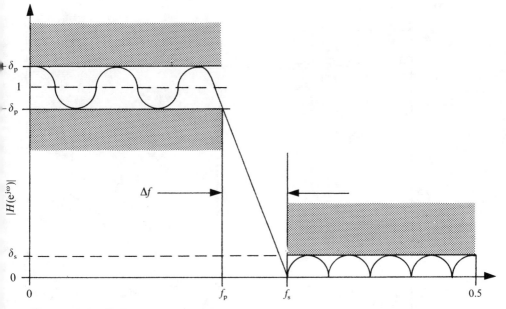

Figure 4.9 Tolerance scheme for low-pass filter

where f_p and f_s are normalized to the sampling frequency, $R_N F_{sa}$. The number of coefficients N needed to fulfil the constraints of the tolerance scheme of Fig. 4.9 can be estimated by an empirical formula (Rabiner and Gold, 1975) as

$$N = \frac{D_\infty(\delta_p, \delta_s)}{f_s - f_p} - f(\delta_p, \delta_s)(f_s - f_p) + 1 \qquad (4.11)$$

where

$$D_\infty(\delta_p, \delta_s) = \log_{10}\{\delta_s[a_1(\log_{10}\delta_p)^2 + a_2\log_{10}\delta_p + a_3]\} \\ + [a_4(\log_{10}\delta_p)^2 + a_5\log_{10}\delta_p + a_6] \qquad (4.12)$$

$$f(\delta_p, \delta_s) = 11.012 + 0.512(\log_{10}\delta_p - \log_{10}\delta_s) \qquad (4.13)$$

and

$a_1 = 0.00539 \qquad a_2 = 0.07114 \qquad a_3 = -0.4761$
$a_4 = -0.00266 \qquad a_5 = -0.5941 \qquad a_6 = -0.4278$

However, the McClellan, Parks and Rabiner (1973) procedure cannot be used directly to design half-band filters, where the aim is to set alternate coefficients to zero. Mintzer (1982) has demonstrated an iterative procedure to the half-band filter, though this is suboptimal. However, Vaidyanathan and Nguyen (1987) and Ansari (1988) have proposed the 'half-band design trick', where only the non-zero coefficients are calculated. Effectively an impulse response $g(k)$ is

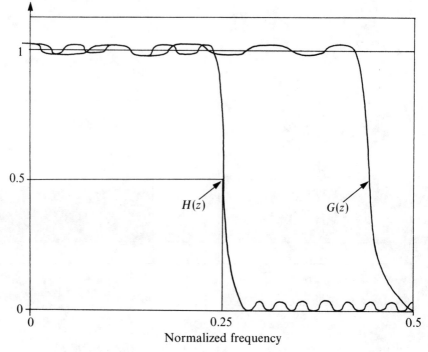

Figure 4.10 Half-band design trick

calculated which is related to the desired half-band impulse response $h(k)$, where $g(k)$ is designed using the McClellan program. Since the expected zero coefficients do not appear in the design, there is no constraint on $g(k)$. The z-domain relationship between $g(k)$ and $h(k)$ is

$$H(z)=0.5[G(z^2)+z^{-(N+0.5)}] \qquad (4.14)$$

$$h(k)=\begin{cases} 0.5g(k/2) & k \text{ even} \\ 0 & k \text{ odd} \neq (N-1)/2 \\ 0.5 & k=(N-1)/2 \end{cases}$$

To illustrate the relationship between $H(z)$ and $G(z)$, Fig. 4.10 shows the corresponding magnitude responses.

4.4 Noise shaping

Information theory (Shannon, 1948) indicates that once a band-limited signal is oversampled, the number of bits required to represent each sample can be reduced without introducing significant signal impairment. One method of achieving a reduction in sample resolution is to requantize the oversampled signal and to use a recursive noise-shaping filter to distribute the resulting error signal and cause it to reside substantially outside the audio band. An extreme example of this process is the delta–sigma modulator (Inose, Yasuda and Murakami, 1962), where oversampling and noise shaping enable the input signal to be reduced to a serial 1 bit code. In Sec. 4.6 a number of ADC and DAC applications is presented including the *bitstream* DAC by Philips (Naus *et al.*, 1987) that use a combination of oversampling and noise shaping where the application is biased towards high-quality audio systems.

A noise-shaping coder consists of a quantizer Q enclosed by a negative feedback loop containing a signal processor $A_N(z)$, as shown in Fig. 4.13(a). Providing a closed-loop system is stable, then it will be shown that the quantizing distortion produced by Q is frequency shaped as an inverse function of $A_N(z)$. Consequently, for a signal that is oversampled so as to create a redundant frequency space, it is possible to redistribute the requantization distortion into this band and

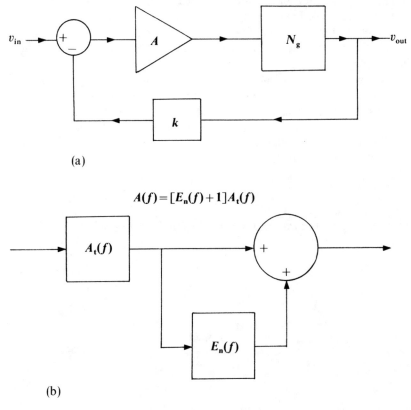

Figure 4.11 (a) Negative feedback amplifier with non-linear output stage N where $N \approx 1$. (b) Error function representation of amplifier distortion process

thus maintain an acceptable in-band SNR. However, an investigation of recursive, noise-shaping coders is complicated by the non-linear behaviour of the feedback loop. Thus, although analyses based upon linear techniques can give good first-order performance predictions, in order to obtain greater insight into the noise-shaping process, computer simulations are required to generate coder responses and hence give an estimate of performance targets in terms of SNR and distortion spectra. This approach offers the advantage of exact system modelling without the limitations imposed by hardware and non-exact mathematical analysis. It also enables the validity of approximate mathematical estimates to be evaluated (Darling, 1987; Darling and Hawksford, 1988; Hawksford, 1985; McCrea, 1987; Steele, 1975, 1980).

To introduce noise shaping it is constructive to consider the analogy with the output-stage distortion reduction process of the analogue feedback amplifier shown in Fig. 4.11(a) where A is the forward path transfer function, N_g the incremental output stage gain (where $N_g = 1 + \Delta N_g, \Delta N_g = 0$ for zero distortion) and k is the feedback factor. In fact, the operation of this amplifier has strong similarities with the heavily oversampled and noise-shaped ADC to be described in Sec. 4.6 and therefore forms a useful and probably more familiar introductory model. To mathematically describe the *noise-shaped* error produced by this simple recursive system an error function $E_n(f)$ is introduced and represented diagrammatically in Fig. 4.11(b). The advantage of this function is the concise way in which it represents the error process, where the output of the transfer function $E_n(f)$ forms the system error expressed in exact relation to the primary output signal. The function $E_n(f)$ is evaluated as follows.

The closed-loop transfer function $A(f)$ of the amplifier is

$$A(f) = \frac{AN_g}{1 + kAN_g}$$

where for $N_g = 1$ the *target* transfer function $A_t(f)$ follows, as

$$A_t(f) = \frac{A}{1 + kA}$$

Defining a transfer error function $E_n(f)$ as

$$E_n(f) = \frac{A(f)}{A_t(f)} - 1$$

then

$$E_n(f) = \frac{N_g - 1}{1 + kAN_g} = \frac{\Delta N_g}{1 + kAN_g} \qquad (4.15)$$

Equation (4.15) demonstrates the effect of negative feedback on the reduction in sensitivity to the amplifier output stage gain, where a finite gain error (ΔN_g) is reduced by the factor $\{1 + kAN_g\}$.

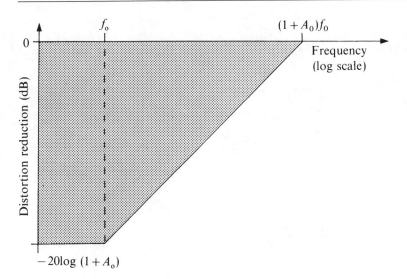

Figure 4.12 Distortion reduction factor for first-order negative feedback amplifier according to Eq. (4.17)

For a first-order amplifier, the loop gain approximates to

$$\{kAN_g\} = \frac{A_o}{1+jf/f_o} \quad (4.16)$$

whereby

$$E_n(f) = \frac{\Delta N_g}{1+A_o}\left\{\frac{1+jf/f_o}{1+jf/[(1+A_o)f_o]}\right\} \quad (4.17)$$

The distortion reduction described by Eq. (4.17) is shown in Fig. 4.12 and can be compared with a basic first-order analogue noise shaper. The result indicates that in the low-frequency region where the loop gain is large, distortion reduction is high while at high frequency, where A is curtailed to maintain closed-loop stability, correspondingly less reduction occurs. Consequently the selection of the loop transfer function strongly influences the error spectrum, where the aim is to maximize loop gain within the constraints of closed-loop stability.

The noise-shaping coder of Fig. 4.13 can be analysed in a similar manner where the quantizer is represented as $Q(z)$ and $A_N(z)$ forms the forward-path processor. To estimate the error in this system, an error function $E_N(z)$ is determined, where

$$E_N(z) = \frac{Q(z)-1}{1+A_N(z)Q(z)} \quad (4.18)$$

expressing

$$Q(z) = 1 + q_n \quad (4.19)$$

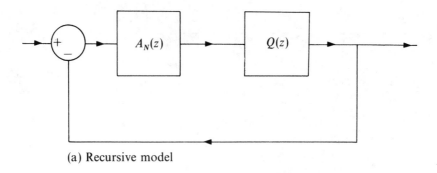

(a) Recursive model

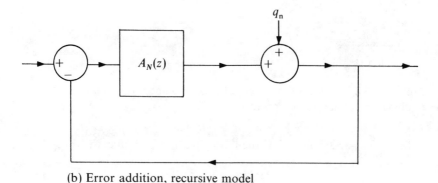

(b) Error addition, recursive model

Figure 4.13 Noise-shaping coder

where q_n represents the quantizer non-linearity. Then to an approximation

$$E_N(z) \approx \frac{q_n}{1 + A_N(z)} \qquad (4.20)$$

Since the transfer function $A_N(z)$ of the filter determines the noise-shaping characteristics, it is constructive to investigate its properties. The aim is to make $A_N(z)$ large at low frequencies and small at high frequencies, thus attempting to bias the quantization distortion into the higher frequency regions. This objective can be met by forming $A_N(z)$ from a cascade of N digital integrators, as shown in Fig. 4.14(a) and (b), where analysis shows that

$$A_N(z) = \frac{z^{-1}}{1 - z^{-1}} + \frac{z^{-1}}{(1 - z^{-1})^2} + \cdots + \frac{z^{-1}}{(1 - z^{-1})^N}$$

Applying the geometric series expansion

$$\frac{1 - x^N}{1 - x} = \sum_{r=0}^{N-1} x^r$$

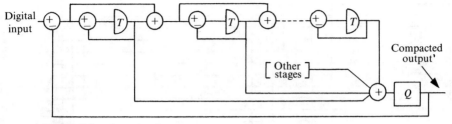

(a) DSM Nth-order noise shaper

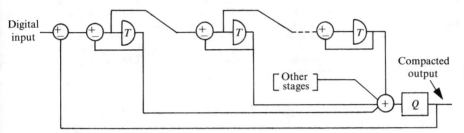

(b) Simplified noise-shaping architecture

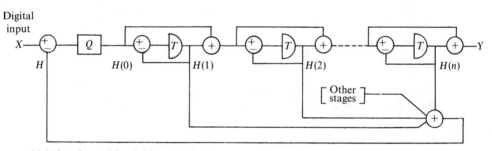

(c) Nth-order multilevel delta modulator

Figure 4.14 Nth-order noise-shaping coder

then

$$A_N(z) = \left(\frac{z}{z-1}\right)^N - 1 \quad (4.21)$$

This result has been demonstrated (Tewksbury and Hallock, 1978) to be an optimal noise-shaping filter and is composed of N cascaded integrators and $(N-1)$ zeros to achieve a first-order response at high frequency and also facilitate closed-loop stability. In Fig. 4.15 a family of frequency response curves of $|A_N(z)|$ is shown for $N=1$ to 6.

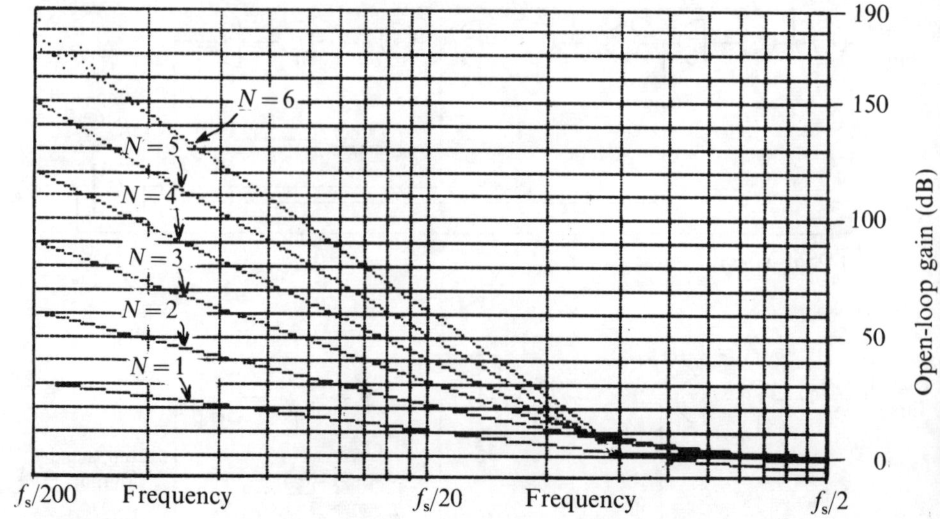

Figure 4.15 Family of open-loop frequency response curves for $N=1$ to 6 noise shaper

The error function $E_N(z)$ can now be calculated by substituting for $A_N(z)$ in Eq. (4.18), whereby

$$E_N(z) \approx q_n \left[\frac{z-1}{z}\right]^N = q_n D_N(z) \quad (4.22)$$

This result shows that the quantizer distortion q_n is frequency shaped by a noise-shaping function $D_N(z)$, where

$$D_N(z) = \left[\frac{z-1}{z}\right]^N \quad (4.23)$$

which has the corresponding amplitude frequency response

$$|D_N(f)| = 2^N \sin^N\left(\frac{\pi f}{f_s}\right) \quad (4.24)$$

The function $D_N(z)$ gives a measure of the anticipated noise shaping of the quantizer distortion, where a family of transfer functions of $|D_N(f)|$ for $N=1$ to 6 is shown in Fig. 4.16. The curves reveal two frequencies of interest:

$$\left|D_N\left(\frac{f_s}{6}\right)\right| = 1 \quad (4.25a)$$

$$D_N\left(\frac{f_s}{2}\right) = 2^N \quad (4.25b)$$

WAVEFORM CODING—DIGITAL AUDIO

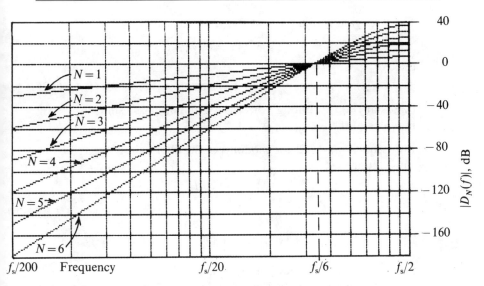

Figure 4.16 Family of $|D_N(f)|$ for $N=1$ to 6 noise shaper

The results show that all curves for any N are coincident at $f=f_s/6$ hertz and that in the range

$$\frac{f_s}{6} < f \leq \frac{f_s}{2}$$

the distortion resulting from q_n is amplified, while distortion reduction occurs only for $f<f_s/6$ hertz.

The increase in quantization distortion with a maximum gain at $f=f_s/2$ hertz has a direct implication on the amplitude range of the quantizer, where there must be sufficient levels to prevent amplitude saturation. Indeed, computer simulation has demonstrated the importance of a non-saturating quantizer for filter orders >2, where saturation in association with a high-order filter instigates entry into divergent limit-cycle instability. Simulation (Darling, 1987; Darling and Hawksford, 1988; Hawksford, 1985; McCrea, 1987) has also revealed that quantizer saturation is the mechanism that prevents the successful operation of delta–sigma modulation with a filter order >2. Hence in all the noise-shaping examples discussed in this section, a non-saturating quantizer is assumed enabling $N>2$, providing that the appropriate feedforward paths are included with the integrators to enable stable loop behaviour.

The following observations of the N-stage integrator, noise shaper should be noted:

1   The filter uses only addition; there are no multiplication coefficients. The algorithm is therefore efficient and simplifies hardware realization.

2. The filter gain is minimum at $f = f_s/2$, where noting

$$z = e^{j2\pi f/f_s} = e^{j\pi} = -1$$

then

$$A_N(-1) = 0.5^N - 1 \qquad (4.26)$$

that is at $f = f_s/2$, although there is a π radian phase shift, the modulus of the filter gain is <1 for all N.

3. For frequency $f \ll f_s/2$, the modulus of the filter gain is derived by substituting

$$z \approx 1 + j2\pi f/f_s,$$

that is

$$|A_N(z)|_{f \ll f_s/2} \approx \left[j2\pi \frac{f}{f_s} \right]^{-N} \qquad (4.27)$$

that is the low-frequency gain is equivalent to N cascaded integrators with individual gain-bandwidth products f_s hertz.

The noise-shaper topology of Fig. 4.14(a) and (b) can be transformed into a system similar to a delta modulator (Inose and Yasuda, 1963; Inose, Yasuda and Murakami, 1962), where the N integrators and associated feedforward paths now appear in the feedback path. The input signal X is assumed to be sampled and quantized to integer values and the quantizer Q has a matched mid-tread, unit-step transfer characteristic.

Observing the notation of Fig. 4.14(c), then the output sequence Y and feedback signal H are expressed as

$$Y = \sum_{r=0}^{N} H(r) = H(0) + \sum_{r=1}^{N} H(r)$$

$$H = \sum_{r=1}^{N} H(r)$$

Since signals assume only integer values

$$H(0) = X - H$$

where on substitution for $H(0)$ and H

$$X = Y$$

This result shows that although the reconfigured system exhibits no useful noise-shaping advantage, it is totally transparent to integer data and represents a fundamental system characteristic.

4.4.1 Computer simulation

To demonstrate the coding performance of an Nth-order oversampled noise shaper, a computer program (McCrea, 1987) using a transient

WAVEFORM CODING—DIGITAL AUDIO

analysis procedure was written to enable exact time-domain simulation and to enable accurate SNRs to be evaluated. The input parameters of the program specified the oversampling ratio R_N, filter order N, and input amplitude and frequency, where sinusoidal inputs were generally used. The program assigned a unit quantum to the quantizer and an input level of 0 dB corresponded to $\sqrt{2}$ units.

In order to eliminate non-synchronous errors and linear distortion from corrupting the running average of the SNR, two parallel coders were run side by side, together with two identical reconstruction filters. However, one coder included a unit-quantum quantizer while the second used full computer precision. At each sample, both signal and instantaneous error were squared and summed with earlier values to enable an accurate running SNR estimate.

In Fig. 4.17(a), four time-domain plots corresponding to $R_N = 200$ and $N = 1$ to 4 are shown for an input level of 0 dB and frequency 20 kHz; also the resulting quantizer activity is presented by histograms in Fig. 4.17(b), revealing the required minimum quantizer range for each filter order. The time-domain results demonstrate high-frequency quantizer activity that increases as N is increased, where for $N = 4$ an almost random raster is apparent. In Fig. 4.18 a family of curves is shown of the SNR versus I/P signal level for $R_N = 25, 50, 100, 200$ and $N = 1$ to 5, while Fig. 4.17(c) shows two computed time-domain sequences for a 2 kHz input of -120 dB to reveal low-level coding performance, where $N = 4$ is seen to offer an acceptable resolution while $N = 3$ exhibits significant noise.

To test the noise-shaping properties of the coder, Fig. 4.19 presents computed Fourier transforms of the output waveforms for $N = 1$ to 4, where the characteristics are similar to those in Fig. 4.16. Although the output rasters appear random, the spectra show structured activity exhibiting the desired noise-shaping properties. The high-level, noise-shaper activity therefore forms an intelligent dither which in Sec. 4.6 is shown to be of advantage in digital-to-analogue conversion.

4.4.2 Linear analysis

The distortion reduction factor $D_N(f)$ described by Eq. (4.24) can be used as the basis of an SNR estimate within the audio band $0 < f < f_s/(2R_N)$, where the output noise power N_a is given as

$$N_a = k_N \int_0^{f_s/(2R_N)} \left[2^N \sin^N\left(\frac{\pi f}{f_s}\right) \right]^2 df \qquad (4.28)$$

where

R_N = oversampling ratio for order N
k_N = a constant for order N

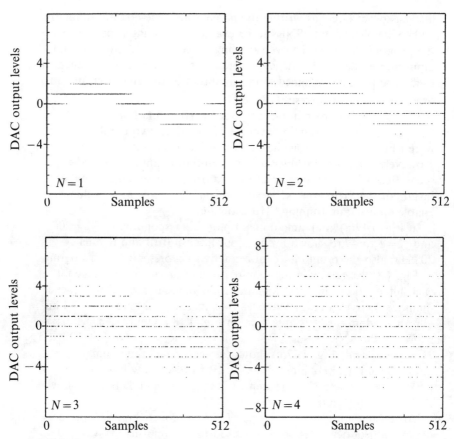

Figure 4.17a Noise-shaper time-domain outputs for $N=1, 2, 3, 4$, input 0 dB (corresponding to $\pm\sqrt{2}$ of an output quantum)

For large R_N, $\sin(\pi f/f_s) \approx \pi f/f_s$, whereby

$$N_a = \frac{\{k_n f_s\}}{2(2N+1)R_N}\left(\frac{\pi}{R_N}\right)^{2N} \qquad (4.29)$$

The term $\{k_n f_s\}$ is related to the total noise power generated by the noise shaper within the band 0 to f_s hertz, where the noise-shaping characteristics follow the curves shown in Fig. 4.16 derived from $|D_N(f)|$. The total output noise power N_t can be estimated from a knowledge of the histograms shown in Fig. 4.17(b), where the simulation reveals a quantizer range approximating to $\pm 2^{N-1}$, which corresponds closely to the product of quantization error ± 0.5 and the distortion reduction factor $D_N(f_s/2)$ described by Eq. (4.25b). To evaluate N_t for a given order, the corresponding histogram is approximated here to a linear distribution, where in Fig. 4.20 an idealized example is shown for $N=3$.

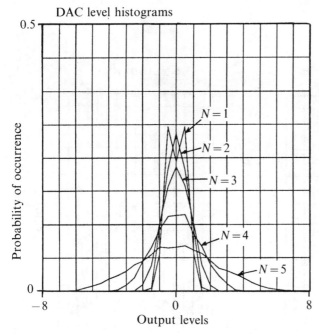

Figure 4.17b Histogram of noise output for $N=1$ to 5

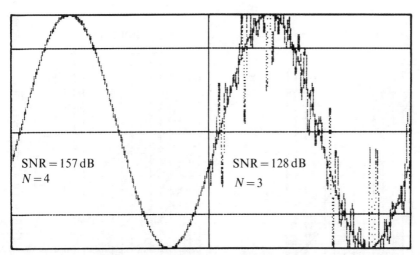

I/P signal -120 dB, $f=2$ kHz, sampling rate $=8$ MHz
3 dB bandwidth of O/P filter $=20$ kHz

Figure 4.17c Low level (-120 dB) time-domain coding of noise shaper for $N=3, 4$

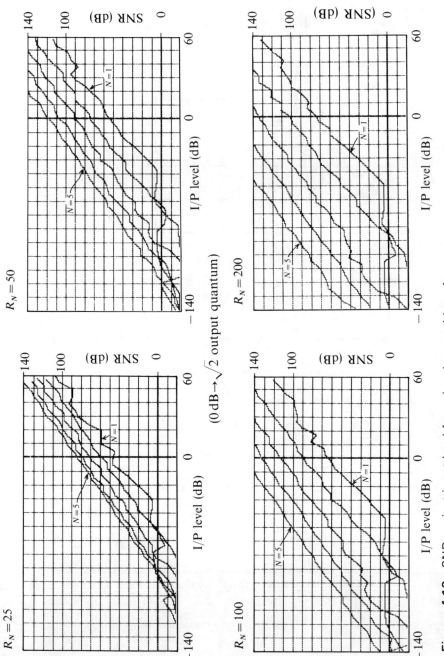

Figure 4.18 SNR against input level for noise shaper with perfect DAC, $N=1$ to 5

WAVEFORM CODING—DIGITAL AUDIO 127

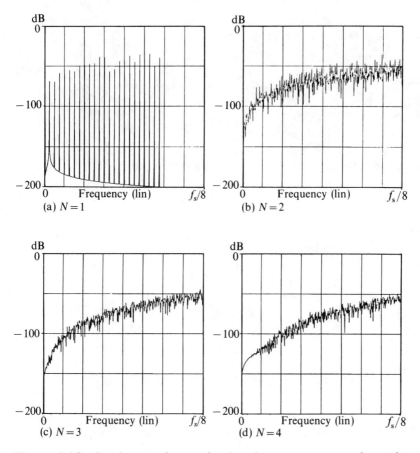

Figure 4.19 Fourier transforms of noise-shaper output waveforms for $N=1$ to 4 (ideal DAC)

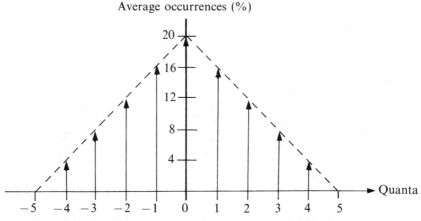

Figure 4.20 Idealized histogram of quantizer activity for $N=3$

Generalizing to order N, the total power N_t can be evaluated from the triangular histogram approximation as

$$N_t = 2 \sum_{r=1}^{\gamma} r^2 \left[\frac{\gamma + 1 - r}{(\gamma + 1)^2} \right] \qquad (4.30)$$

where

$$\gamma = 2^{N-1} \qquad (4.31)$$

The mean spectral density k_n is computed by equating the area $\{k_n f_s\}$ to the area under the curve $\{N_t/f_s\}|D_N(f)|^2$ integrated over $0 < f < f_s$, that is

$$\{k_n f_s\} = \frac{N_t}{f_s} \int_0^{f_s} \left[2^N \sin^N \left(\frac{\pi f}{f_s} \right) \right]^2 df$$

Observing the spectral symmetry about $f_s/2$ Hz (see Fig. 4.16) and using the recurrence formula for

$$\int_0^{\pi/2} \sin^n x \, dx \qquad \text{for even } n$$

then

$$\{k_n f_s\} = 2^{2N} N_t \prod_{r=1}^{2N} \left(\frac{2r-1}{2r} \right)$$

whereby substituting for N_t and $\{k_n f_s\}$, N_a (Eq. (4.29)) is given by

Table 4.1 Comparison of theoretical and computed SNR for noise-shaping filter computed with 20 kHz input signal at -20 dB (ref $\sqrt{2}$ amplitude)

R_N	N	SNR (Eq. (4.32))	Computer simulation, dB
50	1	50	49
50	2	67	68
50	3	83	87
50	4	97	103
100	1	59	53
100	2	82	82
100	3	104	115
100	4	124	134
200	1	68	62
200	2	98	98
200	3	125	130
200	4	152	156

$$N_a = \frac{1}{(2N+1)R_N} \left(\frac{2\pi}{R_N}\right)^{2N} \sum_{r=1}^{\gamma} r^2 \left[\frac{\gamma+1-r}{(\gamma+1)^2}\right] \prod_{r=1}^{2N} \left(\frac{2r-1}{2r}\right) \quad (4.32)$$

To give a measure of the accuracy of the expression for N_a, Table 4.1 compares simulated against estimated SNR for a range of order N and oversampling ratio R_N with reference to a signal of 0 dB, where the computed results were generated using an input signal of 20 kHz at -20 dB (0 dB corresponds to a sine wave of amplitude $\sqrt{2}$ output quanta). The simulation revealed that for low-order loops ($N=1, 2$) quantization noise varied considerably with level (see Fig. 4.18), while for higher orders ($N=3, 4$) the noise was almost level independent, indicating decorrelation of signal and quantization distortion. Equation (4.32) is shown to slightly underestimate the SNR, though this is due mainly to the linear approximation of the histogram distribution which, as Fig. 4.17(b) reveals, there is some curvature making the theoretical noise estimate slightly high.

4.5 Application of mild oversampling in ADC and DAC

Section 4.2 introduced the principal advantages of oversampling which are of direct relevance to ADC systems. In ADC mild oversampling has been employed in a range of equipment that includes digital processors and RDAT machines where the key advantages are:

1 Reduction in complexity of analogue antialiasing filters where low-order filters are viable (e.g. 2, 3 order).

2 Digital filters now perform high-order, antialiasing filter function with the associated advantages:
   (a) Low in-band amplitude ripple.
   (b) High rate of attenuation.
   (c) High-stop band attenuation.
   (d) Linear phase.
   (e) Impulse response synchronized to system clock.
   (f) All filters exhibit identical responses.
   (g) Aperture correction for sample and hold can be included.
   (h) Compensation for analogue filters can also be included.

3 Noise averaging can increase ADC resolution.

4 Allows incorporation of ADC devices with enhanced sampling rate—amplitude resolution products.

A typical system configuration is shown in Fig. 4.21, where the input signal is serially processed by a low-order analogue filter, a sample-and-hold network (SH), the ADC, decimation filter (van der Kam, 1986)

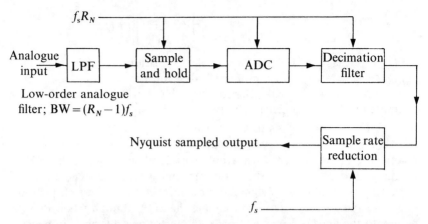

Figure 4.21 Analogue-to-digital conversion using mild oversampling

and subsampler. Although no formal noise shaping is included in this scheme, there is a degree of noise averaging which yields a mild enhancement to ADC resolution as the quantization noise power is spread over a band $R_N f_s/2$ hertz. The resolution of the oversampled scheme has an upper bound N_x bits that is related to the ADC resolution N_y bit and oversampling ratio R_N as

$$N_x = N_y + 0.5 \log_2 R_N \qquad (4.33)$$

The above result suggests, for example, that if $R_N = 4$, a 1 bit improvement in resolution results. However, it should be noted that Eq. (4.33) assumes that the ADC quantization noise power ($\Delta^2/12$) is uniformly distributed across the frequency band 0 to $R_N f_s/2$ hertz, allowing the decimation filter to remove noise from the band:

$$\frac{f_s}{2} < f < \frac{R_N f_s}{2} \qquad \text{hertz}$$

Should non-optimal dither be used (Vanderkooy and Lipshitz, 1984) or insufficient high-frequency dither be present, then quantization noise in general will be non-uniform and concentrated at lower frequencies; consequently, the predicted increase in resolution will not result. A similar argument also reveals that for large oversampling ratios, Eq. (4.33) is inappropriate where the resolution advantage reaches a ceiling.

The performance of the sample-and-hold network and ADC may also become a limiting factor at higher oversampling ratios, where distortion increases due to finite acquisition time and conversion time, which then act to reduce the overall system resolution. It may therefore be concluded that for the ADC of Fig. 4.21 most of the system's advantage is realized with mild oversampling (e.g. $\times 2 \rightarrow \times 4$), particularly when viewed in relation to modern ADC resolution.

WAVEFORM CODING—DIGITAL AUDIO

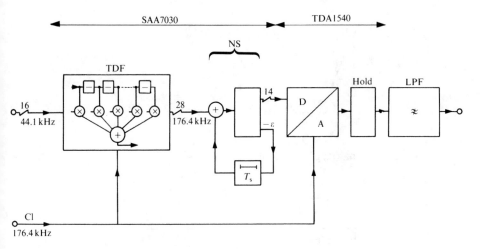

Figure 4.22 A ×4 oversampling filter and first-order noise shaper (after Goedhart, van de Plassche and Stikvoort, 1982)

However, the advantages of a digital antialiasing filter (or decimation filter) are considerable, particularly with the reduction in analogue circuit complexity and the resulting improvements in phase linearity and associated transient performance.

Oversampling has also been applied successfully in DAC systems and has been featured from the earliest Philips compact disc players (Goedhart, van de Plassche and Stikvoort, 1982), as shown in Fig. 4.22, and consists of a ×4 interpolation filter, a first-order noise shaper and DAC running at 176.4 kHz (for a 44.1 kHz sampling rate). The advantages of this scheme are similar to those cited for the ADC where linear phase and predominantly digital filtering perform the recovery filter function and allow a simple third-order Bessel filter with a 3 dB break frequency at 30 kHz to be used in the analogue domain.

The original Philips digital FIR filter employed 96 coefficients which, taking advantage of the reconfiguration discussed in Sec. 4.3, require only 24 multiplications per Nyquist period. Goedhart, van de Plassche and Stikvoort (1982) report that the filter coefficients are quantized to 12 bits and thus each product has a length of $16+12=28$ bits. Also the numbers have been selected such that a 24 point summation does not exceed 28 bits; consequently requantization occurs after the FIR filter. A feature of the digital filter response is that it includes frequency-domain correction for the aperture distortion arising from the sample-and-hold period of $1/4f_s$ second.

In Fig. 4.23, the process of the Philips DAC using interpolation is shown, together with the spectrum of the $\text{sinc}(\pi f/176\,400)$ function corresponding to the sample-and-hold, where the transmission zero at

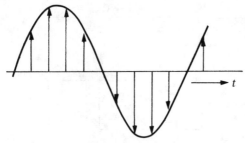

(a) Sampled analogue waveform

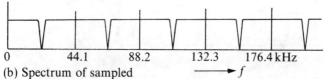

0   44.1   88.2   132.3   176.4 kHz
(b) Spectrum of sampled waveform where $f_s = 44.1$ kHz

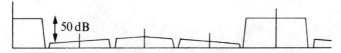

(c) Reduction first, second and third replication after interpolation LPF

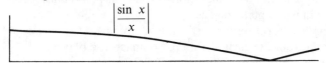

(d) Sine (x) function weighting due to sample and hold on ×4 oversampled signal

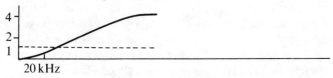

(e) Inherent aperture correction for sample and hold compensation

Figure 4.23 Frequency spectrum of Philips ×4 oversampling filter showing effect of ×4 interpolation filter, sample and hold and corresponding aperture correction (after Goedhart, van de Plassche and Stikvoort, 1982)

176.4 kHz (and subsequent multiples) aids suppression of the replicated audio pass-band centred on this frequency.

An interesting feature of the Philips DAC is the way the requantization process is performed, where advantages from both oversampling and noise shaping are achieved. In the earliest system the DAC offered only a 14 bit resolution (more recently upgraded to 16 bits). However, the data held on compact disc is in 16 bit format; thus Philips took the decision to enhance the resolution by operating the DAC at $4f_s$ and to use the potential advantage of oversampling and noise shaping.

4.5.1 The oversampling advantage

When the oversampled output of the FIR filter is truncated to 14 bits and assuming adequate signal activity and/or digital dither (Lipshitz and Vanderkooy, 1986) to decorrelate distortion from the signal, the requantization distortion is uniformly distributed over the band 0 to 88.2 kHz. Consequently the additional noise power within the audio band of 0 to 22.05 kHz is only 0.25 of the total and represents about a 1 bit enhancement in resolution. However, for this advantage to be fully realized, the DAC must have an accuracy significantly above 14 bits, where in latter conversion systems the DAC has been upgraded to 16 bits, although oversampling and noise shaping is still retained to help reduce residual impairments.

4.5.2 The noise-shaping advantage

The truncation of the 28 bit output word to the 14 bit word required for the (14 bit) DAC is performed by a first-order noise shaper as illustrated in Fig. 4.24(a). Effectively the truncation error is fed back to the FIR output but with a sample delay of (1/176 400) second. The noise shaper is also shown in Fig. 4.24(b) and (c) in two equivalent forms: in Fig. 4.24(b) the system is seen to be similar to error feedback distortion correction (Hawksford, 1981) as used in analogue amplifier systems, while in Fig. 4.24(c) it is seen to be identical to the first-order noise shaper ($N = 1$) described in Sec. 4.4, Fig. 4.13. Goedhart, van de Plassche and Stikvoort in their 1982 paper suggest that the effective reduction in noise offered by the noise shaper is 7 dB and corresponds approximately to 1 bit; however, the noise spectrum is not flat and is approximately proportional to frequency, being correspondingly less at low frequency with little high-frequency advantage, and follows curves similar to Fig. 4.16 for $N = 1$ where Eqs (4.24) and (4.25a) suggest a noise advantage below 176.4/6 kHz. Consequently, a combination of oversampling and noise shaping is seen to exercise the DAC in a way that produces close to 16 bit resolution, which is further aided by the ear's insensitivity to low-level, high-frequency noise (Fletcher and

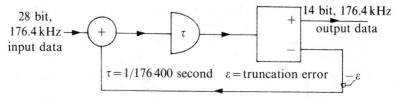

(a) Philips noise shaper (after Goedhardt, van de Plassche and Stikvoort, 1982)

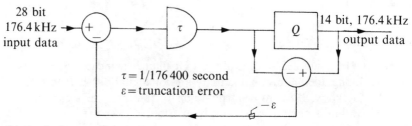

(b) Equivalent model similar to error feedback distortion correction

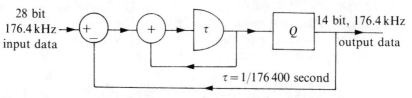

(c) First-order noise shaper derived from (b) and similar to Fig. 4.13 with $N=1$

Figure 4.24 Noise shaper system as used in Philips DAC

Munson, 1937), which makes the rising high-frequency noise of the noise shaper of little concern.

The DAC topology used in the Philips systems uses an ingenious process called *dynamic element matching*. The DAC is formed from a tree structure (shown in Fig. 4.25(a)) of current dividers based upon matched transistor arrays (Goedhart, van de Plassche and Stikvoort, 1982). Although transistor matching gives reasonable current division there are still small errors, as represented by ΔI in Fig. 4.25(b). These errors are reduced by an order of magnitude by arranging an electronic switch to rapidly interchange the output currents, where the switching ideally is infinitely fast and has a 1:1 switching ratio; the switched output currents are then smoothed to reduce residual ripple. The technique offers very accurate current division with both excellent long-term stability and immunity from temperature dependence.

This section has reviewed two complementary examples of ADC and DAC using mild oversampling and, in the case of the DAC, first-order noise shaping. More recently similar systems have emerged using higher oversampling ratios in the range ×8 to ×64, where performance gains in interpolation filter performance are claimed. However, these systems still rely on conventional DACs where a minimal attempt is made to address possible large-scale DAC non-ideality. In the next section, the potential of more powerful oversampling and noise-shaping filters is investigated as a means of decorrelating inherent distortion with signal and of extending the dynamic range to greater than 16 bits.

4.6 Applications of high oversampling in ADC and DAC

The introductory section 4.1 referred to *more natural conversion topologies*. This phraseology intended to imply an almost linear and continuous conversion from analogue-to-digital and digital-to-analogue, with the minimum of discontinuous processing such as sample-and-hold, where the conversion system would resemble an analogue process but with a digital output structure. As an example, the delta–sigma

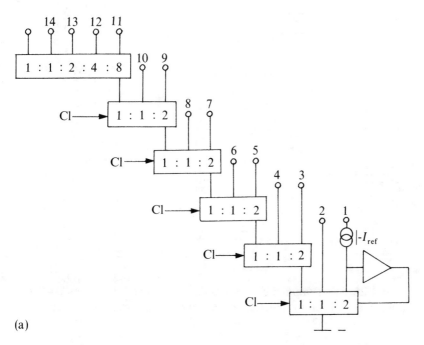

(a)

Figure 4.25(a) DAC tree structure of current dividers (after Goedhart, van de Plassche and Stikvoort, 1982)

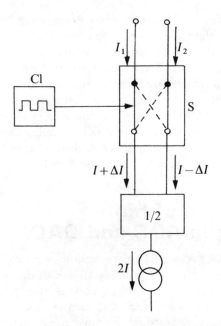
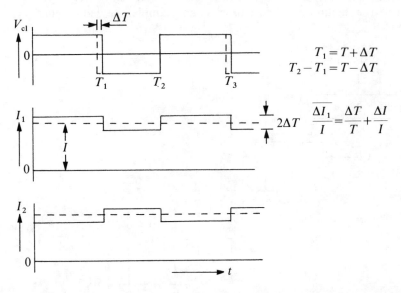

(b)

Figure 4.25(b) Dynamic element matching as used in Philips DAC to compensate for current division errors (after Goedhart, van de Plassche and Stikvoort, 1982)

modulator matches this concept where the digital signal when viewed macroscopically closely resembles the analogue input signal, but when examined microscopically the signal retains a digital structure. Indeed, taking a limiting case, the electron activity in analogue circuitry can be considered as discrete events which, when averaged, result in the more common expression of signal currents (Hawksford, 1983), though this latter viewpoint is very much a philosophical stance.

This section considers three applications of oversampling applied to both ADC and DAC and to PCM-to-1 bit conversion, the latter now more commonly called *bitstream*. All applications follow the same philosophy and use a combination of oversampling and noise shaping together with either decimation or interpolation as described in Secs 4.3 and 4.4.

4.6.1 High-order noise shaping and oversampling in ADC

A system designed by Adams (1986) claims to achieve 18 bit resolution (later updated to 20 bit) using an oversampled converter with a quasi-fourth-order noise-shaping filter configured within the analogue domain using analogue integration and a flash converter for the quantizer. Figure 4.26(a) shows the conceptual noise shaper derived from the digital model of Fig. 4.14(a) together with the front-end coder of the Adams system in Fig. 4.26(b).

The fourth-order system is constructed from a cascade of three virtual earth integrators and a passive integrator together with phase advance compensation to generate the transmission zeros necessary for closed-loop stability. The output signal V_x is related to the effective quantized output signal V_Q and signal V_i by the transfer function

$$V_x = (V_i + 4.03 V_Q) \frac{-(1 + jf/121 \text{ kHz})^3}{(jf/121 \text{ kHz})^2 (jf/437 \text{ kHz})(1 + jf/15.75 \text{ kHz})} \quad (4.34)$$

while a 4 bit flash converter (later upgraded to 6 bit) digitizes the resultant signal. A fast 4 bit DAC is constructed using a CMOS buffer and a parallel ladder network of fifteen $20 \text{ k}\Omega$ precision resistors, which allows the loop to be closed in the analogue domain.

The sampling rate of the front-end coder is set to 6 MHz, and digital output data is presented in a 4 bit format following 15-4 code conversion after the flash converter. A critical factor that affects coding accuracy is the precision of the DAC where effects of level imbalance, of slew rate limiting and of clock jitter result in non-linear area modulation of the reconstructed analogue pulses. Since the closed-loop activity attempts to match the DAC output to the input signal, distortion in the DAC is directly related to distortion in the digital output code. However, distortion in the flash converter is less significant as this appears in the forward path of the noise shaper. The

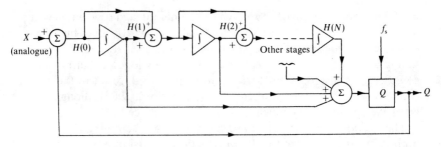

(a) Basic analogue Nth-order noise shaper

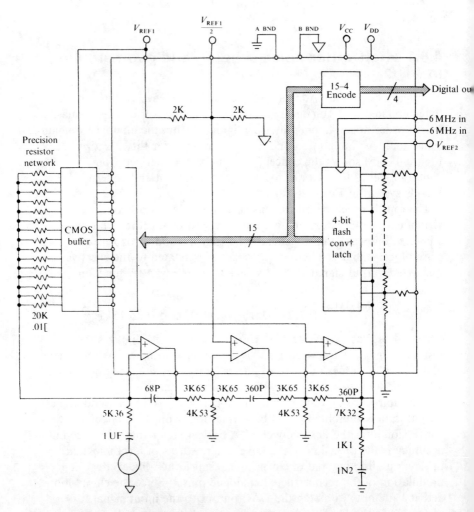

(b) dbx front-end, fourth-order noise-shaped ADC

Figure 4.26 Oversampled and noise-shaped ADC front-end coder

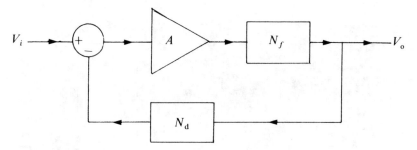

Figure 4.27 Simplified oversampled and noise-shaped coder showing flash converter and DAC non-linearities

effects of the flash converter and DAC non-linearity can be understood using the simplified model of Fig. 4.27, where A is the forward path gain and N_f and N_d the respective non-linear transfer characteristics of flash converter and DAC.

The non-linear transfer function A_N follows as

$$A_N = \frac{AN_f}{1 + AN_f N_d}$$

where for zero non-linearity the optimum transfer function $A_0(N_f=1, N_d=1)$ is

$$A_0 = \frac{A}{1+A}$$

Following the procedure of Sec. 4.4 and defining an error function E_N,

$$E_N = \frac{A_N}{A_0} - 1$$

whereby

$$E_N = \frac{(N_f - 1)}{1 + AN_f N_d} - \frac{AN_f(N_d - 1)}{1 + AN_f N_d} \qquad (4.35)$$

This result demonstrates that, for $AN_f N_d \gg 1$, the forward path error term $(N_f - 1)$ is desensitized while the DAC error term is virtually unprotected. Consequently, the importance of matching the DAC quantization characteristic to that of the flash converter is evident.

The translation of the oversampled signal (4 bit by 6 MHz) is converted to either 48 or 44.1 kHz by a two-stage decimation filter, as shown by the two integrated circuit modules in Fig. 4.28, where the second stage is a half-band filter as discussed in Sec. 4.3.

The Adams converter claims a 104 dB SNR with 0.003 per cent THD at the standard 48 kHz sampling rate (though 44.1 kHz is also available). The decimation filter structure and high oversampling ratio

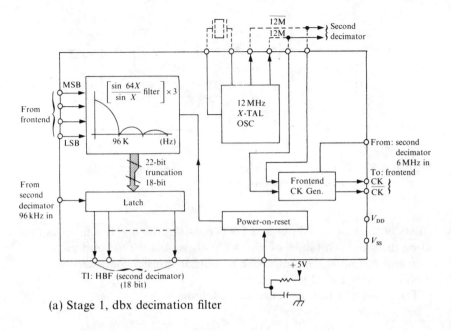

(a) Stage 1, dbx decimation filter

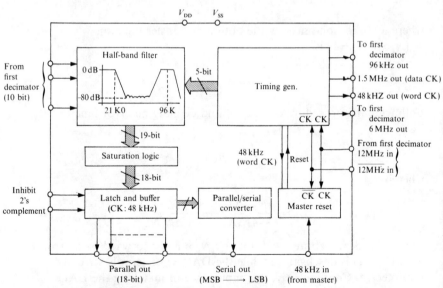

(b) Stage 2, dbx decimation filter

Figure 4.28 Two stages of decimation used in dbx noise-shaped and oversampled ADC

also ensure an effective antialiasing performance with linear phase, while the self-dither of a fourth-order noise shaper aids low-level resolution by decorrelation of quantization distortion. The similarity to delta–sigma modulation is evident, but by relaxing the constraints upon the loop quantizer an order >2 is feasible, yielding significant enhancements in coding performance that are well matched to the needs of high-performance digital audio.

4.6.2 High-order noise shaping and oversampling in DAC

A complementary technique (Hawksford, 1987; McCrea, 1987) to the Adams system can be used to achieve digital-to-analogue conversion, where the basic structure is shown in Fig. 4.29. The scheme employs an interpolation filter (see Sec. 4.3) which translates the Nyquist sampled signal at f_s hertz to an oversampled signal $R_N f_s$ hertz where in this scheme R_N ranges typically from 64 to 256. A noise shaper is then used to reduce the sample resolution from, say, 16 to 4 bits. The performance of this system has already been reviewed in Sec. 4.4 where in Fig. 4.17 the quantizer time-domain activity was shown together with SNR plots in Fig. 4.18. However, these results correspond to an ideal DAC together with an ideal interpolation filter; they do not include typical errors encountered in the hardware.

DAC hardware exhibit a range of non-linear distortion mechanisms that includes:

1. Systematic displacement of DAC reconstruction levels to give large-scale curvature in the quantization transfer characteristic.

2. Random perturbations of each reconstructive level, superimposed upon the curvature in mechanism 1.

3. Differential timing error in DAC logic to model glitch distortion.

4. Slew-rate limiting distortion that results in non-linear pulse area modulation as a function of adjacent pulse difference.

5. Clock jitter which modifies the pulse area, particularly where there are large intersample changes.

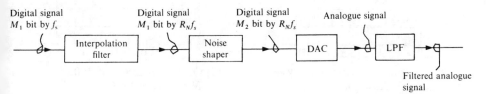

Figure 4.29 Structure of DAC system using high-order noise shaping and oversampling

In a conventional DAC, these non-linearities result in distortion due to correlation between the error waveform and signal. However, a noise-shaping DAC of order ≥4 exhibits a high-level dither that can span the full range of the quantizer. Consequently, all DAC reconstruction levels on average participate in the conversion process, where random-like or chaotic loop activity reduces the correlation of error waveform and signal.

The decorrelation process is demonstrated in the computer-generated spectral plots shown in Fig. 4.30 where an imperfect DAC is used in

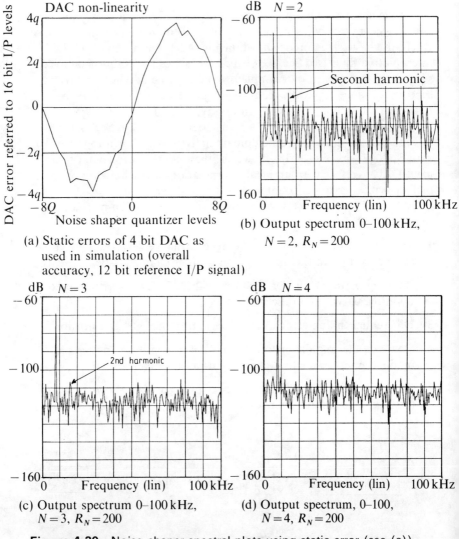

Figure 4.30 Noise-shaper spectral plots using static error (see (a)) for non-ideal DAC, $N=2$ to 4, $R_N=200$

WAVEFORM CODING—DIGITAL AUDIO

association with noise shapers $N=1$ to 4. The results show that as N is increased, there is a marked reduction in non-linear distortion, where for $N \geq 4$, decorrelation is virtually complete. A broad conclusion derived from simulation is that for digital audio applications where, say, $R_N = 200$, $N = 4$, a DAC with 4 bit resolution requires an accuracy ≥ 12 bits and that the slew rate induced and clock jitter dependent distortion should be minimized.

4.6.3 Multilevel to 1 bit conversion

Delta–sigma modulation (Inose, Yasuda and Murakami, 1962) is a means of converting signals into a 1 bit serial code, where traditionally a single-stage process employs either a first- or second-order noise shaping filter together with two-level quantization (Naus et al., 1987). Simulations reveals that filter orders >2 generally lead to instability that is directly a function of the limited quantizer range; however, by relaxing the range constraint high-order noise shaping as described in Sec. 4.4 is feasible. The need for a quantizer with >2 levels is evident by observing the time-domain rasters of Fig. 4.17, where for $N=4$ at least 16 levels are required as predicted by Eq. (4.25b).

However, a high-order noise shaper can be used in PCM-to-1 bit conversion if the process is performed in two stages of level compaction as shown in Fig. 4.31 (Hawksford, 1988). Effectively the first stage

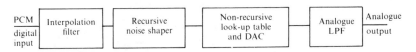

(a) Basic two-stage recursive/non-recursive noise-shaping PCM to 1 bit DAC

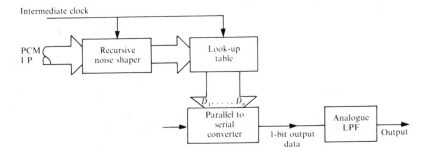

(b) Basic m bit to 1 bit transformer

Figure 4.31 The m bit to 1 bit transformer using recursive and non-recursive noise shaping

consists of an interpolation filter and a noise shaper as already described in this section, but the second stage consists of an open-loop look-up table that translates the multilevel output data to a serial 1 bit format; the level compaction system is, therefore, a combination of recursive noise shaping and non-recursive interpolation.

In performing non-recursive code conversion, codes must be selected that are both efficient yet introduce minimum extra distortion. The parallel-to-serial code conversion is a form of time dispersive coding where a sample defined at an instant in time is translated to a binary sequence spanning the stage 1 noise-shaper sampling period. Although it is possible to select a serial code that matches exactly the multilevel sample, this does not guarantee zero distortion as different codes will display different Fourier transforms. As a simple example, quantized pulse width modulation can form the basis of a code generator where the pulse width is made proportional to the input sample, where Fig. 4.32(a) shows transforms of pulses of varying width. Hence, although the d.c. component can match exactly the input sample, the spectrum is dynamically modulated as a function of pulse width which directly leads to the generation of non-linear distortion. In order to minimize

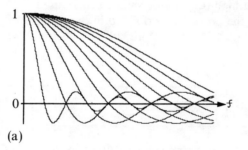

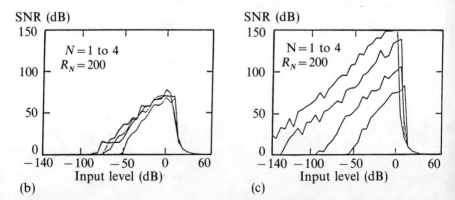

Figure 4.32 (a) $\sin(x)/x$ frequency response weighting as a function of pulse width: a fundamental distortion mechanism in PWM. (b), (c) SNR versus input for code conversion tables 4.2 and 4.3

WAVEFORM CODING—DIGITAL AUDIO

dynamic distortion, codes must be selected that offer near-identical transforms within the audio band, where the following three criteria are identified:

1. The sum of 1 and 0 pulses must average in exact proportion to the input sample.
2. Codes must be time symmetric about the sampling instant to ensure zero dynamic phase modulation.
3. Codes must be selected to offer minimum frequency response deviation within the audio band after normalization by the d.c. term.

To compute the Fourier transform of a code sequence of $2M$ or $2M+1$ (depending on whether an even or odd sequence is selected) the following equations apply:

$$A_e(f) = 2 \sum_{r=1}^{M} a_r \cos\left(\frac{2\pi f}{Mf_s}\right) \text{(even sequence)} \quad (4.36a)$$

$$A_o(f) = a_0 + \sum_{r=1}^{M} a_r \cos\left[\frac{2\pi f}{(M+1)f_s}\right] \text{(odd sequence)} \quad (4.36)$$

However, since the shape of the spectra need only be matched within the audio band where $f \ll f_s$, then, using the approximation $\cos x \approx 1 - x^2/2$,

$$A_e(f) \approx 2 \sum_{r=1}^{M} a_r - \sum_{r=1}^{M} a_r \left(\frac{2\pi f}{Mf_s}\right)^2 \quad (4.37a)$$

$$A_o(f) \approx a_0 + 2 \sum_{r=1}^{M} a_r - \sum_{r=1}^{M} a_r \left[\frac{2\pi f}{(M+1)f_s}\right]^2 \quad (4.37b)$$

whereby the normalized distortion spectra $d_{en}(f), d_{on}(f)$ are

$$d_{en}(f) = \frac{-\sum_{r=1}^{M} a_r \left(\frac{2\pi f}{Mf_s}\right)^2}{2 \sum_{r=1}^{M} a_r} \quad (4.38a)$$

$$d_{on}(f) = \frac{-\sum_{r=1}^{M} a_r \left[\frac{2\pi f}{(M+1)f_s}\right]^2}{a_0 + 2 \sum_{r=1}^{M} a_r} \quad (4.38b)$$

Equations (4.38a) and (4.38b) can be used as a basis for a computer search where, having decided upon the code length M, the available codes are sifted to extract the set with minimum $d_{en}(f), d_{on}(f)$ values. There is a trade-off to be made of the code efficiency against distortion where by increasing the word length a greater variety of available codes emerge but with a corresponding increase in sampling rate.

Table 4.2 Suboptimal code conversion

Level	Output of look-up table
8	1 1 1 1 1 1 1 1 1 1 1 1 1 1 1 1
7	0 1 1 1 1 1 1 1 1 1 1 1 1 1 1 1
6	0 1 1 1 1 1 1 1 1 1 1 1 1 1 0 1
5	0 1 0 1 1 1 1 1 1 1 1 1 1 1 0 1
4	0 1 0 1 1 1 1 1 1 1 1 1 0 1 0 1
3	0 1 0 1 1 1 1 1 1 1 0 1 0 1 0 1
2	0 1 0 1 0 1 1 1 1 1 0 1 0 1 0 1
1	0 1 0 1 0 1 1 1 0 1 0 1 0 1 0 1
0	0 1 0 1 0 1 0 1 0 1 0 1 0 1 0 1
−1	0 1 0 1 0 1 0 0 0 1 0 1 0 1 0 1
−2	0 1 0 1 0 0 0 0 0 1 0 1 0 1 0 1
−3	0 1 0 1 0 0 0 0 0 0 0 1 0 1 0 1
−4	0 1 0 1 0 0 0 0 0 0 0 0 0 1 0 1
−5	0 1 0 0 0 0 0 0 0 0 0 0 0 1 0 1
−6	0 1 0 0 0 0 0 0 0 0 0 0 0 0 0 1
−7	0 0 0 0 0 0 0 0 0 0 0 0 0 0 0 1
−8	0 0 0 0 0 0 0 0 0 0 0 0 0 0 0 0

To illustrate this process, two code sets are presented in Tables 4.2 and 4.3 together with corresponding SNR plots in Fig. 4.32(b) and (c), where the first-stage noise shaper uses the parameters $N=4$ and $R_N=200$.

The Table 4.2 code conversion is suboptimal where the codes are asymmetrical and non-optimum on spectral criteria; consequently there are both dynamic phase and amplitude distortion mechanisms. However, Table 4.3 corresponds to an optimal sifted code set where the corresponding increase in the SNR is shown in Fig. 4.32(b). These

WAVEFORM CODING—DIGITAL AUDIO

Table 4.3 Enhanced code conversion

Level	Output of look-up table
8	1 1 1 1 1 1 1 1 1 1 1 1 1 1 1 1 1 1 1 1 1 1 1 1 1 1 1 1
7	1 1 1 1 1 1 0 1 1 1 1 1 1 1 1 1 1 1 1 1 1 1 0 1 1 1 1 1
6	1 1 1 0 1 1 1 1 1 1 1 0 1 1 1 1 1 1 0 1 1 1 1 1 1 0 1 1 1
5	1 1 1 0 1 1 1 1 0 0 1 1 1 1 1 1 1 1 1 1 0 0 1 1 1 0 1 1 1
4	1 0 1 1 1 1 1 0 0 1 1 1 1 0 1 1 0 1 1 1 1 0 0 1 1 1 1 1 0 1
3	1 1 1 1 0 0 0 0 1 1 1 0 1 1 1 1 1 1 1 0 1 1 1 0 0 0 0 1 1 1 1
2	1 0 1 0 1 1 0 1 1 0 1 1 0 1 1 0 0 1 1 0 1 1 0 1 1 0 1 1 0 1 0 1
1	1 1 1 0 0 0 0 1 0 1 0 0 1 1 1 1 1 1 1 1 0 0 1 0 1 0 0 0 0 1 1 1
0	0 1 1 0 1 0 0 1 1 0 0 1 0 1 1 0 0 1 1 0 1 0 0 1 1 0 0 1 0 1 1 0
−1	0 0 0 1 1 1 1 0 1 0 1 1 0 0 0 0 0 0 0 1 1 0 1 0 1 1 1 1 0 0 0
−2	0 1 0 1 0 0 1 0 0 1 0 0 1 0 0 1 1 0 0 1 0 0 1 0 0 1 0 0 1 0 1 0
−3	0 0 0 0 1 1 1 1 0 0 0 1 0 0 0 0 0 0 0 0 1 0 0 0 1 1 1 1 0 0 0 0
−4	0 1 0 0 0 0 0 1 1 0 0 0 0 0 1 0 0 1 0 0 0 0 0 1 1 0 0 0 0 0 1 0
−5	0 0 0 1 0 0 0 0 1 1 0 0 0 0 0 0 0 0 0 0 0 0 1 1 0 0 0 0 1 0 0 0
−6	0 0 0 1 0 0 0 0 0 0 0 0 1 0 0 0 0 0 0 1 0 0 0 0 0 0 0 0 1 0 0 0
−7	0 0 0 0 0 0 1 0 0 0 0 0 0 0 0 0 0 0 0 0 0 0 0 0 0 0 1 0 0 0 0 0 0
−8	0 0 0 0 0 0 0 0 0 0 0 0 0 0 0 0 0 0 0 0 0 0 0 0 0 0 0 0 0 0 0 0

results clearly demonstrate the importance of minimizing dynamic distortion in the selection of codes, especially if a performance compatible with high-resolution digital audio systems is required.

The strategy presented in this section demonstrates how a two-stage process can be used to perform PCM-to-1 bit DSM code conversion by relaxing the stability constraints imposed by using a quantizer with a limited range. The advantage of the technique is that high oversampling together with high-order noise shaping (e.g. $R_N = 200$, $N = 4$) yields a wide dynamic range while the non-recursive code converter produces a

SPEECH PROCESSING

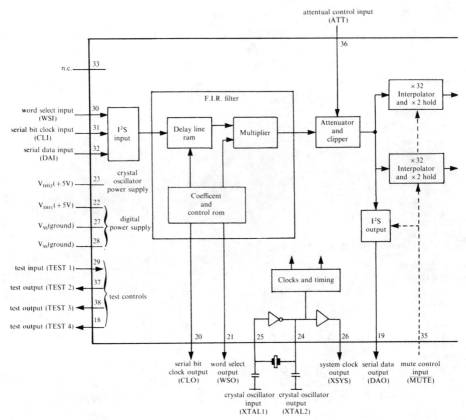

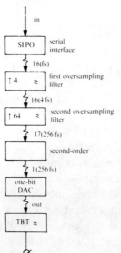

WAVEFORM CODING—DIGITAL AUDIO

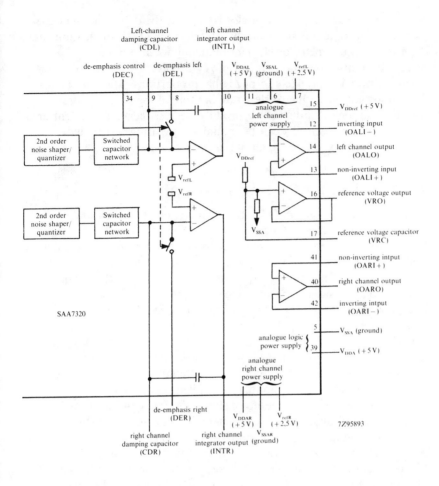

Figure 4.33 Philips 1 bit DAC and flowchart using oversampling and delta–sigma modulation (reproduced courtesy of Philips Semiconductors)

serial signal requiring only a single fast gate at the digital-to-analogue gateway. The system is particularly attractive for DAC applications as there is no requirement for element matching to ensure linear quantization also, the employment of $N \geq 3$ noise shaping reduces idle channel distortion artefacts to negligible proportions. The only major distortions are those due to slew-rate limiting of the output pulse waveform and clock jitter which produce code sequence dependent area modulation, but even this can be minimized by using a switched capacitor network to ensure all pulses have equal area.

4.6.4 CMOS DAC using 1 bit code

A *bitstream* stereo CMOS DAC that exploits oversampling and 1 bit conversion has been introduced by Philips (Naus *et al.*, 1987) (device SAA7320/1), where the basic chip architecture and system flow diagram is shown in Fig. 4.33. Oversampling is performed using a three-stage process: $\times 4$, 128 tap FIR interpolation yielding ± 0.02 dB in-band ripple, $\times 32$ oversampling using linear interpolation and $\times 2$ interpolation using sample-and-hold, resulting in a sampling rate of 11.2896 MHz. The interpolator also includes an internally generated, 176.4 kHz out-of-band dither signal, the latter being included to minimize low-level idle channel noise in the 1 bit converter. The transformation of the output code to 1 bit conversion is performed by a second-order delta–sigma modulator implemented in the digital domain, and is similar to the noise shaper of Fig. 4.14 but using only a two-level quantizer. A characteristic of low-order DSM is that at a low input signal level, idle channel bit patterns are generated with an almost periodic structure which result in objectionable distortion as the error waveform is no longer noise-like; it is to randomize the idle channel noise that digital dither is used prior to the DSM converter. Finally, 1 bit digital-to-analogue conversion is performed using a switched capacitor network, a two-phase clock and a virtual earth, transimpedance operational amplifier as shown in Fig. 4.34, where a third-order Butterworth, analogue reconstruction filter is recommended with a 3 dB break frequency at 60 kHz, as compensation is included within the digital interpolation filter. The switched capacitor network produces constant-charge packets and is therefore more tolerant of clock jitter which would otherwise impair the SNR.

4.7 Conclusion

This chapter has reviewed the principles of noise shaping and oversampling and attempted to present a more unified picture of the relationship between analogue and digital domains through the modelling of oversampling and quantization. Special emphasis was

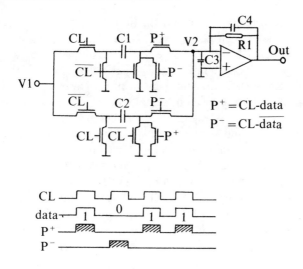

(a) Switched capacitor filter, 1 bit DAC

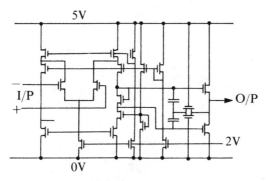

(b) Operational amplifier

Figure 4.34 A 1 bit DAC as used in Philips digital delta–sigma modulator

given to delta–sigma modulation both as an application of noise shaping and as a more natural means of ADC and DAC. However, the relaxation of the stability constraints of DSM using a multilevel quantizer enabled the introduction of higher order noise shapers which also represented a method of matching the amplitude resolution/conversion speed performance of a converter to maximize the throughput of audio information.

The performance bounds of a noise-shaped, oversampled coder were described both by analysis and more accurately by exact computer modelling. Of particular interest with the higher order noise-shaping filter was the intrinsic level of random-like behaviour of the output signal which led directly to useful distortion decorrelation. However, at a philosophical level such performance is an excellent example of a chaotic system (Gleick, 1988), where relatively simple recursive equations together with quantizer non-linearity led to chaotic behaviour where the exact output waveform is extremely sensitive to the slightest perturbation in input signal, system parameters and initial conditions, yet within this apparent chaos there is order such that on band-limiting the output sequences, a structured waveform emerges that can closely approximate to the input excitation.

The advantages of noise shaping and oversampling were seen both as a means of performance optimization of ADC and DAC and as a means for implementing a more predictable conversion system, where the subsequent use of digital filters, dither and companding algorithms can in principle define the channel characteristics but with the minimum impairment from the converter.

References and further reading

References

Adams, R. W. (1984) 'Companded predictive deltamodulation: a low cost conversion technique for digital recording', *Journal of Audio Engineering Society*, vol. 32, no. 9, 659–672.

Adams, R. W. (1986) 'Design and implementation of an audio 18-bit analog-to-digital converter using oversampling techniques', *Journal of Audio Engineering Society*, vol. 34, no. 3, 153–166.

Ansari, R. (1988) 'Satisfying the Haar condition in halfband FIR filter design', *IEEE Transactions on Acoustics, Speech, and Signal Processing*, vol. 36, no. 1, 123–124.

Bellanger, M. G., J. L. Daguet and G. Lepagnol (1974) 'Interpolation, extrapolation and reduction of computation speed in digital filters', *IEEE Transactions on Acoustics, Speech and Signal Processing*, vol. ASSP-22, no. 4, 231–235.

Crochiere, R. E. and L. R. Rabiner (1975) 'Optimum FIR filter implementation for decimation, interpolation and narrow band filtering', *IEEE Transactions on Acoustics, Speech and Signal Processing*, vol. ASSP-23, 444–456.

Crochiere, R. E. and L. R. Rabiner (1983) *Multirate Digital Signal Processing*, Prentice-Hall, Englewood Cliffs, N.J.

Darling, T. F. (1987) 'Oversampled analogue–digital conversion for digital audio systems', M.Phil. Thesis, submitted to University of Essex, England.

Darling, T. F. and M. O. J. Hawksford (1988) 'Oversampled analogue-to-digital conversion for digital audio systems', 85th AES Convention, Los Angeles, Preprint 2740 (H-11).

de Jager, F. (1952) 'Deltamodulation, a method of PCM transmission using 1-unit code', Philips Research Report 7, 442–466.

Deloraine, E. M., S. van Miero and B. Derjavitch (1947–1948) 'Methods et systeme de transmission par impulsions', French Patent 932.140.

Fletcher, H. and W. A. Munson (1937) 'Relation between ioudness and masking', *Journal of Acoustical Society of America*, vol. 9.1, 1.

Gleick, J. (1988) *Chaos: Making a New Science*, William Heinemann, London.
Goedhart, D., R. J. van de Plassche and E. F. Stikvoort (1982) 'Digital to analogue conversion in playing a compact disc', *Philips Technical Review*, vol. 40, no. 6, 174–179.
Goodman, D. J. (1969a) 'Delta-modulation granular quantising noise', *Bell System Technical Journal*, vol. 48, pt 1, 1197.
Goodman, D.J. (1969b) 'The application of delta modulation to analogue-to-PCM encoding', *Bell System Technical Journal*, vol. 48, no. 2, 321–343.
Goodman, D. J. and M. J. Carey (1977) 'Nine digital filters for decimation and interpolation', *IEEE Transactions on Acoustics, Speech, and Signal Processing*, vol. ASSP-25, no. 2, 121–126.
Hansen, V. (1987) 'Design of a multistage decimation-interpolation filter', International Conference on Acoustics, Speech, and Signal Processing, 21.9.1–21.9.4.
Hawksford, M. J. (1977) 'Deltamodulation coder using a parallel realisation', IERE Conference Proceedings, vol. 37, 547–557.
Hawksford, M. J. (1981) 'Distortion correction in audio power amplifiers', *Journal of Audio Engineering Society*, vol. 29, no. 1, 2, 27–30.
Hawksford, M. J. (1983) 'Fuzzy distortion in analog amplifiers: a limit to information transmission?' *Journal of Audio Engineering Society*, vol. 31, no. 10, 745–754.
Hawksford, M. J. (1985) 'Nth-order recursive sigma-ADC machinery at the analogue–digital gateway', 78th AES Convention, Anaheim, Preprint 2248 A-15.
Hawksford, M. J. (1987) 'Oversampling and noise shaping for digital to analogue conversion', Institute of Acoustics, Reproduced Sound 3, 151–175.
Hawksford, M. J. (1988). 'Multi-level to 1 bit transformations for applications in digital to analogue converters using oversampling and noise shaping', *Proceedings of the Institute of Acoustics*, vol. 10, pt. 7, 129–143.
Inose, H. and Y. Yasuda (1963) 'A unity bit coding method by negative feedback', *Proceedings of the IEEE*, vol. 51, 1524.
Inose, H., Y. Yasuda and J. Murakami (1962) 'A telemetering system by code modulation—delta sigma modulation', *IRE Transactions*, vol. PGSET 8, 204.
Lipshitz, S. P. and J. Vanderkooy (1986) 'Digital dither', 81st AES Convention, Preprint no. 2412.
McClellan, J. H., T. W. Parks and L. R. Rabiner (1973) 'A Computer program for designing optimum FIR linear phase digital filters', *IEEE Transactions on Audio and Electroacoustics*, vol. AU-21, no. 6, 506–517. See also *Programs for Digital Signal Processing*, IEEE Press, New York, 1979, 5.1.1–5.1.13.
McCrea, B. A. (1987) 'Simulation of audio digital-to-analogue conversion using noise shaping and oversampling', M.Sc. Dissertation, Department of ESE, University of Essex.
Matsuya, Yasuyuki (1987) 'A 16-bit oversampling A-to-D conversion technology using triple-integration noise shaping', *IEEE Journal of Solid State Circuits*, vol. SC-22, no. 6.
Mintzer, F. (1982) 'On half-band, third-band and Nth-band FIR filters and their design', *IEEE Transactions on Acoustics, Speech, and Signal Processing*, vol. ASSP-30, 734–738.
Naus, P. J., E. C. Dijkmans, E. F. Stikvoort, A. J. McKnight, D. J. Holland and W. Bradinal (1987) 'A CMOS stereo 16-bit D/A converter for digital audio', *IEEE Journal of Solid State Circuits*, vol. SC-22, no. 3.
Peetz, B. *et al.* (1986), 'An 8-bit 250 megasample per second analog-to-digital converter', *IEEE Journal of Solid State Circuits*, vol. SC-21, no. 6.

Philips, N. V. (1949) French Patent specification 987.238.

Rabiner, L. B. and B. Gold (1975) *Theory and Application of Digital Signal Processing*, Prentice-Hall, Englewood Cliffs, N.J.

Ritchie, G. R., J. C. Candy and W. H. Ninke (1974), 'Interpolative digital-to-analog converters', *IEEE Transactions on Communications*, vol. COM-22, 1797–1806.

Shannon, C. E. (1948) 'A mathematical theory of communication', *Bell System Technical Journal*, vol. 27, 379–423, 623–656.

Steele, R. (1975) *Deltamodulation Systems*, Pentech Press, Plymouth, UK.

Steele, R. (1980), 'SNR formula for linear delta modulation with band-limited flat and RC shaped Gaussian signals', *IEEE Transactions on Communications*, vol. COM-28, no. 12.

Sun, M. T. (1986) 'Efficient design of the oversampling filter for digital audio applications', 81st AES Convention, Los Angeles, Preprint no. 2378 (C11).

Tewksbury, S. K. and R. W. Hallock (1978) 'Oversampled linear predictive and noise-shaping coders of order $N > 1$', *IEEE Transactions on Circuits and Systems*, vol. CAS-25-25/7, 437–447.

Vaidyanathan, P. P. and T. Q. Nguyen (1987) 'A "trick" for the design of FIR half-band filters', *IEEE Transactions on Circuits and Systems*, vol. CAS-34, no. 3, 297–300.

van der Kam, J. J. (1986) 'A digital decimating filter for analog-to-digital conversion of hi-fi audio signals', *Philips Technical Review*, vol. 42, 230–238.

Vanderkooy, J. and S. P. Lipshitz (1984) 'Resolution below least significant bit in digital systems with dither', *Journal of Audio Engineering Society*, vol. 32, no. 3, 106–113.

Watkinson, J. (1988) *The Art of Digital Audio*, Focal Press, Butterworths, London.

Further reading

Bars, G. and J. P. Petit (1986) 'High quality sound decoder using noise shaping techniques and digital filters', IEEE Acoustics, Speech, and Signal Processing Workshop, Monk Inn.

Blesser, B. A. (1974) 'An investigation of quantization noise', *Journal of Audio Engineering Society* (Project Notes), vol. 22, 20–22.

Blesser, B. (1978) 'Digitisation of audio: a comprehensive examination of theory, implementation and current practice', *Journal of Audio Engineering Society*, vol. 26, no. 10.

Blesser, B. A. (1982) 'Advanced analog-to-digital conversion and filtering—data conversion', AES Premiere Conference Papers on Digital Audio.

Blesser, B. A. and B. Locanthi (1986) 'The application of narrowband dither operating at the Nyquist frequency in digital systems to provide improved signal-to-noise ratio over conventional dither', 81st AES Convention Preprint no. 2416.

Boser, B. E., K. P. Karmann, H. Martin and B. A. Wooley (1988) 'Simulating and testing oversampled analog-to-digital converters', *IEEE Transactions on CAD*, vol. 7, no. 6, 668.

Candy, J. C. (1985) 'A use of double integration in sigma deltamodulation', *IEEE Transactions on Communications*, vol. COM-33, 249–258.

Candy, J. C. (1986) 'Decimation for sigma delta modulation', *IEEE Transactions on Communications*, vol. COM-34, 72–76.

Candy, J. C. and O. J. Benjamin (1981) 'The structure of quantisation noise from sigma-delta modulation', *IEEE Transactions on Communications*, vol. COM-29/9.

Cattermole, K. W. (1969) *Principles of Pulse Code Modulation*, Illife, London.
Crochiere, R. E. and L. R. Rabiner (1981) 'Interpolation and decimation of digital signals—a tutorial review', *Proceedings of the IEEE*, vol. 69/3, 300–331.
Dalton, C. J. (1971) 'Deltamodulation for sound distribution: a general survey', BBC Research Report no. 1971/12, UDC 621.376.5.
Everard, J. D. (1976) 'Improvements to delta–sigma modulators when used for PCM encoding', *Electronic Letters*, vol. 12/15.
Fielder, L. D. (1987) 'Evaluation of the audible distortion and noise produced by digital audio converters', 82nd AES Convention, London, Preprint 2424 (A-5).
Flood, J. E. and M. J. Hawksford (1971) 'Exact model for deltamodulation processes', *Proceedings of the IEE*, vol. 118, 115.
Freeman, D. M. (1987) 'Slewing distortion in digital-to-analogue conversion', *Journal of Audio Engineering Society*, vol. 25, no. 4, 178–183.
Gibbs, A. J. and L. R. Rabiner (1971) 'Techniques for designing finite-duration impulse-response digital filters', *IEEE Transactions on Communications Technology*, vol. COM-19, 188–195.
Gilchrist, N. H. C. (1980) 'Analogue-to-digital and digital-to-analogue converters for high quality sound', Presented at the 65th AES Convention, London, Preprint no. 1583.
Greefkes, J. A. and F. de Jager (1968) 'Continuous deltamodulation', Philips Research Report, 233.
Halijak, A. and J. S. Tripp (1963) 'A deterministic study of deltamodulation', IRE Internal Convention Report, pt 8, 247.
Hawksford, M. J. (1974) 'Unified theory of digital modulation', *Proceedings of the IEE*, vol. 121/2, 109–115.
Iwersen, J. E. (1969) 'Calculated quantization noise of single-integration deltamodulation coders', *Bell System Technical Journal*, vol. 48, no. 7, 2359.
Jongepier, A. (1980) 'A D/A converter which improves dynamic range by oversampling, slope modulation and slope demodulation', Proceedings of International Conference on Circuits and Systems, ISCAS.
Kasuga, M. (1986) 'An approach to high resolution D/A converters utilizing linear predictive coding', Proceedings of the IEEE Conference on Acoustics, Speech, and Signal Processing, Tokyo.
Koch, R. *et al.* (1986) 'A 12-bit sigma–delta analog-to-digital converter with a 15 MHz clock rate', *IEEE Journal of Solid State Circuits*, vol. SC-21.
Laane, R. R. (1970) 'Measured quantization noise spectrum for single-integration deltamodulation coders', *Bell System Technical Journal*, vol. 49, no. 2, 159.
Lagadec, R. *et al.* (1984) 'Dispersive models for A-to-D and D-to-A conversion systems', Presented at the 75th Convention, Paris, Preprint no. 209.
Lender, A. and M. Kazuch (1961) 'Single-bit deltamodulation system', *Electronics*, 125.
Meyer, J. (1984) 'Anti aliasing filters in digital audio system', *Journal of Audio Engineering Society*, vol. 32, no. 3, 132–137.
Mitra, D. (1977) 'Large amplitude, self-sustained oscillations in difference equations which describe digital filter sections using saturation arithmetic', *IEEE Transactions on Acoustics, Speech, and Signal Processing*, vol. ASSP-25, 134–143.
Mitra, D. (1978) 'The absolute stability of high-order discrete time systems utilizing the saturation non-linearity', *IEEE Transactions in Circuit Systems*, vol. CAS-25.
Mitsuhashi, Y. (1983) 'Mathematical analysis of pulse-width modulation

digital-to-analog converter', *Journal of Audio Engineering Society*, vol. 31, no. 3, 135–138.

Nakamara, Y. and H. Kaneko (1960) 'Deltamodulation encoder', NEC Research and Development, no. 1, 621.376.56–621.382, 45.

Nielson, P. T. (1971) 'On the stability of a double integration delta modulator', *IEEE Transactions on Communications Technology*, 364–366.

Oetken, G., T. W. Parks and H. W. Schuessler (1975) 'New results in the design of digital interpolators', *IEEE Transactions on Acoustics, Speech, and Signal Processing*, vol. ASSP-23, 301–309.

Oppenheim, A. V. and R. W. Schafer (1975) *Digital Signal Processing*, Prentice-Hall, Englewood Cliffs, N.J.

Parks, T. W. and J. H. McClellan (1972) 'Chebyshev approximation for nonrecursive digital filters with linear phase', *IEEE Transactions on Circuit Theory*, vol. CT-19, 189–194.

Rabiner, L. R. (1982) 'Digital techniques for changing the sampling rate of a signal', Collected Papers from the AES Premiere Conference Digital Audio, New York.

Ritchie, G. R. (1977) 'Higher order interpolative analog-to-digital converters', Ph.D. Dissertation, University of Philadelphia, Pa.

Roettcher, U., H. L. Fiedler and G. Zimmer (1986) 'A compatible CMOS-JFET pulse density modulator for interpolative high resolution A/D conversion', *IEEE Journal of Solid State Circuits*, vol. SC-21.

Rorabacher, D. W. (1975) 'Efficient FIR filter design for sample rate reduction or interpolation', Proceedings of 1975 International Symposium on Circuits and Systems.

Schouten, J. F., F. de Jager and J. A. Greefkes (1952) 'Deltamodulation, a new modulation system for telecommunications', *Philips Tech Tidjdschr*, vol. 13, 249, September 1951 (in Dutch); *Philips Technical Review*, vol. 13, 237.

Schouwenaars, H. J., E. C. Dijkmans, B. M. J. Kup and E. J. M. Van Tuijl (1986) 'A monolithic dual 16-bit D/A converter', *IEEE Journal of Solid State Circuits*, vol. SC-21, 424–429.

Schuchman, L. (1962) 'Dither signals and their effect on quantization', *AIEE Transactions on Communications Theory*, vol. COM-12.

Seitzer, D., P. Gunter and N. A. Hamdy (1983) *Electronic Analog-to-Digital Converters*, Wiley, Chichester.

Shively, R. R. (1975) 'On multistage finite impulse response (FIR) filters with decimation', *IEEE Transactions on Acoustics, Speech, and Signal Processing*, vol. ASSP-23, 353–357.

Spang, H. A. and P. M. Schultheiss (1962), 'Reduction of quantizing noise by use of feedback', *IEEE Transactions on Communications Systems*, vol. CS-10, 373–380.

Sripad, A. B. and D. L. Snijder (1977) 'A necessary and sufficient condition for quantization errors to be uniform and white', *IEEE Transactions on Acoustics, Speech, and Signal Processing*, vol. ASSP-25.

Stritek, P. (1987) 'Prospective techniques for improved signal-to-noise ratio in digital audio conversion systems', 82nd AES Convention, Preprint (2477 M-2).

Tewksbury, K. (1976) 'A/D converters for digital filters', Proceedings of the IEEE International Symposium on Circuits and Systems, 602–605.

Urkowitz, H. (1975) 'Parallel realizations of digital interpolation filters for increasing the sampling rate', *IEEE Transactions on Circuits and Systems*, vol. CAS-22, no. 2, 146–154.

Vaidyanathan, P. P. (1986) 'New cascaded lattice structure for FIR filters having extremely low coefficient sensitivity', Proceedings of IEEE

International Conference on Acoustics, Speech, and Signal Processing, Tokyo, Japan, 450–497.
van de Plassche, R. J. (1976) 'Dynamic element matching for high accuracy monolithic D/A converters', *IEEE Journal of Solid State Circuits*, vol. SC-11, 795–800.
van de Plassche, R. J. (1978) 'A sigma–delta modulator as an A/D converter', *IEEE Transactions on Circuits and Systems*, vol. CAS-25, no. 7, 510–514.
van de Plassche, R. J. and E. C. Dijkmans (1982) 'A monolithic 16-bit D/A conversion system for digital audio', AES Premiere Conference Papers on Digital Audio, New York.
van de Plassche, R. J. and D. Goedhart (1979) 'A monolithic 14-bit D/A converter', *IEEE Journal of Solid State Circuits*, vol. SC-14, no. 3.
van de Weg, H. (1953) 'Quantizing noise of a single-integration deltamodulation system with an N-digit code', Philips Research Report 8, 367.
Ward, R. G. (1986) 'Simulation of analogue to digital conversion processes', Ph.D. Dissertation, University College of Swansea.
Wood, K. (1980) 'Analogue-to-digital conversion techniques for audio systems', M. Phil. Dissertation, University of Essex.
Yee, P. W. (1986) 'Noise considerations in high-accuracy A/D converters', *IEEE Journal of Solid State Circuits*, vol. SC-21.

Acknowledgements

The discussion, contributions and computer modelling of noise-shaping and oversampling conversion schemes by audio engineering researchers at Essex University, Timothy Darling, Basil McCrea, Wolfgang Wingerter and Li Mu, are greatly appreciated. Also my thanks go to Chris Rowden for encouragement to publish this text.

5 Parametric coding of speech

CHRIS ROWDEN and STEPHEN HALL

5.1 Parametric coding

The basis for all parametric coding methods is to make use of the redundancy inherent in the speech signal to reduce the amount of information that must be sent. The redundancy arises from two sources: the most obvious is the repetition of waveshapes at a periodic rate, and, secondly, the presence of noise components in some speech sounds, for which reconstruction of the exact waveform is not perceptually important.

We can examine the redundancy by considering four different types of signal: voiced, unvoiced, mixed (in which the first two types overlap) and silence. Firstly, for voiced speech the signal waveform is periodic at a rate corresponding to the glottal pulse frequency. This periodicity may be variable over the duration of a speech segment, and the shape of the periodic wave usually changes gradually from segment to segment. Secondly, for unvoiced speech the signal is like random noise, being produced by the turbulent flow of air at restrictions in the vocal tract. However, the noise does not have a truly flat energy spectrum as would Gaussian noise; the spectrum is shaped by the resonances of the vocal tract, and this gives rise to a small amount of predictability. Thirdly, the two signal types occur together in voiced fricatives and in transitions between other sounds. Finally, neither is present during the silent periods within speech, generally within words before stop consonants, but also longer periods between talkspurts.

In any of these cases much more information is provided by waveform coding than is strictly required from information theoretic analysis, so there is a considerable scope for using parametric coding methods, which reduce the redundancy by separating the excitation component of the speech sound from the spectral envelope component. The excitation is then characterized as either a pulse train for voiced sounds or noise for unvoiced sounds. The spectral envelope, which is

PARAMETRIC CODING OF SPEECH

produced by resonances of the vocal tract, can be characterized by parameters of a frequency selective filter which has the same transfer characteristic as the vocal tract. It is conventional, but perhaps slightly counter-intuitive, to call the filter that models the vocal tract the prediction filter and for a filter that is optimized in the analysis process to be called the inverse filter. In this chapter, the issue is side-stepped and the terms synthesis filter and analysis filter will be used respectively.

The coding advantage arises from the slow rate at which the parameters of both the excitation signal and the filter characteristics change. It is difficult to determine exactly the rate at which the parameters change; it depends on the speaker and on what is being spoken. However, it is comparatively rare for the parameters to be significantly *different* from the values held 5 ms previously, and it is comparatively rare for the parameters to be *similar* to those held 30 ms previously. Thus the updating rate, or *segmental rate* $1/t_s$, is usually selected such that the nominal segment duration t_s is in the range $5\,\text{ms} < t_s < 30\,\text{ms}$.

5.2 An overview of linear predictive coding

A linear predictor can be used to estimate the value of the next sample of a signal, based upon a linear combination of the most recent sample values. An outline of the technique was given in Chapter 2 and is summarized in Fig. 5.1.

Delay elements remember the p most recent values of the sample sequence, s_n, and these values are then linearly combined according to

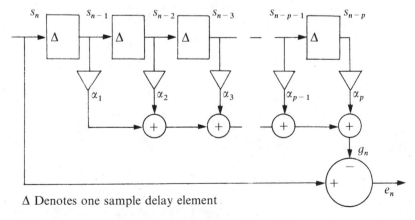

Δ Denotes one sample delay element

Figure 5.1 Linear predictive analysis of a speech signal

the weightings α_1 to α_p to produce a prediction of the value of the next sample g_n

$$g_n = \sum_{k=1}^{p} \alpha_k s_{n-k}$$

By subtracting the predicted signal, g_n, from the current value we obtain a residual (or error) signal, e_n

$$e_n = s_n - \sum_{k=1}^{p} \alpha_k s_{n-k} \tag{5.1}$$

The best coding advantage is obtained when the weightings α_k are chosen such that the residual signal is minimized over the analysis frame during which the characteristics are reasonably stable.

If the speech signal is passed through the analysis filter with prediction coefficients set to the optimum values, the residual sequence needs fewer bits of information to achieve a satisfactory signal-to-noise ratio. The prediction parameters α_k need to be quantized and coded only at the rather slow segmental rate. Examination of the residual signal sequence shows that it consists of either a series of impulses, each coinciding with the start of a glottal pulse, or of random noise, or a mixture of impulses and noise, or silence—corresponding to the four different types of speech signal. Such a filtering process is often referred to as *spectral flattening* or *spectral whitening*, because regardless of the spectral shape of the original speech signal, the spectrum of either type of residual sequence is substantially flat. Spectral flatness is a property that is also exhibited by 'white' noise.

For the task of reproducing a replica of the original speech, we must synthesize the reconstructed speech sequence x_n from the prediction parameters a_k and an excitation sequence i_n. The form of reconstructor that corresponds to the analysis chosen earlier is a purely recursive (all-pole) digital filter, which is a special case of the infinite impulse response (IIR) digital filter. An example is shown in Fig. 5.2, in which the signal h_n is a linear combination of past outputs x_{n-k}:

$$h_n = \sum_{k=1}^{p} a_k x_{n-k} \tag{5.2}$$

to which is added the excitation i_n, and the reconstructed speech x_n is described by the equation

$$x_n = i_n + h_n$$

Thus from Eq. (5.2) we have

$$i_n = x_n - \sum_{k=1}^{p} a_k x_{n-k} \tag{5.3}$$

The similarity between Eq. (5.1), which relates the residual signal to the original speech signal, and Eq. (5.3), which relates the excitation signal

PARAMETRIC CODING OF SPEECH

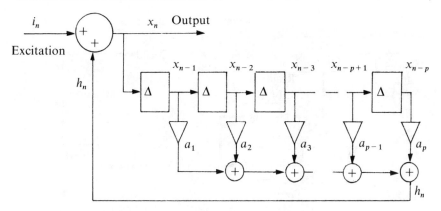

Figure 5.2 Linear predictive synthesis of speech

to the reconstructed speech output, arises because the forms of the analysis and synthesis filters were chosen to be of compatible form; they are *inverse filters*. In the z domain, the form of the analysis filter defines the relation between the sequence of predicted signals $G(z)$ to the sequence of speech input samples $S(z)$:

$$G(z) = S(z) \sum_{k=1}^{p} \alpha_k z^{-k}$$

and the error sequence $E(z)$ is

$$E(z) = S(z) - G(z) = S(z)\left(1 - \sum_{k=1}^{p} \alpha_k z^{-k}\right)$$

In the synthesis filter the feedback sequence $H(z)$ is related to the speech output sequence $X(z)$ by the transfer function

$$\frac{H(z)}{X(z)} = \sum_{k=1}^{p} a_k z^{-k}$$

To create a matching analyser–synthesizer pair of filters the coefficients a_k of the synthesis filter should be periodically updated to the latest values of α_k. The transfer function of the synthesis filter is

$$R(z) = \frac{X(z)}{I(z)} = \frac{1}{1 - \sum_{k=1}^{p} a_k z^{-k}} \tag{5.4}$$

The transfer function $R(z)$ relates the reconstructed speech signal $X(z)$ to the excitation $I(z)$, which corresponds to the source filter model of speech (described in Chapter 1), in the arrangement shown in Fig. 5.3.

In addition to the analysis and synthesis processes, speech coding applications require that the filter parameters and the residual signal must both be quantized and coded. These coded signals are transmitted

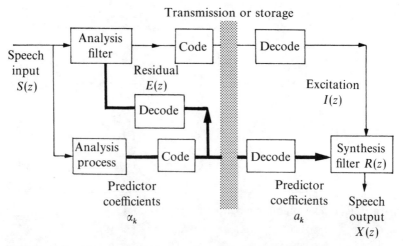

Figure 5.3 Linear predictive coding for transmission or storage of speech

over a channel or stored in some appropriate memory system. In the decoder, the residual signal is decoded and passed to the input of the synthesis filter, the parameters of which are updated regularly by the decoded predictor coefficients. It is worthy of note that the input to the analysis filter is the original speech signal, and the parameters are a *locally decoded* version of the same coded parameters that are supplied to the decoder. By this means the parameters supplied to the analysis and synthesis filters are exactly inverse, as both are subject to the identical quantization and coding processes.

5.3 Linear predictive analysis

The analysis process shown in block form in Fig. 5.3 must find values of α_k that minimize the residual over a period of time. The most commonly used method is to compute the mean square of the error E over a sequence of signals m, which for the moment is not specified. From the expression for the residual given in Eq. (5.1), we have

$$E = \frac{1}{m}\sum_m e_m^2 = \frac{1}{m}\sum_m \left[s_m - \sum_{k=1}^{p} \alpha_k s_{m-k} \right]^2$$

The minimum value of E can be found by the usual method of equating to zero each of the partial derivatives with respect to α_i:

$$\frac{\delta E}{\delta \alpha_i} = 0 \quad \text{for } 1 \le i \le p$$

PARAMETRIC CODING OF SPEECH

which gives a set of linear equations

$$\sum_{k=1}^{p} \alpha_k \sum_m s_{m-i} s_{m-k} = \sum_m s_{m-i} s_m \qquad (5.5)$$

with $1 \leq i \leq p$. This describes a set of p equations in p unknowns, which are $\alpha_1, \ldots, \alpha_p$. The solution of that set of equations to yield values for $\alpha_1, \ldots, \alpha_p$ is the equivalent of inverting a $p \times p$ matrix. When that is done the variance of the error over the sequence of signals m can be evaluated from

$$E_{\min} = \sum_m s_m^2 + \sum_{k=1}^{p} \alpha_k \sum_m s_m s_{m-k} \qquad (5.6)$$

which is sometimes more conveniently expressed in the form

$$E_{\min} = \sum_{k=1}^{p} \alpha_k \sum_m s_m s_{m-k} \qquad \text{where } \alpha_0 = 1 \qquad (5.7)$$

In practical applications it is usual for $10 < p < 26$, but inversion of $p \times p$ matrices of that order is a very expensive computational process. Two practical methods (covariance and auto-correlation) are described which place some constraints on the generality of Eq. (5.5) and result in more efficient processing. A third (lattice) method is then described which uses a completely different approach to optimizing the analysis filter.

5.3.1 Unwindowed (covariance) method

Direct implementation of Eq. (5.5) requires that samples are taken from outside the nominal frame of N samples $h, \ldots, h+N-1$, as shown in Fig. 5.4 (Atal and Hanauer, 1971). However, that requires just a small extension to the nominal length of the frame, and a minor overlap with

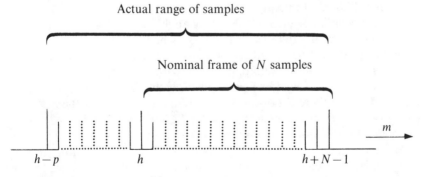

Figure 5.4 The range of samples for the unwindowed (co-variance) method

the preceding frame so that m can take the value down to $h-p$. The set of linear equations becomes

$$\sum_{k=1}^{p} \alpha_k \sum_{m=h}^{h+N-1} s_{m-1} s_{m-k} = \sum_{m=h}^{h+N-1} s_{m-i} s_m$$

with $1 \leq i \leq p$. This set of equations can be stated in terms of the covariance of the speech samples:

$$\sum_{k=1}^{p} \alpha_k \phi(i,k) = \phi(i,0) \tag{5.8}$$

where covariance $\phi(i,k)$ is defined as

$$\phi(i,k) = \frac{1}{N} \sum_{m=h}^{h+N-1} s_{m-i} s_{m-k} \tag{5.9}$$

Now it is clear from the definition (5.9) that $\phi(i,k) = \phi(k,i)$ so the covariance matrix defined by Eq. (5.8) is symmetric, which can be converted by a method based upon Cholesky decomposition (Atal and Hanauer, 1971) into triangular forms, for which the inversion is easier than general matrix inversion. By way of an example, an expansion of Eq. (5.8) for which $p=4$ produces a 4×4 covariance matrix as shown by

$$\begin{bmatrix} \phi(1,1) & \phi(2,1) & \phi(3,1) & \phi(4,1) \\ \phi(2,1) & \phi(2,2) & \phi(3,2) & \phi(4,2) \\ \phi(3,1) & \phi(3,2) & \phi(3,3) & \phi(4,3) \\ \phi(4,1) & \phi(4,2) & \phi(4,3) & \phi(4,4) \end{bmatrix} \begin{bmatrix} \alpha_1 \\ \alpha_2 \\ \alpha_3 \\ \alpha_4 \end{bmatrix} = \begin{bmatrix} \phi(1,0) \\ \phi(2,0) \\ \phi(3,0) \\ \phi(4,0) \end{bmatrix} \tag{5.10}$$

The unwindowed (covariance) method is not much used on speech as Cholesky decomposition only works properly for positive definite matrices, which is not strictly true for the covariance matrix of speech signals. Although it apparently does not window the speech signal itself, the covariance method effectively applies a rectangular window to the residual signal. However, because no windowing is applied to the speech signal itself the length of an analysis frame can be short, about 100 samples, and fully overlapping frames are not needed, as would be the case if a cosine window were used. A further disadvantage of the method is that there is no simple condition for stability on the predictor coefficients α_k, and it does produce unstable filter configurations in some circumstances. A procedure has been described (Atal and Hanauer, 1971) for identifying poles outside the unit circle and replacing them by poles inside the unit circle; if the distance from the unit circle is the same, then the amplitude function in the frequency domain is unchanged.

PARAMETRIC CODING OF SPEECH

5.3.2 Windowed (auto-correlation) method

This method is described as the auto-correlation method in most texts, but the important feature is not the normalization which makes auto-covariance into auto-correlation, although this is often performed, but that useful reduction in computation is achieved by the application of a window function to the speech samples. From inspection of the covariance matrix defined by Eq. (5.10), we can see that the term $\phi(i+1, k+1)$ is on the same diagonal as $\phi(i, k)$, and the two are related by the expansion of $\phi(i+1, k+1)$ as shown:

$$\phi(i+1, k+1) = \phi(i, k) + s_{h+N-1-i} s_{h+N-1-k} - s_{h-i} s_{h-k}$$

so the difference between the two terms is formed from two product terms taken from the extreme ends of the frame of samples. Now there is a form of matrix, a Toeplitz form, in which the elements on each diagonal are equal, and for which an efficient method for inversion exists. We therefore need to reduce these product terms to zero to make the matrix of Toeplitz form. This can be achieved if $N \gg p$ and if we apply a cosine window to the frame. The product terms will then receive very low weighting at the edge of the window function, and the difference between $\phi(i+1, k+1)$ and $\phi(i, k)$ will be so small that it can be neglected, and we can assume that the elements on each diagonal are equal. Therefore the application of a cosine window gives a symmetric, Toeplitz matrix. We can conveniently express the covariance matrix in terms of the auto-covariance function $C(i-k)$ or, as it is often called in speech processing literature, the auto-correlation function $R(i-k)$

$$C(i-k) = R(i-k) = \frac{1}{N} \sum_{m=h}^{h+N-1} ss_{i-k}$$

In most speech processing literature, the auto-correlation function defined in Chapter 2 is called the normalized auto-correlation function $r(i-k)$:

$$r(i-k) = \frac{R(i-k)}{R(0)}$$

Using the auto-correlation function, the least square error condition can be expressed as

$$\begin{bmatrix} R(0) & R(1) & R(2) & R(3) \\ R(1) & R(0) & R(1) & R(2) \\ R(2) & R(1) & R(0) & R(1) \\ R(3) & R(2) & R(1) & R(0) \end{bmatrix} \begin{bmatrix} \alpha_1 \\ \alpha_2 \\ \alpha_3 \\ \alpha_4 \end{bmatrix} = \begin{bmatrix} R(0) \\ R(1) \\ R(2) \\ R(3) \end{bmatrix} \quad (5.11)$$

The solution of Eq. (5.11) for α_k can be achieved by an algorithm defined by Levinson (1949), with a particularly efficient recursive procedure devised by Durbin (1960). In applications that use fixed-point arithmetic, an equivalent form of Eq. (5.11) in terms of $r(i-k)$ gives the best numerical accuracy of computation, because all $r(i-k)$ values are normalized to lie between $+1$ and -1.

Durbin's recursive procedure applied to Levinson's method

Define:
    Error term E
    Reflection coefficients k_j [partial correlation (PARCOR) coefficients]
    Predictor coefficients $a_j(m)$ for which $1 \leq j \leq p$ and $1 \leq m \leq j$

$E = R(0)$
$k_1 = R(1)/E$
$a_1(1) = k_1$

Repeat for j varying from 2 to p:

$$k_j = \frac{R(j) - \sum_{l=1}^{j-1} a_{j-1}(l) R(j-l)}{E}$$

$a_j(j) = k_j$

    Repeat for m varying from 1 to $j-1$:
        $a_j(m) = a_{j-i}(m) - k_j a_{j-1}(j-m)$
        $E = E(1 - k_{j-1}^2)$

Repeat for j varying from 1 to p:
    $\alpha_i = a_p(j)$

Durbin's procedure also allows the reflection coefficients to be calculated from the auto-correlation coefficients, and an error term is evaluated for each iteration of the procedure. If the error term reduces below an acceptable threshold value then the iteration could be truncated and succeeding reflection coeffients equated to zero.

If data space is at a premium the coefficients $a_j(m)$ can overwrite the previous values $a_{j-i}(m)$, as it is only the final set of which are used as the predictor coefficients, as shown in the final evaluation of α_i in the procedure. The auto-correlation method requires quite large frames of speech data because the windowing reduces the significance of events at the ends of the frame. Overlapping data frames must be employed to ensure that all speech events influence the coded signal. It can be shown that the stability of the synthesis filter can be guaranteed if $-1 < k_j < 1$, so it is relatively simple to ensure a stable coder.

PARAMETRIC CODING OF SPEECH

The minimum residual is given in terms of the auto-correlation coefficients by

$$E_{min} = R(0) - \sum_{k=1}^{p} \alpha_k R(k)$$

which can be computed when the predictor coefficients α_k have been evaluated. The value of E_{min} is a useful measure of the closeness of fit between a speech frame represented by a set of auto-correlation coefficients and a vocal tract filter represented by a set of prediction parameters. It can conveniently be rewritten as

$$E_{min} = \sum_{k=0}^{p} \alpha_k R(k) \quad \text{where } \alpha_0 = 1 \quad (5.12)$$

5.3.3 Lattice analysis method

The lattice method of LPC analysis does not require either the auto-correlation coefficients or the co-variance to be evaluated. Partial correlation (PARCOR) coefficients k_i are computed for a sequence of forward and backward predictors e_i and b_i. The coefficients k_i can be used directly as reflection coefficients of an analysis filter with concatenated stages, each of which is a single lattice section as shown in Fig. 5.5.

The signal flow for each stage of the analysis filter can be represented by a matrix equation:

$$\begin{bmatrix} e_i \\ b_i \end{bmatrix} = \begin{bmatrix} 1 & -k_i z^{-1} \\ -k_i & z^{-1} \end{bmatrix} \begin{bmatrix} e_{i-1} \\ b_{i-1} \end{bmatrix} \quad (5.13)$$

The analysis method requires that, for each stage in the lattice, the coefficient k_i is calculated from

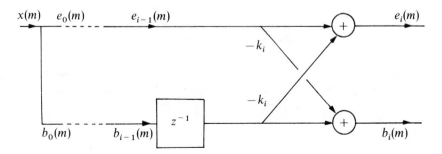

Figure 5.5 Single stage of lattice filter for analysis

$$k_i = \frac{\sum_{m=0}^{N-1} e_{i-1}(m) b_{i-1}(m-1)}{\left[\sum_{m=0}^{N-1} e_{i-1}(m)^2 \sum_{m=0}^{N-1} b_{i-1}(m-1)^2\right]^{1/2}} \quad (5.14)$$

The summations of product and square terms is carried out over an N-sized frame of the variables e and b. The value of N is usually in the range 100 to 200, so it can be seen that *for each stage of the lattice* the evaluation of the numerator corresponds to N multiply and add operations, and for the denominator $2N$ multiply and add operations, a further multiplication and a square root operation are necessary. This represents a considerable computation effort, and also quite large storage area for the intermediate results.

The coefficients k_1 are first calculated from Eq. (5.14) by setting the sequence of e_0 and b_0 variables to be equal to the sequence of input speech values s_0:

$$e_0(m) = b_0(m) = s_0(m)$$

The values of the sequences $e_1(m)$ and $b_1(m)$ can then be evaluated by Eq. (5.13). Next k_2 can be evaluated from Eq. (5.14) and then $e_2(m)$ and $b_2(m)$ again from Eq. (5.13). This process is iterated for as many stages as required. As with the Durbin procedure, the residual is calculated for each iteration, in this case represented by the sequence $e_i(m)$, and early truncation of the iterative procedure can be chosen if the residual reaches an acceptably low level.

5.3.4 Comparison of LPC analysis methods

The three methods described above have strengths and weaknesses as summarized in Table 5.1. Were it not for the conditional stability of the covariance method, the advantage of not requiring a window function

Table 5.1 Comparison of LPC analysis methods

	Unwindowed (co-variance)	Windowed (auto-correlation)	Lattice
Stability	Conditional	Unconditional, but subject to wordlength	Unconditional
Framelength/samples	100	250	100–200
Window	No	Yes	No
Data space	Low	Medium	High
Computation	Medium	Medium	High

would commend it for most applications. Apart from occasional instabilities, it produces good quality of speech. The autocorrelation method requires a longer data frame to allow for the windowing and overlapping frames. This produces a stable synthesis filter, but at a cost of some loss of clarity produced by the windowing process. The lattice method produces stable, clear speech, it can be used with a frame length anywhere between 100 to 200 but the computation cost is quite high, the square root function in particular being computationally expensive.

5.4 Synthesis method for LPC speech

The arrangement shown in Fig. 5.2 for the synthesis of speech shows a purely recursive IIR-type filter in canonical form. It is not practically suitable as a synthesis filter as the condition for stability for such realizations is defined in terms of the poles being within the unit circle in the z plane. The computation involved in evaluating the poles, testing for stability, moving poles, and recomputing prediction coefficients is unattractive. For a lattice filter for which the characteristics are determined by the reflection coefficients k_i for each stage, there are much simpler conditions for stability, namely that $-1 < k_i < 1$, for all k_i. Reflection coefficients can be simply derived from the predictor coefficients by an iterative process; indeed the example of the Durbin procedure in Sec. 5.3.2 included the calculation of the reflection coefficients in addition to the predictor coefficients. The lattice filter for synthesis is the inverse to the analysis lattice filter; the signal flow is similar in that there is a forward predictor and a backward predictor at each stage, as shown in Fig. 5.6.

The signal flow can be represented by the matrix equation

$$\begin{bmatrix} e_i \\ b_i \end{bmatrix} = \begin{bmatrix} 1 & -k_i z^{-1} \\ -k_i & z^{-1} \end{bmatrix} \begin{bmatrix} e_{i-1} \\ b_{i-1} \end{bmatrix}$$

Lattice sections are concatenated as shown in Fig. 5.7, which shows a filter with two complex poles. From the figure it can be seen that

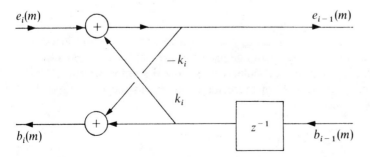

Figure 5.6 Single stage of lattice filter for all-pole synthesis

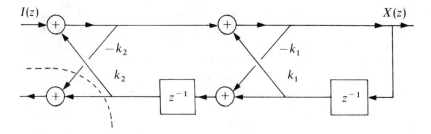

Figure 5.7 A complete two-stage lattice filter for synthesis

$$H(z) = \frac{X(z)}{I(z)} = \frac{1}{1 - k_1(1-k_2)z^{-1} - k_2 z^{-2}} \qquad (5.15)$$

which corresponds to an IIR filter with a two-pole transfer function:

$$H(z) = \frac{1}{1 - a_1 z^{-1} - a_2 z^{-2}}$$

By comparison of the two forms it can be seen that $k_2 = a_2$ and $k_1 = a_1/(1 - a_2)$.

As a general rule, the number of stages required for the lattice filter is two plus twice the number of poles n_p in the filter transfer function. Using a rule of thumb that the speech signal has one complex pole for each kilohertz of bandwidth, we can calculate that for 10 kHz sampled speech, the Nyquist bandwidth is 5 kHz, so $n_p = 5$ and a 12-stage LPC lattice would be sufficient. In practice, extra poles might be added to provide a closer match between the synthesis filter and the speech spectrum. The use of more stages also allows spectral zeros to be modelled, as it can be shown that a zero can be approximated by an infinite number of poles:

$$1 - az^{-1} = \frac{1}{\sum_{n=0}^{\infty} b_n z^{-n}}$$

If $|a| < 1$, then the coefficients b_n rapidly progress to zero, and just a few terms of the infinite series can adequately represent a zero. Zeros do occur in speech signals, particularly in nasalized sounds, and so all-pole LPC lattices may have 20 or more stages to provide accurate modelling of the speech spectrum.

5.5 Coding the predictor parameters

In this section, scalar and vector quantization of the predictor parameters are described. In most applications the predictor parameters must be quantized and coded at a regular rate. Although some applications of parametric coding can make use of the variable information rate of speech, variable bit-rate coding of speech is not yet widely used because the small savings in bandwidth are outweighed by the increased complexity of codecs.

5.5.1 Scalar quantization of the predictor parameters

Direct scalar (i.e. PCM) quantization of the predictor coefficients of the analysis filter is not recommended, because a relatively high accuracy (8 to 12 bits per coefficient) is generally required to ensure stability of the synthesis filter in the decoder. If scalar quantization is to be used, a better approach is to use the reflection coefficients. They are preferred because if the values are constrained within the range +1 to −1 they offer unconditional stability of the synthesis filter. Also a non-linear transformation can be applied to the reflection coefficients before quantization. This provides the same advantage as log-law companding in PCM, and it is found that fewer bits per coefficient are sufficient to avoid round-off error affecting stability. Sensitivity to coefficient round-off error is non-uniform through the lattice. Early stages require 5 or 6 bits per coefficient, later stages only 3 bits per coefficient.

5.5.2 Vector quantization of the predictor coefficients

The aim of any quantization scheme is to map from the sample space onto the symbol space in such a way that the quantization error is within acceptable limits and the bit rate as low as possible. Vector quantization (VQ) (see Fig. 5.8) extends the process of scalar quantization by taking a block or vector of samples $\mathbf{x}$ and seeking from a codebook of stored vectors $\mathbf{y}$ the most appropriate vector as measured by some distortion measure $d(\mathbf{x}, \mathbf{y})$ (Buzo et al., 1980). The nature of the stored vector $\mathbf{y}$ is crucial to the efficiency of the system. Firstly, the vectors must be stored in a gain-normalized form, so the vectors are insensitive to the amplitude of the frame being compared with codebook entries. Secondly, the codebook entries must be stored in a form that minimizes the amount of computation needed to compute a distance between the codebook entry and the incoming vector. This is especially important as a number of such comparisons will normally be required before the optimum vector is identified.

The output of the vector quantizer is the code $\mathbf{U}_n$, which takes the

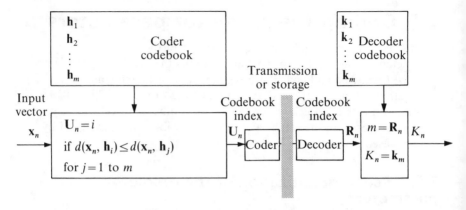

Figure 5.8 Algorithm for vector quantization of filter parameters

value of the index i of the nearest vector $\mathbf{y}_i$ of the m such vectors stored in the codebook. $\mathbf{U}_n$ is then sent on the channel instead of the vector $\mathbf{x}_n$ itself. Considerable reduction of the data rate of a speech signal can be achieved by this technique, although it is accompanied by a reduction of speech quality. It is typically possible to obtain an order of magnitude reduction in the bit rate required for the predictor coefficients.

The complexity of the scheme makes it much more difficult to implement than scalar quantization. Part of this complexity is the need to search through the codebook to find the best matching vector, during the operation of the coder. A *full search* of a 1024 vector codebook will require 1024 comparisons to be made before the best match is identified. The number of comparisons that need to be made can be reduced by organizing it in such a way that a *binary search* is possible (Buzo et al., 1980). The codebook is organized as a binary tree structure, with two vectors at the first level, and the initial comparison selects between one half of the codebook and the other. There are four vectors at the second level allowing a second comparison to select between the two quarters of the selected half of the codebook. This continues until the tenth level when the final comparison identifies the one codebook entry that best matches the incoming vector. The binary search requires 20 comparisons to be made, two at each level, compared to 1024. There is a cost in that the codebook is almost twice the size and quantization distortion is slightly increased. The codebook search has to be done at the coder, but *not* at the decoder where the codebook reference is a simple indexing procedure, so the use of vector quantization leads to asymmetric coding systems. The asymmetry means that the advantages of the scheme are greater where there are fewer encoders than decoders (e.g. broadcasting) or where the encoder does not need to run in real time (e.g. speech storage).

The other part of the complexity occurs once during the design or commissioning of a system. It is the generation of the codebook, which is a crucial stage of designing a VQ system. It is generally based upon the analysis of a large training sequence of speech samples which have been converted into the vector format. Because of the complexity and expense of VQ, most systems in which it is used will be speaker independent. The training sequence must therefore include examples of speech from a wide range of speakers, both male, female and of differing ages. In binary coding systems, the maximum coding advantage is obtained when the codebook size, m, is an integer power of 2, so the codebook vectors are generated by an iterative process of dividing the set of training vectors into two subsets of equal size, and then using the centroid of each subset as the starting point for the next iteration. After M iterations the centroids of the 2^M subsets will be the vectors for the codebook (Linde, Buzo and Gray, 1980).

VQ can be applied to PCM speech (Gersho and Cuperman, 1983), but the best advantage is given when used in conjunction with LPC methods, in which case the auto-correlation coefficients can be used as the input vector. In this application the most commonly used distortion measure is that described by Itakura and Saito (1972), d_{IS}:

$$d_{IS}(X, H) = \int_{-\pi}^{\pi} \left[\frac{X(e^{-j\phi})}{H(e^{-j\phi})} \right]^2 - \ln\left[\frac{X(e^{-j\phi})}{H(e^{-j\phi})} \right]^2 - 1 \frac{d\phi}{2\pi}$$

which can be expressed as (Buzo et al., 1980)

$$d_{IS}(X, H) = \frac{\alpha}{\sigma^2} + \ln(\sigma^2) - \ln(\alpha_0) - 1$$

in which the last two terms are constant, σ is the gain of the speech model and α is the residual energy which results from the (theoretical) application of the speech frame to the codebook model H. It is theoretical because α can be evaluated from

$$\alpha = \sum_n r_m(n) r_a(n)$$

where r_m and r_a are the auto-correlation functions for the codebook model and the speech frame. The method has the advantage of being a perceptually relevant measure of distance, which is readily computed from the auto-correlation coefficients. Thus the coder codebook stores a set of gain-normalized auto-correlation coefficients and the appropriate codeword is selected by distortion measure comparisons made directly from them, and without the need to calculate predictor parameters. The decoder codebook stores the equivalent reflection coefficients which can be used directly in the synthesis filter. The reflection coefficients are effectively precomputed from the auto-correlation coefficients, in the sense that they are calculated once only at the time of codebook generation.

5.6 Coding the residual sequence

There are many approaches to the coding of the residual sequence, giving better or worse approximations to the residual signal after the original speech is passed through the optimum analysis filter.

5.6.1 Monopulse coding

Atal and Hanauer (1971) characterized the residual signal by either a periodic train of impulses during voiced speech or by random noise during unvoiced speech. This has since become known as the *monopulse coding* technique. The actual noise signal need not be coded, as a noise signal can be generated in the decoder without loss of information, but of course the power or gain of the residual must be coded. This approach typically requires:

Voiced/unvoiced flag	1 bit
Pitch period	6 bits (64 discrete values, non-linear progression)
Gain or power	5 bits (magnitude of impulse or noise)

If these 12 bits are transmitted every 20 ms, then the data rate associated with the residual signal is 600 bit/s.

There are practical problems with the monopulse approach. It is difficult to determine the pitch period of speech with a high degree of reliability, particularly for child and female voices, where the fundamental pitch becomes confused with the lower formants. Also there are some speech sounds that require a mixture of voiced and unvoiced excitation, which is not easily achieved. Nevertheless, monopulse coding is quite appropriate for communication where speech quality lower than 'toll quality' is acceptable. Typical frame lengths are 15 to 25 ms, and by using coding techniques that represent zero energy or silent frames by a special short code, the total bit rate for residual and filter parameters can be in the range 1.2 to 2.4 kbit/s. The LPC vocoder is also usable as a compact method of storing speech messages within digital systems. There are large-scale applications that have prompted the design of special purpose integrated circuits implementing the decoder function. An example is the Texas Instruments TMS5220, which achieves further compression of the bit rate by coding 10 reflection coefficients for voiced speech frames but only 6 for unvoiced frames.

5.6.2 Multipulse coding

In an attempt to improve upon the subjective performance of the monopulse technique, techniques have been developed for the representation of the residual by a multiplicity of pulses (Atal and Remde, 1982), so-called *multipulse coding*. The optimum positioning of the pulses is determined for each analysis frame, relative to the frame itself, not synchronized to pitch events. This removes the need for accurate pitch period estimation, although it considerably increases the computation needed in the coder. The position and amplitude of the pulses must be optimized by an iterative process, and then that information must be quantized and coded. The bit rate for this method is often in the region of 8 to 10 kbit/s. A recently proposed variant of the method uses fixed positions for an increased number of pulses per frame (Natvig, 1988). This scheme offers a significant reduction in the complexity of the coder, although the bit rate is much the same.

5.6.3 Residual excited linear prediction (RELP)

This technique allows the residual to be coded by conventional waveform coding techniques such as PCM, DPCM or DM. If the bandwidth of the residual is the same as for the original speech, the coding advantage comes from the reduced variance of the residual, needing fewer bits to achieve a given signal-to-noise ratio. Very natural sounding speech can be produced, although this increases the total bit rate to about 16 kbit/s.

Another approach shown in Fig. 5.9 is to band-limit the residual signal and apply a waveform coding technique to it. This form of residual coding exploits the fact that the nature of the speech excitation spectrum is relatively frequency independent. For voiced excitation it consists of a fundamental and a sequence of harmonics which occur at regular intervals and of gradually reducing amplitudes. For unvoiced excitation the spectrum is essentially uniform, representing random noise. These observations mean that if the high-frequency part of the spectrum is removed by a process of *down-sampling*, it can subsequently be recreated with a reasonable degree of accuracy. The down-sampling consists of band-limiting the signal and then sampling at a lower rate, so it can be coded, by a scalar quantization scheme, at a lower bit rate. In the decoder the missing band of frequencies can be regenerated by means of *spectral folding*, which produces a mirror image of the lower band into the upper band. This can simply be achieved by interpolating zero-valued samples between each baseband sample and band-limiting to the original bandwidth of the residual signal. Coding of the residual at about 8 kbit/s can produce speech of good overall quality, although in some cases background tones are audible. This is probably due to the fact that the harmonic structure of the speech is interrupted at the folding frequency (Makhoul and Berouti, 1979).

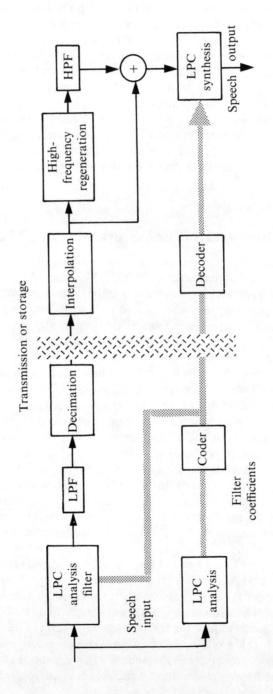

Figure 5.9 Residual excited LPC coder–decoder

5.6.4 Code excited linear prediction

Vector quantization of the residual produces advantages in the same way as when used for the predictor coefficients, but again at the cost of a considerable increase in complexity. Using 2 kbit/s vector quantization of the residual, LPC speech which is virtually indistinguishable from the original has been obtained (Schroeder and Atal, 1985). That was considered as the state of the art in residual coding, well ahead of the ability to implement in practical coding systems, because the codebook search occupied almost 125 seconds of CPU times on a Cray-1 supercomputer to process 1 second of speech signal.

5.6.5 A practical LPC speech codec

An LPC-based speech coder is currently being standardized by the Conference of European Post and Telecommunication Administrations for use in the European mobile radio network (Natvig, 1988). In this context, bandwidth efficiency is a paramount concern, and the delay introduced by LPC analysis and the possibility of slightly degraded speech quality are of lesser importance. The proposed coder operates with a basic bit rate of 13 kbit/s, although the addition of error-protection coding, required because of the high error rates on mobile radio channels, brings this up to 16 kbit/s. It uses an analysis frame of 20 ms, an eighth-order lattice-type predictor and codes the residual in terms of a sequence of regularly spaced variable amplitude pulses. It also uses a long-term predictor, which attempts to remove long-term correlations in the input signal which correspond to the pitch pulses, just as the short-term predictor removes short-term correlations due to vocal tract resonances.

Although final testing of this coder is not complete at the time of writing, it achieves a subjective speech quality which is at least 'fair' to 'good' for bit error rates up to 1 per cent, which is considerably better than can be achieved for analogue transmission with similar bandwidth constraints (Natvig, 1988).

5.7 Interaction between coding of filter parameters and coding of the residual

The linear predictive coding method possesses a very unusual characteristic in that imperfections in determining the prediction filter can be overcome in the coding of the residual signal. The performance of the overall system can therefore be much better than its least accurate part. This arises from the use of digital filtering techniques, in which a synthesis filter can be created which is an *exact* inverse of the analysis filter. Therefore, whatever the limitations of the filter pair might be, the relationship between the original speech and the residual is determined by the analysis filter, and the relationship between the excitation (decoded residual) and the reconstructed speech is determined by the synthesis filter. Since the filters are exactly inverse, the fidelity of the overall system is primarily determined by the fidelity with which the residual is replicated at the decoder.

The argument would be exact in the case of a time-stationary signal for which no change in filter parameters occurs. This is a rather uninteresting application of LPC, and clearly for a speech signal the filter parameters are updated at the segmental rate, which will produce some extra impairment. This is, however, a secondary consideration, and becomes progressively less important as the updating period is reduced towards the practical minimum around 10 ms.

The advantage of having a good prediction filter rather than a poor one is that it produces a residual spectrally flat signal of low energy, which allows the use of a reduced bit-rate coding scheme without undue quantization noise being introduced.

5.8 Sub-band coding: a hybrid between parametric and waveform coding

We have seen various coding schemes that derive a coding advantage from some general characteristics of speech signals. Some of these characteristics are more marked in one part of the frequency spectrum than in other parts, and this feature is used in sub-band coders. The sub-band coder is one in which the input signal is divided by a filter bank into a number of distinct frequency bands, which are then coded independently (Crochiere, Webber and Flanagan, 1976). The technique utilizes properties of the speech signal in a less direct way than LPC-based techniques. It is a hybrid between parametric and waveform coding, like adaptive differential pulse code modulation (ADPCM), which typically uses a short-term and a long-term predictor, introduced in Chapter 3.

The sub-band coder reduces the bandwidth of each sub-band to less than that of the input signal, so the sampling rate for each of the sub-bands can be reduced by a process of down-sampling prior to encoding. At the decoder each sub-band is decoded separately, and then by a process of up-sampling and band filtering is restored to its original place in the frequency spectrum. The components for all of the sub-bands are summed to form the complete decoded speech signal.

The process is illustrated for a two sub-band system in Fig. 5.10. The coding advantage is not as easy to understand intuitively as it is for LPC methods, but it also exploits correlation of the speech signal in the time domain, which is equivalent to spectral 'non-flatness' in the frequency domain (Jayant and Noll, 1984). If the signal-to-quantization noise ratio (SQNR) for each sub-band is constant and the overall spectrum of the input signal is non-flat (signal samples are correlated) then the total noise power in the reconstructed signal in the decoder is less than that of a full-band coder. This is shown schematically in Fig. 5.11. Alternatively a constant noise power for each of the sub-bands can be achieved by allocating fewer bits to those sub-bands that have a lower signal power. This reduces the total bit rate required to achieve a particular overall SQNR. The bandwidth compression attainable from sub-band coding depends on the non-flatness of the signal spectrum and also on the number of the sub-bands into which the speech signal is split.

A potential problem with the sub-band coding technique is that bandpass filters cannot be designed to have infinitely steep transition regions, and so the *aliasing* effect that occurs when each sub-band signal is decimated cannot be neglected. For exactly the same reason *imaging* will occur when the sub-band signals are interpolated and band-filtered in the decoder. Although both of these effects could be reduced by increasing the sampling rate for the sub-bands, this would have the undesirable effect of increasing the overall bit rate of the encoder. Early examples of sub-band coding eased this problem by using non-contiguous bands (Crochiere, Webber and Flanagan, 1976) which could reduce the bit rate from 16 kbit/s to below 9.6 kbit/s, but at the cost of introducing a reverberant quality into the speech signal.

Recent examples of the sub-band technique use *quadrature mirror filters* to overcome these limitations to the bandpass filtering function. The design of the quadrature mirror filter banks (Vaidyanathan, 1987) is such that aliased components produced during the down-sampling stage are exactly cancelled by the image components produced by the up-sampling process. The unwanted components are cancelled in the overall system rather than removed by filtering, so bandpass filters of relatively low order can be used, provided that the filters in encoder and decoder are closely matched. With a lower order filter some degree of overlap between adjacent filter bands is inevitable, and the band edges of the filters occur within the range of speech signal frequencies. If the sub-bands are constrained to have equal bandwidth, which is related to the original bandwidth by a power of 2, the sampling rate changes can be particularly simple. Down-sampling becomes a decimation process (removing samples from a sequence) and up-sampling becomes an interpolation process (adding zero-valued samples to the sequence).

In general, sub-band coders yield similar performance to differential

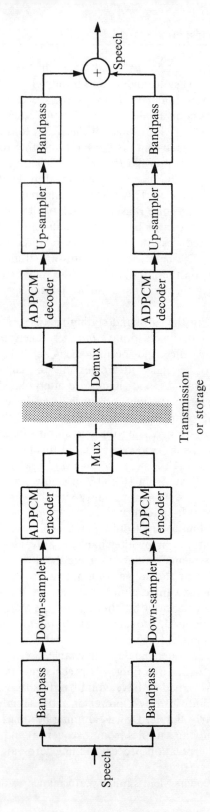

Figure 5.10 Two-band sub-band coder

PARAMETRIC CODING OF SPEECH

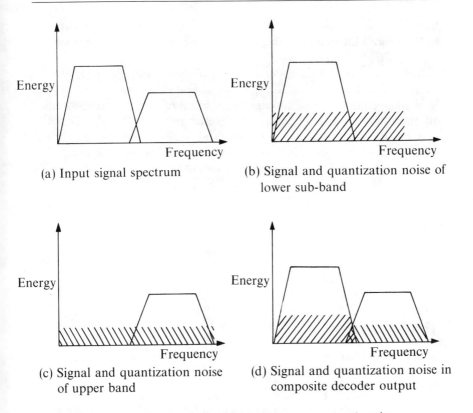

Figure 5.11 Quantization noise distribution in two-band coder

coders of similar complexity (e.g. the ADPCM coder) for bit rates above 16 kbit/s. Below that rate the performance of differential coders degrades rapidly because of quantization error feedback. As sub-band coders restrict the spread of quantization error between the sub-bands, their performance at bit rates below 16 kbit/s is considerably better.

A recent application of sub-band coding principles is the CCITT recommendation G.722 on wideband speech coding (Mermelstein, 1988). Wideband speech is defined as having a bandwidth of 50 to 7 kHz, compared to the conventional narrowband speech with a bandwidth of 300 to 3400 Hz. Apart from the higher quality and intelligibility that it brings to telephony, wideband speech is particularly advantageous to teleconferencing applications, in voice-only and voice-and-video services. The G.722 speech coder splits the input signal into a lower sub-band from 0 to 4 kHz and an upper sub-band from 4 to 8 kHz, using quadrature mirror filters. These sub-band signals are then coded by independent ADPCM coders which produce 6 bits per sample for the low band and 2 bits per sample for the high band. The sample rate for each ADPCM coder is decimated to 8 kHz, so the overall bit rate is 64 kbit/s, so the coder is rate compatible with existing PCM systems.

A feature of the G.722 coder is the possibility of using embedded coding in the lower sub-band. This means that either one or two of the least significant bits in each output code word can be removed, without affecting the tracking of adaptive parameters in the ADPCM encoder and decoder, as it would in a conventional ADPCM system. One or two bits in every eight can then be used for data which is multiplexed into the speech code stream. This is compatible with ISDN standards, in which the overall data rate of 64 kbit/s can be used in three different ways:

Speech	Data
64 kbit/s	—
56 kbit/s	8 kbits
48 kbit/s	16 kbits

References

Atal, B. S. and S. L. Hanauer (1971) 'Speech analysis and synthesis by linear prediction of the speech wave', *Journal of the Acoustic Society of America*, vol. 50, pt 2, 637–655.

Atal, B. S. and J. R. Remde (1982) 'A new model of LPC excitation for producing natural-sounding speech at low bit-rates', IEEE Proceedings of the International Conference on Acoustics, Speech, and Signal Processing, 614–617.

Buzo, A., A. H. Gray, R. M. Gray and J. D. Markel (1980) 'Speech coding based upon vector quantization', *IEEE Transactions on Acoustics, Speech, and Signal Processing*, vol. ASSP-28, no. 5, 562–574.

Crochiere, R. E., S. A. Webber and J. L. Flanagan (1976) 'Digital coding of speech in sub-bands', *Bell System Technical Journal*, vol. 55, 1069–1085.

Durbin, J. (1960) 'The fitting of time series models', *Rev. Inst. Int. Statistics*, vol. 28, 233–243.

Gersho, A. and V. Cuperman (1983), 'Vector quantisation: a pattern matching technique for speech coding', *IEEE Communications Magazine*, Vol. 21, no. 9, 15–21.

Itakura, F. and S. Saito (1972) 'On the optimum quantisation of feature parameters in the PARCOR speech synthesiser', IEEE Conference on Speech Communication and Processing, 434–437.

Jayant, N. S. and P. Noll (1984) *Digital Coding of Waveforms*, Prentice-Hall, Englewood Cliffs, N.J.

Levinson, N. (1949) 'The Wiener RMS (root mean square) error criterion in filter design and prediction', *Journal of Mathematics and Physics*, vol. 25, 261–278.

Linde, Y., A. Buzo and R. M. Gray (1980) 'An algorithm for vector quantiser design', *IEEE Transactions on Communications*, vol. 28, 84–95.

Makhoul, J. and M. Berouti (1979) 'Predictive and residual coding of speech', *Journal of the Acoustical Society of America*, vol. 66, 1633–1641.

Mermelstein, P. (1988) 'G.722, a new CCITT coding standard for digital transmission of wideband audio signals', *IEEE Communications Magazine*, vol. 26, 8–15.

Natvig, J. E. (1988) 'Evaluation of six medium rate coders for the pan-European digital mobile radio system', *IEEE Journal on Selected Areas in Communication*, vol. 6, 324–331.

Schroeder, M. R. and B. S. Atal (1985) 'Stochastic coding of speech signals at very low bit rates: the importance of speech perception', *Speech Communication*, vol. 4, 155–162.

Vaidyanathan, P. P. (1987) 'Quadrature mirror filter banks, M-band extensions and perfect reconstruction techniques, *IEEE Acoustics, Speech, and Signal Processing Magazine*, vol. 4, no. 3, 4–20.

6 Synthesis

CHRIS ROWDEN

6.1 Introduction

Speech has been used for tens of thousands of years as a primary means of providing information, instruction and amusement to people, and so it is natural that speech should be used as an output from information technology machines. Automatic generation of speech messages has been possible since sound recording was developed, and in 1936 the Speaking Clock service was introduced in the telephone network, which used the principle of inserting stored word utterances into a standard phrase. The technology was largely electromechanical, and speech was stored on an optical soundtrack similar to that used in cinematography. Modern counterparts of that system are *copy synthesis* systems which replay a relatively small range of recorded words and phrases. The copy synthesis approach is a stored message system in which word or phrase inserts may be selected from a limited repertoire. The control function for the synthesizer does no more than to select the insertion. However, more elaborate *text-to-speech* systems have been developed which generate a spoken message corresponding to an input of unrestricted text. Such a system needs a sophisticated multilevel control function which can convert any sequence of machine-readable (e.g. ASCII coded) text into an acceptable substitute for a human voice reading it. Essentially it is an expert system for reading aloud, which accesses a knowledge base, held in a dictionary and tables, to relate the written form of the language to the spoken form. Although word insertion and word concatenation systems are still effective solutions to some problems of spoken message generation, this chapter concentrates on the more general technique of text-to-speech synthesis.

Some languages are easier to synthesize from text than others, and British English must be among the most difficult of common languages. (English is such a widely used language that it is necessary to distinguish between British English and, for example, American English.) There are two groups of factors that make British English difficult. The first group are related to the irregular correspondence of the written form of the language, the *orthography*, to the sounds of the spoken form, the *phonology*. The irregularity has arisen because the language has evolved over thousands of years under the successive influence of many other languages. Consider the following words, their pronunciation and their origin:

'cam' /k aa m/† From Dutch
'came' /k ei m/ From Old English, previously from Old Norse
'Cambridge' /k ei m b r I dzh/ English place name, from the late Roman bridge on River Cam
'cameo' /k aa m ee ou/ From Italian

The words 'cam' and 'came' illustrate the well-known 'magic e' rule, in which the letter 'e' placed after a syllable alters the vowel sound, usually by lengthening it or palatizing it. However, 'Cambridge' and 'cameo' contravene the 'magic e' rule. The sound of each phoneme constitutes a segment, and the accuracy with which words can be mapped to phonemes influences the *segmental* quality of synthesized speech.

The second group of factors that makes English a difficult subject for text-to-speech synthesis relates to the *prosody* of speech, which describes the patterns of pitch, stress and rhythm that occur in spoken words and phrases. Prosodic features are not defined explicitly in the written form, and yet most speakers of a language can read a written passage aloud with a prosodic style similar to other speakers of the same dialect. To do this they must be adding some language knowledge to the text that they read. It is knowledge of some general rules of prosody that have been acquired through practice and imitation, rather than having been learnt explicitly. The generation of believable patterns of pitch and rhythm, which span over the many segments in a phrase or sentence, influence the *suprasegmental* quality of synthesized speech.

I have written in Chapter 2 about the imaginative four-level model described by Ian Witten (1986), which provides a framework for his excellent book on speech synthesis. The structure of this chapter is rather different, as it attempts to show the current state of knowledge of text-to-speech synthesis by describing the processing stages in the sequence in which they occur. Essential background material occurs in other chapters, particularly Chapter 1 on production and perception, for the classification of speech sounds as phonemes, and the effect of the articulators on the speech signal; Chapter 2 on analysis, for an introduction to the syntax of natural language; and Chapter 11 on human factors, for a discussion of factors relating to quality of speech and applications of synthesized speech. Throughout the chapter examples are drawn from the MITalk system (Allen, Hunnicutt and Klatt, 1987), the development and implementation of which must be the most fully documented of any text-to-speech synthesis system.

†Unless otherwise indicated the phonetic symbols used in this chapter are as defined in Table 1.1.

6.2 An overview of text-to-speech synthesis

Most systems for converting a textual message to spoken form involve a number of processing stages. The sequential presentation of these processes might imply that the system conforms to a pipeline model of multiprocess systems. Close examination shows that this is not generally appropriate; information that is implicit in the input text is teased out at various stages by reference to the text and to information stored in the computer system about speech and language in general and about speakers in particular.

The first part of this overview briefly describes a possible data structure that could be used by such a sequence of processes. The information generated by each of the processes can be added a piece at a time until the analysis is complete. The second part explores two different ways in which each of the processes could be implemented—by programmed algorithm or by a look-up table.

6.2.1 Example of a data structure for text-to-speech conversion program

The natural unit of input to a text-to-speech system is the sentence. As we shall see various processes in the conversion are sensitive to context, but it is very rare for this context to exceed sentence boundaries. The length of a sentence, in terms of the number of words or letters it contains, is unlimited. This problem of unlimited size of data structures is often solved in computer programs by using a linked list. Each element in the list is a structure that can hold a word-sized string and a set of pointers to other objects that are initialized to null values when they are created. As the program proceeds in converting the input text to eventually producing the large amount of data needed to update the synthesizer control parameters once every 10 ms, various intermediate data forms are produced and added to the data structure (see Fig. 6.1). Firstly, word forms that are not pronounceable are converted into pronounceable words; abbreviations and strings of digits are the most common occurrences, and Sec. 6.3.1 gives some examples of the process. Then the words have to be translated to a sequence of phonemes which indicate the normal pronunciation. Also indicated is whether the syllable of which each phoneme is part can be stressed, and if so at what level. As a preparation for the use of a syntax parser that can group the words into appropriate phrases, the part of speech of each word is determined. Many words can serve as more than one part of speech in a sentence, so this part of the data structure is a set of all *possible* parts of speech. When the parser has operated, the information in the linked list can be augmented by adding appropriate phrase

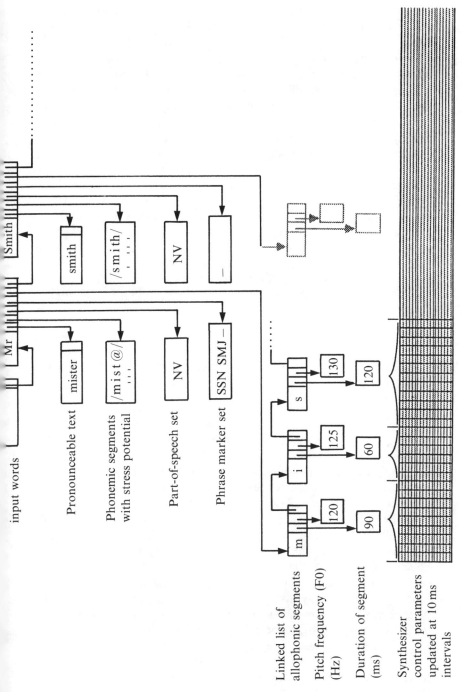

Figure 6.1 Example data structure for text-to-speech conversion

markers to the data structure, and again several markers may coincide so a set of markers is needed. For example, the first word in a sentence could mark both the start of a sentence (SSN) and the start of a major phrase (SMJ).

Further extension of the data structure is implemented with a separate linked list for each word unit in which each element represents an allophonic segment, of which one, two or three may replace a phoneme. Sometimes an allophonic segment may be deleted because it is not required in a particular context. The linked list structure allows for easy addition and deletion of elements. The pitch contour of the synthesized speech and the duration of each segment are placed into the structure at this level, and eventually a stream of control parameters are produced, a set of values for each 10 ms unit of duration.

There are many practical issues that are not defined in this data structure, such as the data types for the various objects. However, the purpose here is not to lay the foundation for a text-to-speech program, but to illustrate the way in which the information needed by the speech synthesizer can be gradually assembled and *added to* an existing data structure rather than replace it. All of the information relating to a particular sentence produced earlier is available to the process at any time.

6.2.2 Implementing processes by algorithm or by table look-up

The translation of a file of text to the spoken form is a deterministic process, in the sense that the same textual input should give rise to the same spoken output every time it appears in the same context. The many ways of implementing a deterministic process fall into two classes: algorithmic solutions and table look-up solutions. If, as is usually the case, a process for text-to-speech conversion is partitioned into a number of subprocesses, then each of them must also be deterministic. For each subprocess the implementation could be either by the algorithmic or by the table look-up method.

Consider one such subprocess, that of converting a word to a sequence of phonemes that represents the sound of the word. A broadly algorithmic approach uses a set of rules to provide a phonetic transcription from the letters of the written words to the phonemes of the spoken word. A set of rules can be devised that can be applied to any word and can provide the correct translation for most words. The irregularity in the English language means that there will be common words that are exceptions to the rules. For these words operation of the rules must be replaced by reference to a small exceptions dictionary, which holds the correct transcription. The table look-up approach uses a much larger dictionary as the primary method of phonetic transcription, which covers the common words in the language. In this case, the exception is a word that is not found in the dictionary, so the look-up process must be augmented by a set of rules that can translate these rarely used words.

It is interesting to note that in either rule-based conversion or table look-up, the dictionary *must be searched first*, as the selection between the two substrategies can only be made by determining whether the target word is included in the dictionary. It will become clear as the chapter progresses that this process of converting letters to phonemes is just one part of a text-to-speech system.

6.3 Textual input-to-speech synthesis systems

A text-to-speech system that is designed to convert from written text to the sounds of speech must be able to deal with the variety of textual forms (abbreviations and strings of digits for example) that do not translate directly into pronounceable words. To translate them requires access to the language knowledge that a human speaker uses when reading aloud. Text that contains these unpronounceable character strings is called *unrestricted text*.

6.3.1 Converting unrestricted text to pronounceable text

Most written text contains abbreviations, symbols and digit strings that do not explicitly define the words that should be spoken. A preprocessor which translates unrestricted text to pronounceable text must refer to the language knowledge that would normally be supplied by a human reader. The following test sentences demonstrate some of the problems; the lines are numbered for reference, but the line numbers are not part of the test passage. In translating from the textual form to a pronounceable form two different techniques are demonstrated. One is *word substitution*, in which an unpronounceable character string is replaced by a group of one or more pronounceable words. The other technique is to replace the character string with a sequence of phoneme descriptors, *phoneme substitution*.

1 Kiteflying is such fun, I'd do it all day!
2 Mr. Smith lives in St. George St.
3 121,040,011 is a large number, £65 000 is a 'city salary', 0444 009123 is a 'phone number.
4 World War II broke out on 3/9/39; will NATO prevent another one?
5 I arrived at 8:00 and stayed till 3:30.

Line 1 *Kiteflying* is not a word that occurs in any dictionary that the author has consulted, but most speakers will pronounce it compatibly with *kite* and *flying*. Both rule-based and dictionary look-up systems fail with invented compound words like this. The use of an apostrophe to show contracted words such as *I'd* is very common; a phonetic substitution would be inserted for such constructions in a rule-based system, whereas most dictionary systems would include such contractions. The temptation to expand to *I would* should be resisted, firstly because it is an unnecessary formalization and secondly because it might be wrong—*I'd* can also mean *I had*. The exclamation mark ! should be recognized as a sentence terminator.

Line 2 *Mr.* should be rewritten as *mister*, or a phonemic substitution could be inserted. *St. George St.* shows an abbreviation which is ambiguous. Placed before a proper noun, *St.* could be rewritten *saint*, but in British English the phonemic substitution /s @ n t/ should be made. Placed after a proper noun, *St.* should be rewritten *street*. The period . that indicates an abbreviation should normally be removed to avoid interpretation as a sentence terminator. However, as it is normal for a sentence that ends with an abbreviation to place just one period after the abbreviation, this sentence gives an example of one period serving two purposes. The sentence termination could easily be missed.

Line 3 Long strings of digits which are grouped into powers of three by commas such as *121,040,011* should be rewritten *one hundred and twenty-one million, forty thousand and eleven*. Where a currency symbol precedes the digit string, the digits are also grouped into powers of three even if there are no commas. *£65 000* should be rewritten *sixty-five thousand pounds*. Telephone numbers present special problems; they are usually pronounced as single digits but grouped in twos or threes as convenient, with repeated digits indicated as such. *0444 009123* could be rewritten *oh four double four, double oh nine, one two three*, but it requires quite a complex algorithm to generate such textual forms without error. Generally an apostrophe ' that appears at a word boundary as in *'phone* rather than within a word can be ignored for pronunciation, although the grouping of words between quotes can significantly influence phrasing and prosody.

Line 4 *World War II* requires a substitution of *two* for the Roman numeral *II*. The sequence of words beginning with capitals provides important information for phrasing and prosody, so a mechanism should be provided for grouping together a sequence of capitalized nouns. Dates can be described in many ways, *3/9/39* would indicate *the third of September nineteen thirty-nine* to a Briton, whereas it means *March ninth nineteen thirty-nine* to an American. To either *NATO* is pronounced /neit@u/. The question mark ? should be recognized as a sentence terminator.

Line 5 Times also have a variety of formats, in this example *8:00* and *3:30* should be rewritten *eight o'clock* and *three thirty*.

6.3.2 Stylistic difference between written passages and spoken passages

There are major stylistic differences between the way people write and the way that they speak, although there are some occasions where the two seem to converge—news broadcasts for example. Most spoken language is learnt informally, usually from parents, siblings and schoolfriends, and one's knowledge of spoken language is continually reinforced and influenced throughout life, usually by examples which are spoken *impromptu* in the presence of the listener. By contrast written language is learnt in a more formal way at school, and the lifetime influences on most people's style of writing comes from printed publications, books, newspapers, magazines, which are usually prepared in isolation from the reader.

The presence of the listener and the isolation of the reader mentioned above are important. While a speaker who can continually look at the listener's face can see whether the message is properly understood, a writer who is isolated from the eventual reader will usually be more careful in expressing ideas, often by using a much more formal style. The main difference between written passages and spoken passages can indeed be summarized by the degree of formality that they display. From the description that emerges from Quirk *et al.* (1972) there are five modes in the spectrum of formality:

1 *Rigid* Using elaborate constructions in situations where it is most important to be unambiguous and definitive. Examples abound in legal and bureaucratic language.

   Notwithstanding prior claims by parties of the first part . . .
   The Minister may determine such issues without recourse to precedent.

2 *Formal* Using constructions in the passive form, and descriptions that are complete enough to avoid potential misunderstandings.

   The remaining ingredients should be mixed slowly until no lumps remain.
   The birds housed at Mr Brown's farm are ducks, geese and chickens, but no pheasants are housed there

3 *Normal*   Normal.

   Mix the flour and butter thoroughly.
   Mr Brown keeps poultry but not game.

4 *Informal*   Using colloquialisms, oblique references and more frequent use of personal pronouns.

   I cream together the flour and butter.
   You know he's got chickens and things, but I haven't seen any pheasant.

5 *Familiar*   In which sentence structures may be incomplete, phrases or words may be abbreviated or slang words used.

   Sort'f mix everything in.
   If it quacks or clucks, he's got some.

When people write it is more probable that they will assume a mode on the formal side of the spectrum; when they speak, it's more likely to be on the informal side. However, a text-to-speech synthesizer cannot be expected to translate between these modes of formality. Therefore preparing text for use in speech synthesis systems requires careful consideration of how it will sound when delivered. Text in formal mode may look quite acceptable when read, but may sound unduly pedantic when spoken, whether by person or machine. On the other hand, if the sounds produced by the synthesizer are less than perfect, the greater redundancy of a more formal mode could increase a listener's understanding.

6.4 Converting words to phonemes (phonetic transcription)

The two ways of implementing this process were briefly reviewed in Sec. 6.2.2. The table look-up approach requires a very large table; in fact a dictionary intended to cover a significant fraction of the words of British English could contain up to 40 000 words. Such a dictionary must contain, for each word entry, the orthographic form, which is usually the field on which the dictionary search is performed; the phonetic transcription of the word, accompanied by accent marks which indicate the potential for primary (') and secondary (") stress on a syllable; and the allowable parts of speech for that word. For example:

   sense: s e n s':   Noun, verb
   ...
   sensitivity: s e n 'sitiv"@ t ee: Noun

The entries can include contextual effects within each word, but not context dependencies that operate beyond word boundaries. For example, if the words 'horse' 'sense' are spoken together, the /s/ at the end of 'horse' and the /s/ at the start of 'sense' become one. In this example an accent mark is shown following the syllable to which it applies.

The process of converting words to phonemes can be comparatively regular in some languages. For instance, the Japanese language has a written form Kana, in which each character corresponds roughly to a syllable, which has an invariant pronunciation, so a completely rule-based conversion system could be relatively uncomplicated. However, for English and many other languages that have evolved from diverse origins, the conversion of letters to phonemes is far from regular and is very context dependent. The sensitivity to context of some common letter-to-phoneme conversions was mischievously highlighted by George Bernard Shaw when he defended the spelling GHOTI for the word 'fish'. He did so by choosing the pronunciation of GH as in 'enough', O as in 'women' and TI as in 'station'.

6.4.1 Rule-based word-to-phoneme conversion

One early example (Ainsworth, 1973) consists of a set of 159 context-sensitive rules for letter-to-phoneme conversion. The rules were derived from the analysis of the most common phonetic translation of each letter and the contextual conditions that accompanied each translation.

For the twenty years from the early sixties to the early eighties, considerable developments were made in speech synthesis for British English by John Holmes and colleagues at the Joint Speech Research Unit at Cheltenham (Holmes, 1984). The JSRU system was originally intended for an input that was a phonemic transcription of an utterance (Holmes, Mattingley and Shearme, 1964) but in later implementations (Golfin, Challener and Millar, 1985) it had developed into a full text-to-speech system which accepts unrestricted text input and is implemented entirely on a single computer card.

In the development and testing of a rule-based system attention is repeatedly focused on the words that are *not* translated correctly. There are two remedies for such a shortcoming:

1 A new context-sensitive rule can be designed to produce the correct translation for the target word, or group of target words, or

2 The word can be entered into the exceptions dictionary with the correct phonetic transcription.

The effectiveness of a rule-based system is very dependent on the order in which the rules are applied. Therefore the design of new rules, as in 1 above, must include careful consideration of where the rule should be placed in relation to existing rules. If the new rule is too general or the placing wrong, it may have side effects on the transcription of other words, possibly beneficial, but possibly detrimental. This means that there is a high development cost for incorporating a new rule, and the cost increases with the number of rules. On the other hand, incorporating an extra word into the exceptions dictionary always has a low development cost. From this line of reasoning it becomes clear why systems that began life with rule-based word-to-phoneme conversion tended to migrate to a greater proportion of dictionary-based conversion during development.

6.4.2 Dictionary look-up

The gradual reduction of memory cost and increase of processor speed also swings the balance towards a system with a more complete dictionary for the word-to-phoneme conversion process. However, there are more important reasons for expecting to find every word in the dictionary:

1. The prosodic quality of synthesized speech can only be improved if the position of phrase boundaries is established by a syntactic parse of the input text, which needs part-of-speech information about each word, and

2. The potential to bear stress of each syllable within a word is extremely irregular; it cannot be satisfactorily generated by rule and so must appear in a dictionary definition.

Therefore the dictionary must be an ordered list of word entries, each of which contains the phonemic translation, the set of possible parts of speech and marking of stress potential. A rather important complication is that more than one part of speech is possible for most words. For example 'lever', 'branch' and 'feed' can all be used as either a noun or a verb, and the word 'but' can be used as any one of seven possible parts of speech.

To cover the English language a dictionary with up to 40 000 word entries might be needed, and the searching strategy for the dictionary must be well designed to keep the computational load low. An inefficient search strategy on a large dictionary will result in greater delay between text input and speech output. A computer program to implement a dictionary search requires a primitive procedure to compare the target field against successive index fields in the dictionary. An example of a suitable procedure is the *strcmp()* function which is in the standard library of the C language.

> int strcmp(s1, s2)
> *s1* and *s2* are pointers to strings; *strcmp()* returns an integer which is less than, equal to or greater than zero, depending on whether string *s1* is lexicographically less than, equal to or greater than string *s2*.

The lexicographical value of a string is an integer formed from the ordinal value of the successive characters in the string. The precise definition of the integer values is not important, as long as comparisons of the form shown below yield the expected results.

strcmp('goat', 'goal') returns a value greater than zero.
strcmp('hat', 'hat') returns zero.
strcmp('hat', 'hate') returns a value less than zero.

In most dictionary search procedures all strings are reduced to one case, so comparisons between upper case and lower case which would give misleading results should never occur.

strcmp('hat', 'Hat') should be avoided.

The value returned from *strcmp()* is particularly useful in that, if the dictionary is lexicographically ordered, it indicates the direction that a search must take to find an index field with the same value as the target field. This makes a binary search possible, for which it can be shown that at most N comparisons are required for a dictionary that contains up to 2^N entries (Knuth, 1973). The example English dictionary with 40 000 word entries would require at most 16 comparisons.

It is often possible to do better than this with a hash function search, in which word entries in the dictionary are each given an arbitrary index number. The entries are arranged numerically in order of the index number of the word. The key to retrieving a word from the dictionary is that the index can be computed from the word by a *hash()* function.

int hash(s)
s is a pointer to a string; *hash(s)* returns an integer value which is easily computed from the ordinal value of the characters in the string.

The integer value returned by *hash(s)* should be distributed as evenly as possible over the range of index values. However, even for a large dictionary it is inconceivable that the hashing function will spread the word entries evenly over the index range without collisions where quite dissimilar words produce the same hash value. A discussion of the problem and methods to solve it appear in Knuth (1973).

The efficiency of the hash search is difficult to reconcile with the idea of grouping similar words together. For instance, the use of affix stripping could be applied to the dictionary look-up approach so that the stems of words are stored in the dictionary. Thus the stem 'interest' may occur with suffixes '-s', '-ed' and '-ing', and some of those forms may occur with the prefix 'dis-' and 'un-'. The compaction of a complete word list by mechanisms of this sort can greatly reduce the word count, but only at the cost of introducing rather complicated rules about affix applicability.

The exceptions that occur in the dictionary look-up process arise when the target word is not found in the dictionary. This might happen for rarely used words, mis-spelt words or foreign words. Because the diversity of words that fall into this category is unlimited, the design of an appropriate set of letter-to-phoneme rules must be a very broad compromise. On one hand a set of rules, which is designed to give the correct pronunciation for all the common words, can hardly be right for all the uncommon ones. On the other hand, a human reader will probably pronounce an unknown word by using the example of similar fragments of known words. The best approach is probably:

> *if* the word contains consonant clusters typical of the language
> *then* carry out a naive letter-to-phoneme translation, grouped into syllables as much as possible
> *else* spell out the word letter by letter.

Clearly a record of the words that are not found in the dictionary would be of considerable interest to the designers and maintainers of text-to-speech systems. If they are appended to a log file, analysis of the contents can determine whether new words should be added to the dictionary.

6.4.3 A rules-dictionary hybrid

Letter-to-phoneme rules have been devised (Elovitz *et al.*, 1976) for American English. In this system there are 329 rules, and their formulation was based on an analysis of the 50 000 word Brown Corpus. It has been computed that 90 per cent of the words in the Corpus are converted correctly by the rules. Most words that are incorrectly converted will not be totally wrong, and the proportion of *phonemes* that are produced correctly was measured at 97 per cent.

An interesting feature of this work (known informally as the Navy rules, as the authors worked for the U.S. Naval Research Laboratory) is that the rules and the exceptions dictionary are combined. The rules contain entries ranging from single letters to complete words. The inclusion of complete words was used sparingly, only for common words. If an extra rule were introduced for each of the 5000 words in the Brown Corpus that is wrongly pronounced, the result would be a system with 5329 rules rather than 329. Reference to such a large rule set would make the conversion process very laborious and slow. Some extracts from the rules are given in Table 6.1, which shows examples that cover a mixture of words, part words and single letters. The rules are grouped alphabetically by the first letter of the match string enclosed by [ ]. Within each letter group the ordering determines the priority of a rule over others, as the first rule for which the conditions

SYNTHESIS

Table 6.1 Extracts from the Navy rules (after Elovitz et al., 1976)

```
GRULE(complete)
[GIV] = /G IH V/                Key to context symbols
   [G]I^ = /G/
[GE]T = /G EH/                  space = start or end of a word
SU[GGES] = /G JH EH S/          ^ = one consonant
[GG] = /G/                      # = one or more vowels
B#[G] = /G/                     + = one front vowel (E I Y)
[G]+ = [JH]
[GREAT] = /GREY T/
#[GH] = //
[G] = /G/

ORULE(extract)
...
[OF]    = /AX V/
...
[OROUGH] = /ER OW/
...
[OUGHT] = /AO T/
[OUGH] = /AH F/
   [OU] = /AW/
...

RRULE(complete)
[RE]^# = /R IY/
[R] = /R/

TRULE(extracts)
...
[THROUGH] = /TH R UW/
[THOSE] = /DH OW Z/
[THOUGH] = /DH OW/
   [THUS] = /DH AH S/
[TH] = /TH/
```

are satisfied is operated. In general a specific match string must be placed before a more general match string. This principle is followed through to the last rule in each set of letter rules, which is always a single letter-to-phoneme translation without context. This ensures that the translation process is completely defined, and there can be no cases where a match is not found.

The rules are implemented by the following procedure:

1. The pointer to the input text is advanced until a letter is encountered.
2. The appropriate set of rules for the letter under the pointer is searched.

3 *For each* rule:

if the input text starting at the pointer matches the string enclosed by [ ]
and both left and right context is satisfied
then the pointer to text is moved past the string enclosed by [ ],
and the phoneme string given by the rule is added to the output.

4 Loop to 1.

The complexity of letter-to-phoneme conversion in the English language is shown by the words 'through', 'thorough', and 'rough'.

'through': a TRULE converts the entire word to the phoneme string /TH R UW/.
'thorough': a TRULE converts [TH] to /TH/, an ORULE converts [OROUGH] to /ER OW/.
'rough': an RRULE converts [R] to /R/, an ORULE converts [OUGH] to /AH F/.

The phoneme descriptors in this example use the letter symbols for IPA symbols that were used in the original source and likewise the pronunciation (particularly for the word 'thorough') is American English.

6.4.4 Phonetic transcription in MITalk

The text analysis modules of the MITalk system (Allen, Hunnicutt and Klatt, 1987), shown diagrammatically in Table 6.2, are designed to accept English text in ASCII coded form and produce a phonetic transcription with appropriate phoneme and allophone descriptors, and markers to indicate stress, and the boundaries of syllable, morph, word phrase and sentence.

The phonetic transcription is achieved by a dictionary look-up process in the module DECOMP; the dictionary provides pronunciation and part-of-speech information for about 12 000 morphs. This is estimated to cover more than 95 per cent of input text, the remainder being converted by letter-to-phoneme rules in the module SOUND1. Morphs were chosen rather than words because the new words that are being coined continually as part of the natural development of language are usually new combinations of existing morphs. Some morphs that are more generally reusable than others are *prefixes* and *suffixes*, which attach to the start or end of a word respectively. A morph that can stand alone as a word is a *root*. The module DECOMP decomposes each word in the standardized input into constituent morphs if possible. To allow for the indefinite number of morphs that may occur the procedure is recursive, and to

SYNTHESIS

Table 6.2 MITalk modules for analysis of text (after Allen, Hunnicutt and Klatt, 1987)

Analysis of text modules
FORMAT—convert symbols to standard form
$3.17 → three dollars seventeen cents Mr & Mrs → mister and missez 1979 → nineteen seventy-nine
DECOMP—phonetic transcription by morphological analysis
scarcity → scarce+ity → SS KK AE RR SS IH TT IY legendary → legend+ary → LL EH JJ EH NN DD AE RR IY
PARSER—phrase level parsing
NOUN GROUP: most of the exercises VERB GROUP: are NOUN GROUP: translations UNCLASSIFIED: .
SOUND 1—letter to sound rules and lexical stress
Letter to sound rules applied to words that are not covered by the morph dictionary, e.g. names of places and people, mis-spelt words. - SS KK AE RR SS IH TT IY → SS KK 'AE RR SS IH TT "IY LL EH JJ EH NN DD AE RR IY → LL 'EH JJ EH NN DD "AE RR IY

discriminate between legal and illegal sequences of morphs DECOMP uses an algorithm based on a finite state machine. The goal of the algorithm is always to find the longest possible morph that can be matched from the right, that is from the end of the word. Clearly once a potential match has been made, the starting point for the next match is determined. If DECOMP cannot match a morph while there are still unmatched characters, then progress can sometimes only be made by accepting a shorter morph at an early stage. In most cases there is more than one possible decomposition, and to arbitrate such conflicts a scoring scheme is provided. A few example scores are given below:

Score	Type	Description
35	DERIV	Derivational suffixes, e.g. 'ness', 'ment', 'ity'
64	INFL	Inflectional suffixes, e.g. 'ing', 'ed', 's'
101 (first)	ROOT	Morph that can appear alone
133 (other)	ROOT	
	ABSOLUTE	Can only precede an INFL suffix

Some aspects of the decomposition of the word 'scarcity' are shown in Fig. 6.2. Starting from the right it is seen that the largest morph 'city' is identified as a ROOT morph scoring 101; this leaves 'scar', which is also identified as a ROOT morph but which scores 133, giving a total of 234. Other possibilities for the second morph 'ar' and 'car' achieve the same score without covering the complete word, and so at this stage 'scar-city' is the preferred decomposition. The next largest suffix that can be applied is 'ity', a DERIV morph scoring 35; in this case there are two morphological mutations which could be applied, as well as simply removing the suffix 'ity'. The lowest scoring of the alternatives gives 'scarce-ity' as the preferred decomposition with a score of 136. Finally the suffix 'y', a DERIV morph scoring 35, could be applied, but the direct application leads to an illegal use of the ABSOLUTE morph 'it', and a morphological mutation gives a score of 170+ for 'scar-cite-y'.

We are left with the decomposition 'scarce-ity', which provides the necessary index terms to the dictionary, which then yields the information that scarcity is a singular noun, formed from the root 'scarce' phonetically described as 1SKE*RS (1 indicates primary stress on that syllable), and derivational vocalic suffix 'ity' with phonetic description *T-Eˆ (*-ˆ are phonetic symbols). The set of allowable parts of speech for the word are determined from the part of speech for the root morph, modified by the suffix morph 'ity'.

Morphological decomposition can yield incorrect results for some words; for example 'colonize' would decompose to 'colon-ize' in preference to the correct 'colony-ize', and 'cobweb' would incorrectly decompose to 'cob-web'. The few examples of this effect can be prevented by entering the complete word as a lexical entry in the morph dictionary. Other words are given a separate lexical entry if the morphological decomposition gives rise to the assignment of an incorrect part of speech.

6.4.5 Phonological recoding in MITalk

The phonetic transcription module produces a sequence of phoneme descriptors which are consistent with the notion of the phoneme as a purely logical unit. The PHONO module replaces some of the phonemes by allophones, which are tokens for the actual sounds of the spoken word.

The rules by which an appropriate allophone is selected are sensitive to context, either within a word or across word boundaries. For example, the plosive release which would normally complete the enunciation of the phoneme /d/ (DD in MITalk notation) is not appropriate for the word 'ladder'. In cases like this the contact between the tongue and the alveolar ridge is fleeting and the sequence of stop and release does not normally occur. The allophone DX represents the

SYNTHESIS

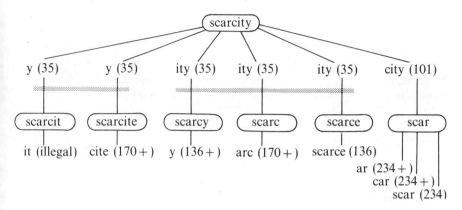

Figure 6.2 Morphological decomposition of the word 'scarcity'

sound of this alveolar tap. Further examples of phonological recording show the effect of context across word boundaries; for instance the determiners 'a' and 'the' would be represented as /@/ and /dh @/ when followed by a word beginning with a non-vowel, but recoded to /ei/ and /dh ee/ when followed by a word beginning with a vowel.

6.5 Concatenation of stored speech units

Synthesized speech is usually generated by a two-stage process. The first of these stages is to define a sequence of speech units, each of which symbolizes the sound of a human utterance. The second stage is to regenerate coded fragments of speech, one for each speech unit, and to join them together in such a way that the joins are not evident. A previous section has looked at the task of converting words to phonemes, but they are not necessarily the best units to join together. A variety of speech units in decreasing order of size are shown in Table 6.3, and most of them have a part to play in solving the synthesis problem. In looking at each of the units it will become clear that larger units are easier to define but more difficult to join together, and the smaller units are harder to define but easier to join together.

The range of speech units, with an example and various measures associated with each, are shown in Table 6.3. The first measure is the quantity of units needed to give a reasonably complete coverage of the whole English language and then appears the average duration in milliseconds, followed by the memory size needed to contain the complete set, assuming that speech fragments are coded by LPC at 2 kbit/s. Finally, there are approximate estimates of the computational complexity of implementing a system for concatenating those units.

Table 6.3 Comparison of speech units for synthesis by concatenation

Speech unit with example	Quantity	Duration, ms	Memory size (LPC 2 kbits/s)	Computational complexity	Speech quality
Word disintegrate	40 000	350	28 Mbit	Low	Low (prosodic) High (segmental)
Morph dis-integr-ate	12 000	250	6 Mbit	—	—
Syllable dis-in-te-grate	5000	200	2 Mbit	Medium	—
Demi-syllable di-is-in-te-gra-ate	3000	100	600 kbit	High	—
Phoneme /d/i/s/i/n/t/e/g/r/ei/t/	40	80	6.4 kbit	Medium	Medium
Diphone /-d/d-i/i-s/s-i/i-n/n-t/ ...	1200	80	192 kbit	High	High

There is not a unique solution to this problem; different applications will be served best by different choices.

The memory requirement gradually reduces for units of word down to the phoneme. This is the range of speech units for which the quantity of units to cover the language, and the average duration of each unit, both decrease steadily. However, at the phoneme the decline of total memory size is thrown sharply into reverse. A diphone is not a subpart of a phoneme because each diphone represents the transition between two phonemes, each of which could be chosen freely from the complete set. Therefore the memory requirement for a complete set of diphone units is much higher than for phonemes.

Computational complexity is rather difficult to establish other than as a rank ordering of the various approaches that are feasible. For concatenation of world-sized units, the computation involves recognizing each word, retrieving the appropriate stored speech unit and presenting the coded parameters to the decoder. None of this represents a complex problem, although for a dictionary of 40 000 words it would undoubtedly be a large problem. What could add to the complexity would be an attempt to reconcile one word-sized unit to the next. In continuous speech there is no gap between successive words, and the co-articulation effects that occur between segments within a word also occur across word boundaries. The extent to which this can be modelled is extremely difficult unless there is information about the phonemic structure of the word. However, phonemic structure is not available in stored word-sized units, so we must assume that this is not a fruitful avenue to explore.

If we look at the computational complexity of storing syllable-sized units, then we have another difficult process of recognizing syllables in the input text; however there are fewer of them than there are words. Syllables in English can be classified into V, CV, VC and CVC types, where C represents a cluster of one or more consonants and V represents a cluster of one or more vowels.

```
I    do    dis  -  crim  -  in  -  ate.
V    CV    CVC     CVC      VC     VC
```

The recognition of valid consonant clusters in a word can be a useful aid to determining syllable boundaries. For example, the word 'straight' is usually accepted as consisting of one syllable of the CVC type, composed of valid consonant clusters [str][ai][ght], each of which occurs commonly in many other words. The word 'thumbnail' could be correctly divided into two syllables by recognition of the consonant cluster [mb] and rejection of the implausible cluster [mbn]. An indication of the potential difficulties is shown in the example sentence above; there is no clear reason why the syllables 'dis-crim' should be preferred to 'di-scrim', or 'disc-rim', or 'discr-im', all of which are formed from valid consonant clusters.

Speech quality is also difficult to quantify and tabulate, but that measure is likely to be in the order given in Table 6.3. Word-sized stored units will have segmental quality which is determined by the coding system. The prosodic quality would depend very critically upon whether a restricted or unrestricted range of messages is to be generated. For restricted message systems such as the speaking clock, in which a selection of prerecorded words are inserted into a prerecorded phrase, a waveform coding approach can be used, provided that each of the word inserts is spoken with a consistent intonation. For unrestricted message systems a coding scheme must be used which can provide continuity of excitation and filter parameters between adjacent stored units. The processing necessary to predict and implement the co-articulation effects on word-sized units is probably beyond present knowledge.

A separate coded segment for each morph presents considerable difficulties. The morph is a lexical unit and there are different pronunciations for the same morph in different contexts. For example, the morph 'cup' is pronounced similarly in words *cupful* and *cuphook* but in *cupboard* it loses the final stop /p/, so it appears not to be a suitable speech unit for concatenation. It has been shown later that the morph is extremely useful in determining phonetic transcription, in which case the problem can be resolved by ensuring that the morph 'cupboard' has priority over the morph 'cup'. The syllable suffers from the same problems as the morph. 'Cup' is a syllable as well as a morph; however, it does have a redeeming feature, because it is the smallest unit at which prosodic effects are identified. There are not many examples of the concatenations of morph, syllable or demi-syllable units, so no comment is made on speech quality.

The phoneme seems to have most to offer on the basis of this survey. There are so few of them, even though most systems augment the 40 phonemes by *allophones* (units that sound similar to a phoneme but differ slightly in length or vowel position), that the set rarely has more than 100 members. They can be joined together without enormous computational effort and the speech quality is reasonably good. The use of the diphone offers a solution to the problems of using phonemes at the cost of more computation.

For the coding of speech units smaller than a word, the use of a parametric coding scheme (described in Chapter 5) makes it easier to join the units together smoothly. This is because the excitation signal and the vocal tract filter parameters, both of which change at the relatively slow segmental rate, can more easily be interpolated when necessary. Linear predictive coding is one such parametric coding scheme, for which simple analysis techniques have been developed for vocoding applications. For use in text-to-speech systems, the analysis is performed only once so more elaborate techniques can be used, as long as they are consistent with relatively simple resynthesis.

6.5.1 Concatenation of stored words

It is possible to concatenate words if care is taken to ensure that the recorded words have intonation and rhythm that are consistent with the eventual use. A potential difficulty with this technique is that it is extremely difficult to add to the vocabulary at a later date. Some reasons for this are that the original speaker may not be available or the voice quality of the speaker may have changed; even the acoustics of the recording studio or the choice of microphone can make the newly recorded words identifiably different when heard in the context of previously recorded utterances.

The speaking clock (Accurist timeline) service available on the British Telecom telephone network is an example of an application for which the very limited range of utterances allows the use of PCM coding of the base phrase with selectable word inserts. There follows an excerpt from the message sequence with the optional word inserts underlined:

At the third stroke the time sponsored by Accurist will be <u>eight</u> <u>thirty</u> <u>two</u> <u>and</u> <u>fifty</u> <u>seconds</u> * * *
At the third stroke the time sponsored by Accurist will be <u>eight</u> <u>thirty</u> <u>three</u> <u>precisely</u> * * *

The word concatenation technique can be used because the repertoire of words, and the sentences that need to be constructed, is precisely defined. There are no foreseeable additions to the range of messages. It is reported (Halliday, 1983) that 8 kHz sampled speech, quantized at 8 bits per sample has been used in this application, the perceived voice quality is good and the original characteristics of the speaker are very clearly identifiable.

Speech output was available for the BBC computer, which was an imaginative development central to the Computer Literacy Project launched in 1981. Provision was made for the Texas Instruments TMS5220 LPC decoder to be added as a speech output device. In the BBC Computer Speech System, the coded speech is stored in word-sized units each of which is LPC coded. Despite the use of LPC, the voice is quite clearly recognizable as Kenneth Kendall, who became famous as a BBC newsreader. However, there is no easy way of reconciling one speech unit to the next, so sentence-length utterances are delivered as isolated words which results in a very jerky style of speech. The range of messages that can be constructed from approximately 165 words or part-words included in the standard ROM is quite limited, although the vocabulary could be extended by the addition of further ROMs.

6.5.2 Concatenation of stored syllables and demi-syllables

The syllable appears to be a convenient unit for storage, as it is large enough to reduce the number of joins that are necessary and yet small enough to make each stored unit reusable in a number of word contexts. There is often a choice as to where to separate one syllable from the next; whenever possible a stop consonant should be placed at the start of a syllable. This allows the join to be made at the silence preceding the stop. A seamless join between the first and second syllables of 'dis - crim - in - ate' would be easier to achieve than it would be for 'disc - rim - in - ate'.

There are no noteworthy attempts at syllable concatenation for the English language, but there are better prospects for the Japanese language; one form of the spoken language uses about 100 syllables of the CV type. These syllables have been used for the synthesis of speech from stored cepstral parameters (Imai and Abe, 1980). The use of demi-syllables reduces some of the problems that occur with syllables, and several examples have been reported of the concatenation of demi-syllables (Browman, 1980; Dettweiler and Hess, 1985).

6.5.3 Concatenation of stored phonemes

In storing phoneme-sized units a considerable advantage can be gained by coding the pitch information separately from the vocal tract filter model. Thus any of the parametric coding schemes allows continuity of the pitch information from one stored segment to the next. A hardware example of a phoneme synthesizer is the Votrax SC01, which is no longer in production. Although some attempt was made to provide continuity of pitch and formant values from one segment to the next, each segment has a duration that is fixed in relation to the others. These durations are only sometimes correct. A brief description of the SC01 is given later in Sec. 6.7.2.

6.5.4 Concatenation of stored diphones

One of the problems associated with storing speech segments corresponding to phonemes is that in many phonemes the character of the sound is produced from the transitions that occur at the beginning and end of the segment. The characteristic sound that these transitions give could be lost from both segments if they are smoothed together. If there is a steady part of a phoneme, it occurs near the centre, and so it would seem less destructive to join one segment to the next at the centre of a phoneme. This technique requires that the coded segments correspond to a *diphone*, a speech segment that spans the end of one phoneme and the beginning of another. If we choose the first half of the diphone from the N phonemes in British English, there are $N-1$ ways to choose the second, and so there are $N(N-1)$ (approximately N^2) possible diphones. It has been said that of the 1600 potential diphones in British English ($N \approx 40$) as many as 400 do not occur. However, construction of a diphone dictionary for the French language (Stella, 1985) showed that although some phoneme combinations did not occur within words, they could occur at word boundaries, and so N^2 was the approximate number of diphones. The extent of co-articulation at diphone level is very important, and it has been evident from early examples of this technique (Dixon and Maxey, 1968) that the determination of a valid library of diphones is important and that the final selection of diphones is context dependent.

6.6 Prosody

The criticism most commonly made of synthesized speech is that it does not recreate the rhythm and intonation (prosody) characteristic of a human speaker. The result is that even though the segmental quality is good, and words can be correctly distinguished, the delivery is flat and monotonous, and the listener soon becomes either inattentive or irritated. For synthesized speech to be acceptable in any application that is more exacting than a novelty item, the generation of prosody must be improved. The prosody of speech is determined by:

1 *Pitch movements* These are short-term variations of pitch (F0) which are aligned to stressed syllables, superimposed on long-term variations which indicate phrase or sentence type.

2 *Duration* The length of individual segments is altered to be compatible with the context in which they occur, and also to indicate emphasis.

3 A third factor, *loudness*, is the least important of the factors; it can create dramatic impact, but has little contribution to the synthesis of natural sounding speech.

Looking in isolation at a list of the words in a language, for example in a dictionary listing, we would expect to see accent marks, which show the *potential* for stress. When the word is spoken in the context of a sentence the *actual* stress applied may be different from the accent marks to provide contrast or reinforcement.

6.6.1 The incidence of stress

An early model for the generation of prosody was based on the distinction between function words and content words. A *function word* serves to define the relationship between the other words, *content words*, which carry information content. The usually accepted list of function words includes the following types, with two examples of each:

Type	*Examples*
Determiners	the, some
Parts of the verb *to be*	am, was
Prepositions	to, over
Conjunctions	and, but
Personal pronouns	I, his
Auxiliaries	might, can

In this model function words do not to bear primary stress in declarative sentences, so segments are typically of short duration, and voiced segments show small pitch variation.

A problem with division into function and content words is that function words *can* bear stress in some circumstances. For example, the word 'can' in the previous sentence is stressed because a contrast is being made, and yet 'can' appears in our list of function words. The classification is generally regarded as a poor basis for the generation of prosody, because there is no satisfactory definition of function words. More recent work does not seek to make a rigid classification, and the analysis of spoken English from a prepared talk (Altenberg, 1987) showed, for example, that determiners were unstressed in 89 per cent of instances and nouns were unstressed in 8 per cent of instances. The proportion of unstressed instances for all other parts of speech lay between those proportions, which were the extreme values.

6.6.2 Pitch movements

The pitch or fundamental frequency (F0) of voiced speech is related to the pressure difference across the vocal cords, and consequently to the rate of air flow through them. Most speakers will draw a good breath before starting the utterance of a sentence, and with high pressure in the lungs the flow of air will be strong and consequently the pitch of voiced speech raised. As the sentence continues the reserves of air are reduced, the pressure in the lungs is reduced and the pitch of the voice falls as the flow rate of air through the vocal cords decreases. Thus the baseline of pitch observed in the utterance of a sentence usually starts at a high level and decreases over the length of the sentence. Temporary replenishments of air supply may occur with snatched breaths taken during the utterance of a long sentence.

The speech produced between drawing one breath and the next, often characterized by the falling baseline, is called a *breath group* (also called a major phrase). In some examples of written text, commas are used to indicate where a pause for breath is expected. Superimposed on the falling baseline are local variations of pitch that serve to accentuate stressed syllables. These local variations occur for several reasons. Most obviously are the conscious efforts to increase air flow to emphasize a syllable, but also there are incidental reductions or cessations of air flow related to the articulation of fricatives or stops. In producing these sounds the constriction of the vocal tract, either partial or complete, with the increase of pressure above the vocal cords reduces the pressure difference across the cords and the pitch falls momentarily.

Within a breath group there are *tone groups* (also called a minor phrase or *tone unit*), which may correspond roughly to a phrase defining an entity, a noun phrase, or an action, a verb phrase. The tone group is the basic unit of prosody, whose character is formed by the pattern of stressed syllables within it. A very simple algorithm for determining the extent of a tone group is to insert boundaries at every point that has a content word to the left and a function word to the right, subject to the restriction that a tone group must contain at least one content word.

On fundamentals like the cost of manufacturing in America
   F C   C |F C |F C |F C
compared to elsewhere, the dollar is already undervalued.
   C |F C |F C |F C   C

SYNTHESIS

Table 6.4 Pitch patterns for sentences

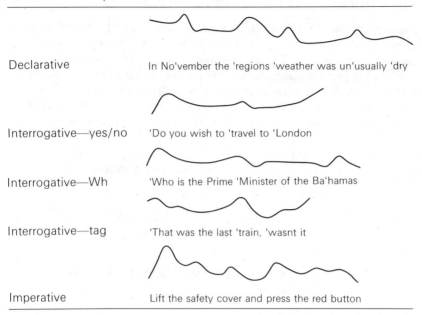

Declarative	In No'vember the 'regions 'weather was un'usually 'dry
Interrogative—yes/no	'Do you wish to 'travel to 'London
Interrogative—Wh	'Who is the Prime 'Minister of the Ba'hamas
Interrogative—tag	'That was the last 'train, 'wasnt it
Imperative	Lift the safety cover and press the red button

The example shows that this algorithm produces rather short tone groups. They can be amalgamated by a scheme (Young and Fallside, 1980) that gives a weighting to every stressed syllable within a tone group, and for every group with a total below a given threshold, it is amalgamated with its lightest neighbour. A better determination of tone group boundaries is to parse the input text, which also establishes the breath group boundaries.

There is a very wide variety of pitch patterns applied to the utterance of a sentence, depending on the sentence type, its length and whether particular words receive stress for emphasis or for contrast. The main types of sentence, each with a range of characteristic pitch patterns, are shown in Table 6.4. Important features in the pitch patterns are aligned to the *head syllable*, which is the first stressed syllable, and the *nuclear syllable*, which is the last stressed syllable in a sentence.

Short sentences are uttered as one breath group. Long sentences may be divided into a number of breath groups, and a comma is sometimes used in the text to indicate a breath group boundary. Confusion can arise because a comma can also be used to separate items in a list. It is often difficult to differentiate between the two uses of the comma, even when they are correctly used. In text-to-speech systems it is prudent to give punctuation within sentences a rather low priority in an algorithm for generating prosodic markers.

6.6.3 Phrase level parsing in MITalk

The module DECOMP shown in Table 6.2 contains a part-of-speech processor which computes a part-of-speech set for each word, based on the part-of-speech set for each constituent morph. The algorithm employs the general principle that the part of speech of morphs at the right of a word dominates others. However, the mechanism is not a simple one, as will be shown in the following examples. Most parts of speech are modified by a list of properties, and some properties transfer from the morph to the word, some are lost and others are transformed. For instance, the word 'titles' decomposes to 'title-s', and the suffix morph 's' transforms the part of speech for the complete word:

'title' NOUN(NUM SING) → 'title-s' NOUN(NUM PL)

Contrast that with the decomposition of 'entitles', in which the last morph operates on the property only after the first morph has transformed the part of speech:

'title' NOUN(NUM SING) → 'en-title' VERB(PL TR)(SING TR) → 'en-title-s' VERB(SING TR)

A similar complexity occurs with the suffix 'er'; when applied to a noun, adjective or adverb, it does not change the part of speech, but applied to a verb it transforms it to a noun:

'fast' ADV → 'fast-er' ADV
'friend' NOUN(NUM SING) → 'friend-ly' ADJ → 'friend-li-er' ADJ
'friend' NOUN(NUM SING) → 'be-friend' VERB → 'be-friend-er' NOUN(NUM SING)
'rock' NOUN(NUM PL) VERB → 'rock-er' NOUN(NUM SING)

The determination of phrase boundaries is achieved by the module PARSER, which operates on the basis of an augmented transition network (ATN) showing allowable structures for a noun group and for a verb group. The input for the parser consists of the sequence of words, together with the morph decomposition and part-of-speech information on both the word and morphs. This is usually sufficient to allow a surface structure to be determined at a phrase level. The parser does not attempt to provide a deep structure parse at the sentence level, as this is very difficult, and for many sentences of unlimited text it would be impossible, given the limited time and computational effort available in a text-to-speech system.

The purpose of the parser is twofold: it creates a sequence of nodes corresponding to phrases of the text and it defines a part of speech for each word within the node and for the node itself. If a word cannot be incorporated into a node corresponding to a phrase, then a node is created for that word and a set of parts of speech is given for that node. The sentence:

Kate noticed the dog's reflection.

is grouped into four nodes:

'kate' NOUN GROUP(NUM SING) — node 1
  'kate' NOUN(NUM SING)
'noticed' VERB GROUP(SING TR)(PL TR) — node 2
  'notice' VERB NOUN→'notice-ed' VERBEN(SING TR)(PL TR)
'the dog's reflection' NOUN GROUP(NUM SING) — node 3
  'the' ART(DEF TR)
  'dog' NOUN(NUM SING) VERB → 'dog's' NOUN(POSS TR)
  'reflect' VERB → 'reflect-ion' NOUN(NUM SING)
'.' UNCLASSIFIED — node 4
  '.' End Punctuation Mark

6.6.4 Fundamental frequency (F0) algorithm in MITalk

The algorithm for defining F0 in MITalk is based on a method described by O'Shaughnessy (1976, 1977). This method defines a high-level system which is based on sentence type, clause contour, phrase contour and word contour, which is augmented by a low-level system based on phoneme type, lexical stress and the number of syllables per word. Pitch contours at the high level are modelled on tune A for declarative sentences and tune B for yes/no questions. In the MITalk system the parser does not establish clause boundaries, so clause contours are not implemented separately from the sentence tune; also a third tune has been added for Wh questions.

The part of speech of each word which was determined by the parser is used to assess the level of pitch variation to be applied to stressed syllables. Some examples of the potential to affect the F0 contour for various parts of speech is shown in Table 6.5. Levels below 6 correspond roughly to function words, and levels of 6 and above to content words.

6.6.5 Segmental duration

The duration of speech segments, which includes phonemes and allophones, is extremely variable between instances. English is a stress-timed language, in which stressed syllables tend to occur on a regular time frame and unstressed syllables between them are reduced in length as necessary. Therefore the duration of unstressed syllables is in approximately inverse proportion to the quantity between stressed syllables. This factor can be determined from the input text *if the stressed syllables are identified*. By contrast, in French or Italian, which are syllable-timed languages, the duration of segments is much less variable.

Table 6.5 MITalk modules for synthesis of speech (after Allen, Hunnicutt and Klatt, 1987)

Synthesis of speech modules
PHONOL—phonological recording
DH AH → DH IY   selection of allophones according to context AH → EY LL AE DD ER → LL AE DX ER   alveolar tap

PROSOD—timing (segmental duration)		
segment	minimum duration	inherent duration
IH	40	135
AE	80	230
RR	30	80
OY	150	280
	(DUR = MIN + (INH − MIN)*PRCNT/100)	

F0ARG—fundamental frequency F0
potential to affect F0 contour 0 article ... 3 personal pronoun ... 7 noun ... 10 quantifier ...

PHONET—phonetic targets and continuation smoothing					
segment	F1	F2	F3	B1	B2	B3
AH	620	1220	2550	80	50	140
IH	400	1800	2570	50	100	140
FF	340	1100	2080	200	120	150
SS	320	1390	2530	200	80	200

Smoothing templates

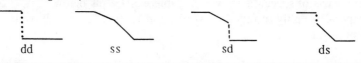

    dd   ss   sd   ds

CWTRAN—parameter conversion
Phonetic parameters converted to control coefficients

COEWAV—waveform generation
5 ms update of coefficients to synthesizer hardware

Klatt (1976) performed an analysis of the duration of phonetic segments of American English which was subsequently used in the MITalk system. The analysis resulted in the determination of an inherent duration, INHDUR, and a minimum duration, MINDUR, for each segment; some example values appear in Table 6.1. He also investigated the factors that influence the duration, and determined which of them were perceptually important. The influence of those factors is described in a set of rules (Klatt, 1979) which determines the duration of speech segments. The length of the segment, DUR, is determined by the relation

DUR ← MINDUR + ((INHDUR − MINDUR)*PRCNT)/100

where a variable PRCNT is created for each segment, initialized to 100, and then modified by the relation

PRCNT ← (PRCNT*PRCNT1)/100

with the appropriate PRCNT1 for each of the rules that applies to that segment. A few example rules follow to indicate the type of factors and the PRCNT1 value:

- Syllabic segments (vowels and syllabic consonants) are shortened by PRCNT1 = 60.
- A phrase-final postvocalic liquid or nasal is lengthened by PRCNT1 = 140.
- Syllabic segments in a polysyllabic word are shortened by PRCNT1 = 80.
- Consonants in a non-word-initial position are shortened by PRCNT1 = 85.

6.7 Electronic analogues of the vocal tract

The most successful approaches to speech synthesis use an electronic analogue of the vocal tract which allows the speech segments to be characterized by a parametric coding method. Chapter 5 explains how parametric coding allows speech to be generated by updating parameters at a relatively slow rate. Generally between 10 and 20 parameters that determine the vocal tract filter are updated at a fixed rate, which may be somewhere between once every 5 ms to once every 20 ms.

One approach is to create an articulatory model, which is a direct analogue of the vocal tract area as a function of its length. The area function models the shape of the vocal tract and the various constrictions between the tract walls and the articulators. A less direct approach is to model the spectrum of the transmission characteristics of the vocal tract. In either case the model of the vocal tract is excited by a periodic signal representing the glottal pulse and a noise signal representing the frication of turbulent air flow.

6.7.1 Articulatory models

This approach seeks to recreate the effect of the exact position of the speech articulators (velum, tongue, teeth, lips, etc.) on the produced speech sounds. The advantage of this approach is that many co-articulation effects can be predicted by the dynamics of the articulators. The phoneme realization of one segment is influenced by the targets for the next segment, and the influence is greater as the segmental rate is increased. In a rather narrow sense the lattice filter realization of the vocal tract can be regarded as an articulatory model, because it represents the area function of the vocal tract, which is determined by the position of the articulators. It is a limited analogy because the constant delay units that are used in the lattice filter represent a constant displacement, and so the area function can only be changed at a number of fixed locations. Greater precision in locating discontinuities in the area function requires lattice filters of higher order.

Speech sounds and the interaction between successive sounds can be described in terms of movement of the articulators, in the parameters of the model. The effect of greater or lesser precision in the location of the articulators that occurs as a result of changes in the speaking rate could be determined automatically from the model (Fig. 6.3). Despite

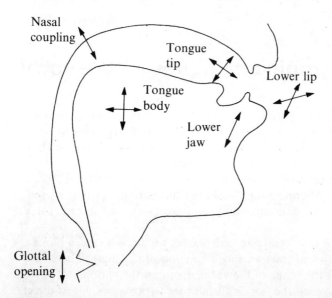

Figure 6.3 A model of human articulation (after Coker, 1976)

SYNTHESIS

extensive investigations of the possibilities, the approach has not yet been developed into a practical speech synthesizer. Further comment on the topic is given by Linggard (1985).

6.7.2 Spectrum models

In this technique the spectrum of the transmission characteristics of the vocal tract is modelled by a transmission filter with multiple resonators (see Fig. 6.4). Each resonator produces a peak in the spectrum which is identified as a formant. In early examples of the method the resonators were connected in cascade. For voiced sounds, the cumulative effect of the 'skirts' of each resonator happens to give approximately correct amplitudes for successive formants in the combined spectrum. Thus individual control of the resonator amplitudes is not needed. This is quite convenient as it is impossible to control the amplitude of one formant without affecting the amplitude of all the others. However, for fricatives and plosive bursts, where the excitation source is above the larynx, the sounds produced can be unrealistic as there is no separate control of the amplitude of the resonators. In voiced fricatives, there are two distinct sources of excitation, which are produced at different places in the vocal tract, so the setting of the resonators must be a compromise.

There are many examples of successful synthesizers using the cascade model including the Votrax SC-01 speech synthesizer, a single-chip solid-state device. Although now out of production, the SC-01 was used in many low-cost speech synthesis applications during the eighties. The input to the system comprises: a 6 bit code which identifies which of the 64 phonemes is to be synthesized and a 2 bit code to control pitch of voiced excitation. Each of the phonemes has its own natural duration, so no timing information is required; a request signal is generated when the next phoneme code is needed. The pitch control provides just four steps, which allows for a natural variation of pitch over the length of a sentence, the best results being obtained if the pitch is altered during unvoiced segments of speech. This control of pitch is too coarse to provide for pitch movements within a voiced syllable, which are needed to give stress to a syllable.

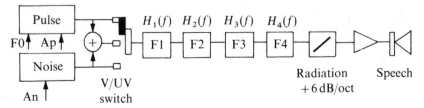

Figure 6.4 Synthesizer based on a cascade resonator model of the vocal tract

Parallel resonator model

When the resonators are connected in parallel the amplitude and frequency of the formants can be controlled individually. A general model of this arrangement is shown in Fig. 6.5.

This approach offers improved flexibility to model speech sounds, but at the cost of increasing the number of control parameters necessary to drive the synthesizer. However, it does bring with it a serious problem. The algebraic addition of signal components is only correct for signals of identical phase relationship. Each of the resonators has a phase response which has a leading phase (positive angle with respect to the excitation) below the resonance frequency and a lagging phase (negative) above the resonance frequency. This means that if an identical signal is applied to the input of two resonators, for a range of frequencies between two resonances, the phase difference between signal components can approach 180° and they will be subtracted rather than added. This effect, shown in Fig. 6.6, will add a spectral zero to the synthesized waveform, which can be clearly heard. One solution to the problem is to alternate the phase of successive resonators in order of centre frequency as in the Holmes parallel formant model (Holmes, 1973). The problem and another solution to it are also discussed by Klatt (1980).

In the general parallel mode, purely voiced or purely unvoiced sounds can both be modelled well, but to model voiced-to-unvoiced transitions requires agile control of filter parameters. As with the cascade organization, the two distinct places of excitation present in voiced fricatives would require a compromise setting of the resonators. The compromise is considerably reduced in the parallel formant synthesizer developed at the Joint Speech Research Unit (Holmes, 1973) in which the voiced/unvoiced selector is replicated for each of the

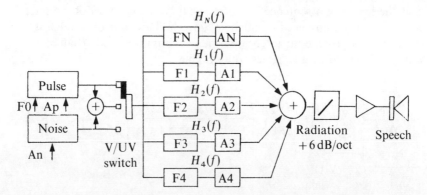

Figure 6.5 Synthesizer based on a parallel resonator model of the vocal tract

SYNTHESIS

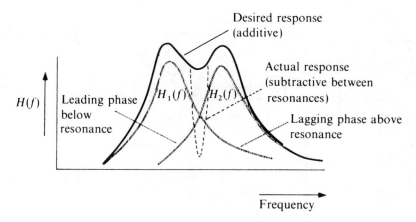

Figure 6.6 The effect of phase on the parallel addition of resonator outputs

parallel signal paths. There are many more recent implementations of the Holmes model, including a single-chip solid-state circuit, implemented on the versatile British Telecom FAD chip.

Parallel and cascade resonator model

A good, if expensive, solution to the either/or argument is to have both, and the Klatt synthesizer model (Klatt, 1980) takes this approach (Fig. 6.7). At the cost of additional complexity and the need for considerably more parameters to be stored for each speech unit, the advantages of both cascade and parallel connections are achieved.

It seems that the original reason for the combined approach was to offer flexibility to experimenters in speech signal generation, for which application the model was originally designed. Indeed the model was first implemented by a FORTRAN program which was not implemented as a real-time process. More recent implementation of the model as a real-time process on a DSP has shown that although the control parameters are more numerous they are not unmanageably so, and the model is capable of producing very good results.

6.7.3 Speech synthesis in MITalk

The modules for the synthesis of speech are shown diagrammatically in Table 6.5. The phonetic transcription, with markers to indicate stress, and syntactic boundaries that are produced by the analysis phase are used as input to the synthesis phase, which produces a timing framework into which the pitch contour and vocal tract model control parameters are placed.

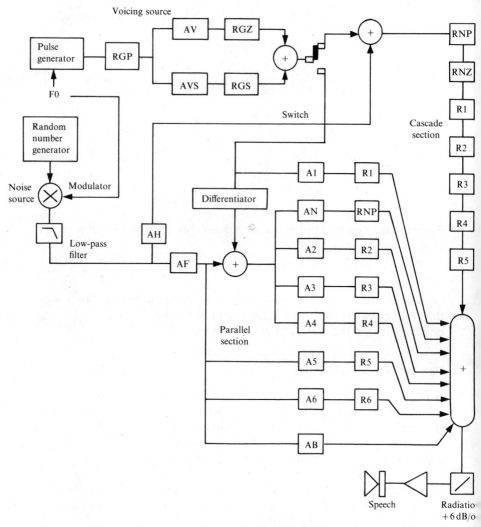

Figure 6.7 The Klatt cascade/parallel formant synthesizer (after Klatt, 1980)

The module PHONET shown in Table 6.5 shows target values for the centre frequency and bandwidths of the first three formants for the production of selected speech segments. If we consider the sequence of control parameters for the vocal tract model then at the centre of each segment the frequency and bandwidth should be close to the indicated target. In the transition between one segment and the next there are a number of possibilities; the value of each parameter may or may not be influenced by the adjacent segment. The case where neither segment has any influence on the other is illustrated by the discontinuous template dd in Table 6.5 and where both segments are influenced by the other by template ss. The other examples sd and ds show the templates for cases where the influence operates in one direction but not the other.

The phonetic parameters, suitably smoothed where necessary, are generated as a set of values for each tick of the update clock, which in MITalk occurs at 5 ms intervals. The formant resonators for F1, F2 and F3, etc., with bandwidths B1, B2 and B3, etc., are modelled by digital filters, operating at the speech sampling frequency f_S, as shown in the example of Fig. 6.8(a), which is the recursive digital filter for a typical resonator. The module CWTRAN produces control coefficients, for example the filter coefficients A, B and C appropriate to the values of f_B, the bandwidth, and f_R, the resonant frequency, from the expressions shown in the figure. Figure 6.8(b) shows the transversal filter which is used to implement the antiresonator digital filter that produces a transmission zero. Finally, the module COEWAV implements all of the digital filters and the generation of appropriate excitation waveforms to produce a synthetic speech waveform.

6.8 Summary

The generation of synthetic speech from text involves many stages of processing which gradually build an acoustic description of a speech signal. The end results can be judged by whether it is:

1   Effective in conveying information,
2   Believable as a human voice and
3   Pleasant to listen to for long periods.

At the time of writing it must be concluded that there is still work to be done in improving each of these factors.

The effectiveness in conveying information is crucially dependent on the segmental quality, the precision with which speech sounds are produced. Vowels are rarely misheard, but plosives, liquids and nasals are frequently misheard in synthetic speech. Many potential errors can be corrected by the listener using contextual clues, and the redundancy of speech acts beneficially in this regard.

The question as to whether the synthesized voice is believable is a complicated one. It is important that the voice is not dull, monotonous and mechanical as some early synthesized voices were, but in some sense it may be important for a listener to realize that the voice is *not* human and that what is spoken is produced by artificial intelligence rather than human intelligence. Prosodic quality is the main contribution to this factor.

It must be admitted that very few text-to-speech systems are pleasant to listen to for extended periods of time. However, this is true also of human speakers who do not vary the intonation and rhythm of speech

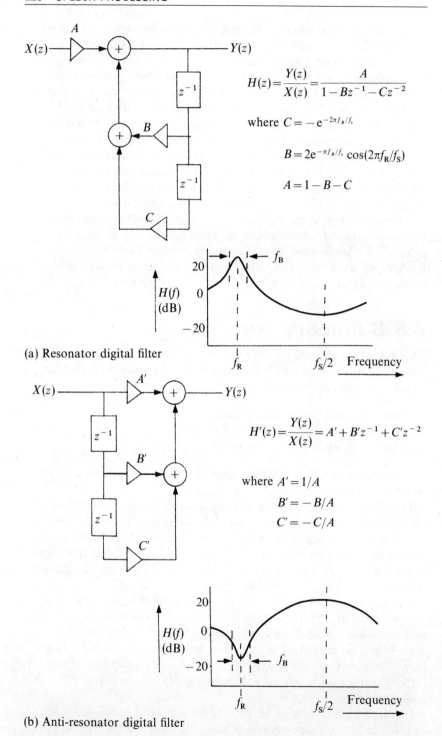

(a) Resonator digital filter

(b) Anti-resonator digital filter

Figure 6.8 Digital filters for resonators and antiresonators

in the way that an actor might. It is probable that as the art of artificially generating prosody improves, greater variation of pitch and duration will be possible. The introduction of some variation in the segmental quality may also improve this factor, and also the believability mentioned above. A large component of the effort put into speech recognition is to overcome the variability that occurs between different utterances of the same word by the same speaker; perhaps some randomness should be introduced into the generation of speech.

From early word-substitution systems with a limited vocabulary, speech synthesis has progressed to elaborate rule-based systems that allow unlimited vocabulary. Vocal tract models have been developed that improve the segmental quality of speech, but only if the control template for each segment is adequately described and matched to its neighbours. The relationship between the surface structure of written text and prosodic patterns is becoming understood, and parsers that can determine that structure can run in real time. However, the irregularity and gradual development of language will ensure that those who write the dictionaries will always be in work.

References

Ainsworth, W. A. (1973) 'A system for converting English text into speech', *IEEE Transactions on Audio and Electroacoustics*, vol. AU-21, 288–290.

Allen, J., M. S. Hunnicutt and D. Klatt (1987) *From Text to Speech: The MITalk System*, Cambridge University Press, Cambridge.

Altenberg, B. (1987) *Prosodic Patterns in Spoken English*, Lund University Press, Sweden.

Browman, C. (1980) 'Rules for demi-syllable synthesis using LINGUA language interpreter', Proceedings of International Conference on Acoustics, and Signal Processing, IEEE.

Coker, C. H. (1976) 'A model of articulatory dynamics and control', *Proceedings of the IEEE*, vol. 64, no. 4, 452–460.

Dettweiler, H. and W. Hess (1985) 'Concatenation rules for demi-syllable speech synthesis', *Acustica*, vol. 57, 268–283.

Dixon, N. R. and H. D. Maxey (1968) 'Terminal analog synthesis of continuous speech using the diphone method of segment assembly', *IEEE Transactions on Audio and Electroacoustics*, vol. AU-16, 40–50.

Elovitz, H. S., R. Johnson, A. McHugh and J. E. Shore (1976) 'Letter-to-sound rules for automatic translation of English text to phonetics', *IEEE Transactions on Acoustics, Speech, and Signal Processing*, vol. ASSP-24, no. 6.

Golfin, N. G., P. Challener and P. C. Millar (1985) 'A single card text-to-speech synthesiser based on the JSRU vocal tract model and text-to-speech rules', Proceedings of International Conference on Digital Processing of Signals in Communications, IERE, Publication no. 62.

Halliday, J. B. (1983) 'The new British Telecom speaking clock', *British Telecommunications Engineering*, vol. 2, part 3, 192–197.

Holmes, J. N. (1973) 'The influence of glottal waveform on the naturalness of speech from a parallel formant synthesiser', *IEEE Transactions on Audio and Electroacoustics*, vol. AU-21, 298–305.

Holmes, J. N. (1984) 'Formant synthesisers: cascade or parallel?', *Speech Communication*, vol. 2, 251–273.

Holmes, J. N., I. G. Mattingley and J. N. Shearme (1964) 'Speech synthesis by rule', *Language and Speech*, vol. 7, 127–143.

Imai, S. and Y. Abe (1980) 'Cepstral synthesis of Japanese from CV syllable parameters', Proceedings of International Conference on Acoustics, Speech, and Signal Processing, IEEE.

Klatt, D. H. (1976) Linguistic uses of segmental duration in English: acoustic and perceptual evidence', *Journal of the Acoustical Society of America*, vol. 59, 1208–1221.

Klatt, D. H. (1979) 'Synthesis by rule of segmental durations in English sentences' appears in *Frontiers of Speech Communication Research*, B. Lindblom and S. Ohman (eds), Academic Press, New York.

Klatt, D. H. (1980) 'Software for a cascade/parallel formant synthesiser', *Journal of the Acoustical Society of America*, vol. 67, no. 3, 971–995.

Knuth, D. E. (1973) *The Art of Computer Programming*, vol. 3, *Sorting and Searching*, 506–549, Addison-Wesley, Reading, Mass.

Linggard, R. (1975) *Electronic Synthesis of Speech*, Cambridge University Press, Cambridge.

O'Shaughnessy, D. (1976) 'Fundamental frequency by rule for a text-to-speech system', Ph.D. Thesis, Massachusetts Institute of Technology, Cambridge, Mass.

O'Shaughnessy, D. (1977) 'Fundamental frequency by rule for a text-to-speech system', *Proceedings of the International Conference on Acoustics, Speech and Signal Processing, IEEE*.

Quirk, R., S. Greenbaum, G. Leech and J. Svartvik (1972) 'Varieties of English and classes of varieties' in *A Grammar of Contemporary English*, 13–32, Longman, London.

Stella, M. (1985) 'Speech synthesis' in *Computer Speech Processing*, F. Fallside and W. A. Woods (eds), Prentice-Hall International (UK), London.

Witten, I. H. (1986) *Making Computers Talk: An Introduction to Speech Synthesis*, Prentice-Hall, Englewood Cliffs, NJ.

Young, S. J. and F. Fallside (1980) 'Synthesis by rule of prosodic features in word concatenation synthesis', *International Journal of Man–Machine Studies*, vol. 12, 241–258.

7 Recognition—the stochastic modelling approach

ROGER MOORE

7.1 Introduction

Throughout the sixties and seventies, research in the field of 'automatic speech recognition' (ASR) was dominated by two competing approaches: one based on 'whole-word' pattern matching and the other based on the explicit representation of 'speech knowledge' in the form of phonetic and linguistic rules. The pattern-matching paradigm was always acknowledged as a rather shallow approach but, by the end of the seventies the exploitation of powerful mathematical search algorithms meant that it achieved some considerable degree of practical and commercial success. On the other hand, despite offering a more intellectually satisfying solution to the problem, the knowledge-based approach consistently failed to demonstrate superior recognition performance.

More recently, the eighties have seen the emergence of a more *integrated* approach to 'speech pattern processing' which is based on a better understanding of the problems involved. In particular, the increasing use of statistical techniques and the growing awareness of the importance of 'structured statistical modelling' has put automatic speech recognition on the road to advanced performance systems.

This chapter outlines the difficulties that are faced by researchers who are concerned with designing effective automatic speech recognition devices, and the two initial approaches—pattern matching and knowledge engineering—are described briefly. The concept of 'speech pattern modelling' is introduced and it is shown how the stochastic modelling paradigm (based on hidden Markov models—HMMs) has led to much higher performance systems through the integration of the pattern-matching and knowledge-based approaches.

The chapter concludes with an indication of the contemporary issues in automatic speech recognition research.

7.1.1 The speech signal

Human speech is a low-data-rate low-dimensional acoustic signal which, in comparison with a visual image, might at first sight appear to be quite easily decodable. However, speech is not the result of the passive illumination of some aspect of the physical world; rather it is a man-made medium for the communication of thoughts and ideas from one human brain to another. The features, objects and relationships embedded in a speech signal are abstract representations of concepts invented by the human mind and encoded by the human vocal apparatus into a sequential acoustic pattern. For these reasons it is profoundly difficult to describe the 'objects' present in an acoustic scene.

These difficulties manifest themselves as a number of key problem areas which must be overcome by any successful automatic speech recognition machine.

Firstly, speech signals tend to be *continuous*. Fluent speech is a continuous changing sound pattern with no explicit markers to indicate the end of one sound and the beginning of another and with no obvious boundaries between one word and the next. In particular, there are no regular pauses between one word and another in an utterance. For example, Fig. 7.1 shows a speech spectrogram for the phrase 'we were away a year ago'. The only pause in this sentence is the 'stop gap' arising from the /g/ sound in 'ago'; the rest is a continuously changing sound pattern with each word merging smoothly into the next.

Secondly, speech signals are highly *variable*. One person's voice is quite different to another's, either due to differing linguistic backgrounds (different dialects and accents) or simply due to a physical disparity between their vocal systems. For example, the word for the digit 'one' is obviously pronounced differently in different languages (French: 'un', German: 'eins', Dutch: 'een', Italian: 'uno', Danish: 'en'),

'we were away a year ago'

Figure 7.1 Speech spectrogram of the phrase 'we were away a year ago'

RECOGNITION—THE STOCHASTIC MODELLING APPROACH

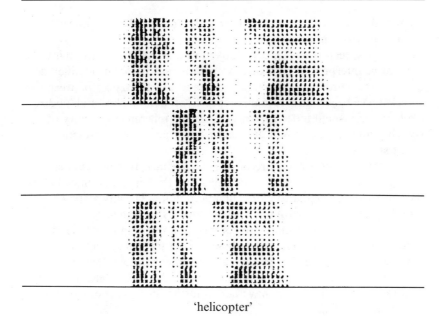

'helicopter'

Figure 7.2 Three versions of the word 'helicopter' spoken by a single talker

but even in English there are geographical variations in the pronunciation of the vowel sound ('one' as in 'sun' versus 'one' as in 'wan'). Also, there are clear differences between the voices of a man, a woman and a child which arise, in part, from differences in physical size.

Also, even the voice of a single individual exhibits variations under different conditions such as while whispering or shouting, or while suffering a head cold. In fact, it is virtually impossible for a talker to say the same word or phrase in exactly the same way on two different occasions. For example, Fig. 7.2 shows spectrograms for three repetitions of the word 'helicopter' spoken by the same talker in quick succession. Note that although the resulting acoustic patterns are visually similar, they are by no means identical.

Another source of variability arises from the continuous nature of speech outlined above. Since words and sounds flow smoothly one into another, the beginnings and ends can affect each other. For example, the phrase 'bread and butter' would be most likely to be pronounced 'bre'm butter' if spoken quickly! In this case, the word 'and' has been severely 'reduced' leaving only a residual /n/ sound ('bread'n butter') which has then been 'assimilated' with the place of articulation of the /b/ in 'butter' ('bread'm butter') which in turn leads to the same effect for the /d/ in 'bread' ('breb'm butter'), the /b/ then being reduced completely ('bre'm butter').

Thirdly, speech is *ambiguous*. Spoken utterances often have many alternative interpretations. For example, there is no acoustic difference between the words 'to', 'two' and 'too', and the differences between 'wreck a nice beach' and 'recognize speech' or 'new display' and 'nudist play' are very small indeed. Hence, given the inherent variability of speech patterns, these small distinctions may not even be present in fluent speech.

Fourthly, speech signals are often *contaminated*. It is only under very tightly controlled laboratory conditions that speech is guaranteed to be the only acoustic signal to enter the microphone. Usually a speech signal occurs in an environment where there is some degree of reverberation or where there are competing acoustic noises (including other voices), and in some cases these interfering signals and noises may be stronger than the speech signal of interest. Also, the speech may have passed through a communications channel (a telephone line or a radio link) which may have added further complications such as distortion and delay.

Finally, speech is highly *complex*. Language and speech are intimately related and speech is only a small component of a complicated symbolic signalling system for the communication of thoughts and ideas between human beings. Speech is the carrier of a message—the meaning; it is not the message itself. Human-to-human dialogue is replete with ritualistic behaviour which is designed for effective communication; the person who says 'hello, how are you' is very rarely interested in the health of the listener!

7.1.2 Implications for automatic speech recognition

For an automatic speech recognition system to be successful, it must therefore overcome a significant number of the problems identified above; it must have a clear strategy for either surmounting or avoiding each of the five problem areas. A successful automatic speech recognizer must therefore be able to recognize words even when they are embedded in a continuous utterance. Algorithms are thus needed which give rise to an effective 'segmentation' of a continuous pattern into its relevant component parts—typically a string of words.

A speech recognizer must also exploit 'similarities' between patterns rather than rely on precise repetition of the same information on different occasions. This implies that it is necessary to define suitable 'metrics' for measuring the similarity (or distance) between different patterns. However, such measures must have properties that are relevant to the task of recognition; for example different patterns for the same word should be measured as being more similar than those for words that are different.

A recognizer must be able to make use of 'context' in order to resolve alternative explanations which arise when the measured similarities are insufficient to discriminate between objects such as words. This means that the identity of a part of a continuous pattern may not be able to be decided independently of the identity of the other parts.

Finally, an automatic speech recognizer must be able to accommodate interfering signals and noise, and it must provide a rich interface into higher-level semantic interpretation processes if it is to provide a capability over and above that of a simple voice-activated button.

7.2 Early approaches to automatic speech recognition

Faced with the problems and algorithmic difficulties outlined above, automatic speech recognition might appear to be an almost unattainable goal. However, by concentrating on a reduced specification and by tackling the problems in a scientific and staged manner, it has been possible to make considerable progress in understanding the precise nature of the problems and in the development of relevant and practical solutions.

However, this has not always been the case. Some of the early work, while interesting in the context of a review of different approaches to automatic speech recognition, tended to be either overambitious about the achievements that could realistically be expected to be realized or somewhat naive with regard to the real difficulties that were being tackled.

Early attempts can thus be categorized into one of two main approaches. In the fifties and sixties the main approach was based on simple principles of 'pattern matching' which in the seventies gave way to a 'knowledge engineering' or rule-based approach. Only towards the end of the seventies was there a growing awareness of the need to integrate these two approaches and move towards a clear and scientific methodology for tackling the problems of automatic speech recognition—a move that ultimately led to a maturation of ideas and algorithms which are now beginning to provide powerful and exploitable solutions.

The following three sections review some of these early approaches to automatic speech recognition.

7.2.1 Pattern matching

The basic principle underlying almost all early speech recognition equipment is illustrated in Fig. 7.3. Such systems employ two modes of operation: a 'training mode' in which example speech patterns (usually words) are stored as reference 'templates' and a 'recognition mode' in which incoming speech patterns are compared with each reference pattern in turn. The identity of the reference pattern that is most similar to the input pattern determines the result.

In this scheme the acoustic pattern of a speech signal typically consisted of a sequence of vectors which had been derived from the speech waveform using some form of 'preprocessing'. For example, it

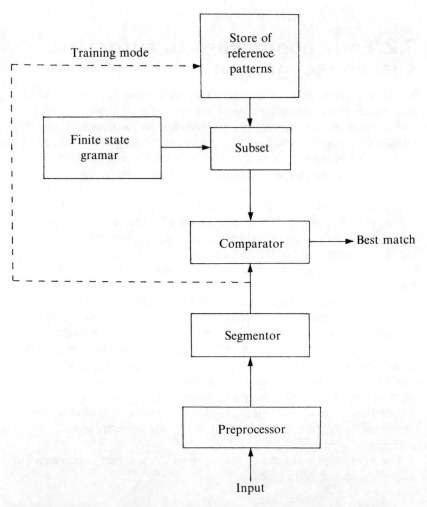

Figure 7.3 Structure underlying early pattern-matching automatic speech recognition systems

was common to perform a frequency analysis by means of an FFT or a filter bank in order to produce vectors that corresponded to the short-time power spectrum of the signal. A 'segmentor' would then divide the continuous signal into discrete pattern segments.

The difficulty associated with attempting to segment a speech signal has been discussed above and early workers found that it could only be made to function reliably for speech that had been spoken in the quiet and with the talker making explicit pauses between each word. This scheme has been used successfully for 'isolated word recognition' and the entire process has been termed 'whole-word pattern matching'.

The key to the success of this approach is the comparison process, and a technique called 'linear time normalization' was commonly used in order to overcome variability in the duration of spoken words. In this situation, the lengths of the patterns were 'time normalized' to a standard duration by lengthening (or shortening) the patterns the appropriate amount by using a fixed expansion (or compression) of the time-scale uniformly over the entire pattern. Comparison of the resulting fixed-length patterns could then be achieved by the calculation of a simple 'Euclidean distance' between the elements in one pattern and the corresponding elements in the other pattern.

It was also possible to raise the performance of such recognizers by using a simple finite-state grammar to limit the choice of words at each point in a spoken sentence (Alter, 1968). This meant that a recognizer could operate in an application that called for a vocabulary that was considerably larger than the maximum number of reference templates that the recognizer could accommodate.

These techniques began to be commercialized by the late sixties, but the performance was only barely acceptable, even for vocabularies as small as ten words. Also, since a user had to provide an example of every word in the vocabulary, such systems would only work well for that person's voice. As a consequence they were termed 'speaker-dependent' speech recognition systems.

Early research workers also attempted to use the pattern-matching scheme for large vocabulary continuous speech recognition by using the segmentor to divide the signal, not into words but into smaller 'phoneme'-sized segments (Reddy, 1967). The idea was that, by transcribing a speech signal into the spoken equivalent of alphabetic symbols, it would be possible to recognize any word simply by looking it up in a suitable 'pronouncing dictionary' (see Fig. 7.4). However, researchers ran into great difficulties, mainly arising from their rather naive assumptions about the supposed invariance of acoustic-phonetic patterns; the approach was severely criticized for not exploiting so-called 'speech knowledge'.

In fact, towards the end of the sixties, researchers began to suspect that there was not sufficient information in the acoustic signal to recognize speech in a 'bottom-up' data-driven manner, and that

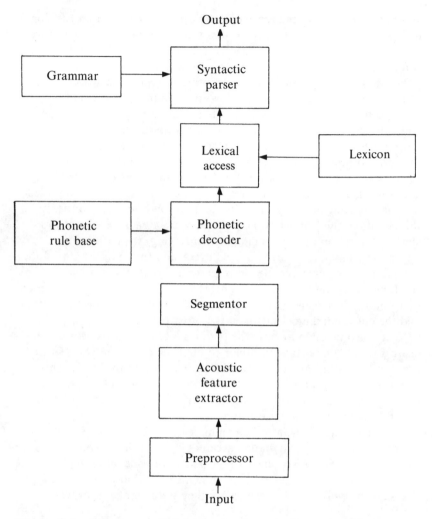

Figure 7.4 Structure underlying early phonetically motivated automatic speech recognition systems

continuous speech recognition could only be achieved by bringing to bear the same type of high-level knowledge that a human being clearly uses (Pierce, 1969). Also, there was a growing feeling that it was more important for a speech system to perform the correct actions as the result of a spoken request than it was for it to determine exactly what words had been spoken.

As a result of all this, the pattern-matching approach was abandoned in favour of a knowledge engineering approach which placed the emphasis on speech *understanding* rather than on speech *recognition*.

7.2.2 Knowledge engineering

The knowledge-based approach, popular in the early seventies, was based on techniques from the field of artificial intelligence (AI), which was then newly emerging. These techniques were applied to traditional concepts from the disciplines of phonetics and linguistics about how speech signals were organized. The key principle was to exploit 'speech knowledge' through its explicit use within a rule-based framework aimed at deriving an interpretation that would be suitable for the purposes of understanding the semantic content of the signal (Newell et al., 1973).

The knowledge was to be taken from different 'knowledge sources' at traditional levels of abstraction from the signal. Typically these might have included acoustic knowledge (concerning the physical signal), phonetic knowledge (concerning the sounds of the language), lexical knowledge (concerning the words of the language), syntactic knowledge (concerning the grammar), semantic knowledge (concerning the meaning) and pragmatic knowledge (concerning the task/application domain).

In order to integrate these different knowledge sources, such 'speech understanding systems' (SUS) typically employed quite complex organizational structures (Reddy and Erman, 1975). Figure 7.5 illustrates a range of different architectures.

Several speech understanding systems were developed during the seventies (Klatt, 1977) yet, despite the high level of available funding, none exhibited an acceptable level of performance. Their failure was due to a number of reasons. Firstly, the complex system architectures tended to operate very slowly (in many times real time). This meant that the behaviour of such systems was difficult to analyse and, more importantly, it was very difficult to optimize their performance by making suitable adjustments. Secondly, many speech understanding systems performed poorly because more attention was given to the higher linguistic levels than to the acoustic-phonetic front-ends. It was possible for such systems to output well-formed meaningful sentences as the result of a noise (such as a cough) entering the microphone! Thirdly, the approach did not take account of the possibilities for using example speech data as a reliable source of speech knowledge; they did not employ a suitable training mode.

7.2.3 Integrated approach

One speech understanding system did work well. The system was called 'HARPY' (Lowerre, 1976) and it was based on the idea that syntactic, lexical and phonetic knowledge could be compiled into a single data structure in the form of an 'integrated network' (Baker, 1974) (see Fig. 7.6). Recognition was then achieved by using a mathematical search

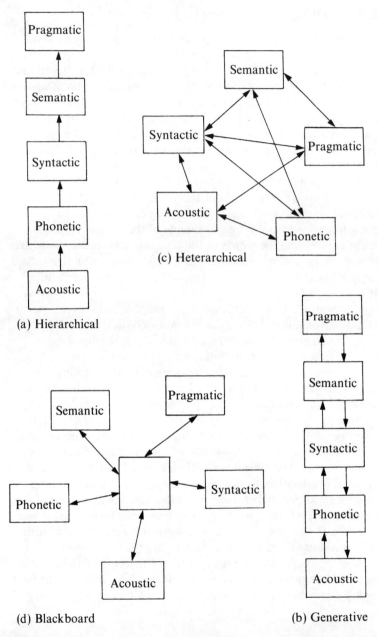

Figure 7.5 Alternative architectures for a speech understanding system

RECOGNITION—THE STOCHASTIC MODELLING APPROACH

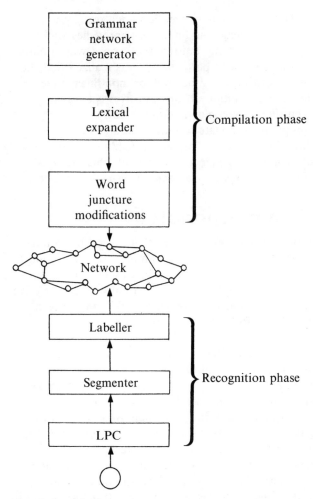

Figure 7.6 Structure of the HARPY continuous speech recognition system

technique called 'dynamic programming' (Bellman, 1957; Vintsyuk, 1968) (also known as the 'Viterbi algorithm') to find a route through the network that provided the best explanation of the input speech.

HARPY was speaker dependent (that is it was speaker trained), but it was highly successful in that it was able to demonstrate (in real time) the recognition of continuous speech using a 1000 word vocabulary with an acceptable level of recognition accuracy. This achievement can be attributed to two factors; firstly, the use of dynamic programming enabled HARPY to avoid segmenting the acoustic signal into phonetic segments *before* the recognition process, and, secondly, the grammar imposed very tight constraints.

As a result of HARPY, mainstream automatic speech recognition research moved away from phonetic and linguistic approaches and back to whole-word pattern matching, but this time armed with a more sophisticated comparison process. In particular, it was realized that the dynamic programming algorithm could be used for 'non-linear time normalization' and for the recognition of *connected* words (words without pauses between them) (Sakoe and Chiba, 1978).

This new process (popular in the late seventies) became known as 'dynamic time warping' (DTW) and it has been a highly successful technique in terms of raising performance to a level at which serious commercialization of automatic speech recognition systems could begin.

7.3 Towards a methodology

By the end of the seventies it had become clear that several important lessons had been learnt; the implementation of an intuitively satisfying knowledge-based approach was much more difficult than it had at first been thought and the use of a simple rule-based formalism was insufficiently powerful to overcome the problems exhibited by speech signals. Also, although the performance of connected word recognizers based on the integrated dynamic time-warping approach had been impressive, it was still found to be lower than that needed for many applications. In both cases, the difficulties arose from the inherent variability of speech. As a consequence, there was a growing awareness of the need to introduce *statistics* into the recognition process.

Thus in the eighties automatic speech recognition (and automatic speech generation—speech synthesis) reached something of a watershed in respect of its further development. Progress in speech synthesis appeared to be moving towards the use of recognition-type techniques to analyse quantities of natural speech to produce a more realistic output (Bridle and Ralls, 1985), and highly organized structures such as those used for many years in speech synthesis seemed to be relevant to recognition. What was missing was a clear methodology which, following the lessons learnt in the sixties and seventies, would be capable of unifying all of these ideas into a single research paradigm.

It therefore became apparent that in order to achieve high accuracy, many talker, large vocabulary speech recognition in a harsh environment, and high quality, high intelligibility, variable talker speech synthesis, it was necessary to try to establish a central *theory* of 'speech pattern processing'—a theory that would be mathematically rigorous, computationally tractable and could make effective use of available information about the structure and use of human speech and language.

The goal in automatic speech recognition research thus became to develop such a theory and to provide a formal foundation for the integration of different sources of speech knowledge and for the process of moving automatically between the different levels of description of speech patterns—from the lowest level *physical* description to the highest level *abstract* description—in other words, to achieve the appropriate 'signal-to-symbol' transformations.

7.3.1 The need for a formalism

In order to achieve these goals researchers found it necessary to focus on the following key concepts:

- The prior *information* (knowledge) that is available in the form of descriptive statements about speech patterns, linguistic structures and the relationships between different levels of description, in addition to corpora of recorded speech material and linguistic data,
- The *representation* (encoding) of speech knowledge and information derived from actual speech data and
- The *computation* (algorithms) that must be executed in order to achieve the required transformations.

Of these three facets of speech pattern processing, the key issue was the nature of the encoding. Evidence from the seventies showed that it was easy to propose intermediate representations which were consistent with various types of prior information, or alternatively which were compatible with efficient algorithms. The challenge was to satisfy both at once. What was required was a mathematical and scientific 'formalism' for encoding *a priori* information in a computationally useful form, a formalism that could exploit regularities in the data (the patterning), generalize from 'seen' data to 'unseen' data (for recognition) and be driven by some efficiency criterion to use the minimum information to achieve its goals.

These arguments led naturally to an approach to speech pattern processing that was founded on information theory and on 'speech pattern modelling' in which information about speech and speech patterns is encoded in a suitable model and appropriate algorithms are used to compute the output of the model for a specified input (for recognition or synthesis).

7.3.2 Structured statistical modelling

In general the problem of modelling physical systems can be decomposed into four subproblems (Mendel, 1973):

1. *Representation* The type of model (static or dynamic, linear or non-linear, deterministic or stochastic, discrete or continuous, etc.).
2. *Measurement* Which physical properties should be measured and how.

3 *Estimation* The determination of those physical quantities that cannot be measured from those that can.

4 *Validation* Demonstration of confidence in the model.

The research issues in speech pattern modelling are therefore concerned with the modelling paradigms (knowledge representation), the representation of acoustic data, the definition of a 'good' interpretation or output, the search strategies for finding a good interpretation (or output) of a model given some input data, methods for model construction and parameter estimation, and performance assessment procedures.

In practice there are two sources of speech knowledge that need to be integrated for high-performance automatic speech recognition: descriptive information arising from the insight of an 'expert' and real data in the form of example speech patterns. It is therefore important to exploit the *structural* constraints suggested by an expert to get somewhere near a solution and then to use *stochastic* behaviour to fill in the gaps in the expert knowledge as a form of 'generalization' (interpolation). It is also important to utilize 'automatic learning' to alter the structural constraints and the stochastic detail in the face of hard evidence from example data.

Of course virtually any recognition scheme will do well given sufficient training, but in reality there is never enough data. In principle, one could train a speech recognizer on all of the possible sentences in a language, but this is obviously completely impractical. However, sentences are constructed from sequences of words and the total number of words in a language is smaller than the total number of sentences, so it would clearly be more efficient to train a machine on each of the words. This is feasible but it could still require an enormous amount of training material (especially if each word is to be spoken several times by each talker). Of course within words there is another level of patterning since words are made up of syllables, and there are fewer syllables in a language than there are words—and syllables are made up of phonemes, etc. Therefore, the methodological trick that needs to be mastered is to fold these different levels of patterning into the model structures, thereby getting the best estimates for the least data—maximum generalization.

7.4 Speech pattern modelling

In order to implement the methodological approach described above, it is necessary to choose a suitable statistical modelling paradigm and to derive algorithms for assessing and optimizing the quality of the models with respect to example data. In other words, what is needed is a trainable maximum likelihood or 'Bayesian classifier', but with the

RECOGNITION—THE STOCHASTIC MODELLING APPROACH

additional property that it should be capable of processing variable-length rather than fixed-length patterns.

From an information theoretic point of view, this means that it is necessary to be able to compute the probability $P(W|O)$ of occurrence of a sequence of segments (say words) W, given an observation sequence (the acoustic evidence) O (Jelinek, 1979; Jelinek, Mercer and Bahl, 1982). It is rather difficult to compute $P(W|O)$ directly; however using Bayes' rule it is possible to express this probability in terms of two *generative* models:

$$P(W|O) = \frac{P(O|W)P(W)}{P(O)}$$

where $P(O|W)$ is based on an 'acoustic model' and $P(W)$ is based on a 'language model'. $P(O)$ may be regarded as a constant. Therefore, in order to construct a statistically based automatic speech recognizer, it is first necessary to choose suitable statistically based generative modelling paradigms.

At the present time the most computationally useful modelling approach is based on stochastic 'automata' in which speech knowledge may be expressed in the structure and parameters of a probabilistic 'finite-state machine' (Baker, 1975). The advantages of such an approach are that (for certain types of model) the likelihood of the model generating the observed data may be found easily using an optimal search procedure such as dynamic programming and that powerful parameter reestimation algorithms exist which are guaranteed to increase the likelihood of a model generating a particular set of observation data.

There are many different classes of automata, each reflecting models of different complexity. One of the simplest is a deterministic finite-state machine, or first-order Markov process, which simply provides an explicit statement of which item can follow which in an observation sequence. Each state is uniquely associated with a particular observation, and the transitions from state to state specify allowable observation sequences. Such a process is often used to provide a basic language model in which an automaton is used to express a 'grammar' or word 'syntax' (the model structure being derived manually for each specified task domain).

A probabilistic language model which is often used to capture word order information in a more general way than a simple finite-state syntax is a 'bi-gram'. This is a non-deterministic first-order Markov automaton which models the probability of one word (or grammatical item) following another. The 'tri-gram' is the equivalent second-order model. Probabilistic language modelling using these techniques has been investigated extensively during the eighties (Jelinek, 1985).

A particularly interesting class of automata are 'hidden Markov models' (HMMs). These are also non-deterministic first-order Markov processes, but each state is not uniquely associated with a particular observation. In fact the term 'hidden' derives from this property in that, because the model is entirely probabilistic, it is generally not possible to determine uniquely the state from which any given observation vector was drawn. The underlying state sequence responsible for generating a given observation sequence is thus hidden from the observer.

All these models can be viewed as being generative because one can imagine choosing an initial state (randomly according to the initial-state probabilities), outputing a vector (randomly according to the state output probability density function) and then moving to another state (randomly according to the state transition probabilities), outputing another vector and so on. The result would be the generation of a particular sequence of output vectors (and its associated probability of occurrence).

In principle such models can generate (and therefore assign a probability to) any output sequence. It is thus a particularly interesting formalism for employing within the Bayesian approach described above since such a model would be able to assign a probability to any observation sequence.

It should be clear from the foregoing that these modelling paradigms are related to 'formal grammars' (Hopcroft and Ullman, 1969). A first-order Markov model is formally equivalent to a 'regular grammar' and a hidden Markov model is equivalent to a 'stochastic regular grammar'. It is also interesting to note that there is an even more general form of hidden Markov model that is equivalent to a 'stochastic context-free grammar'.

7.4.1 Hidden Markov models

The principle of hidden Markov modelling is based on modelling speech patterns as a sequence of observation vectors derived from a probabilistic function of a first-order Markov chain (Levinson, Rabiner and Sondhi, 1983a, 1983b; Rabiner and Juang, 1986). The states in such a model are connected by probabilistic transitions and each state is identified with an appropriate output probability density function. It is also possible to define an equivalent formulation in which the outputs are associated with the transitions.

Basic formulation

Let a hidden Markov model M consist of an underlying Markov chain $A = [a_{j,i}: i,j = 1, \ldots, I]$ and a set of probability density functions $b_1, \ldots, b_I$. The probability of a transition from state j to state i is given by $a_{j,i}$ and if A is 'upper triangular' ($a_{j,i} = 0$ for $j > i$) then the model is termed 'left–right'.

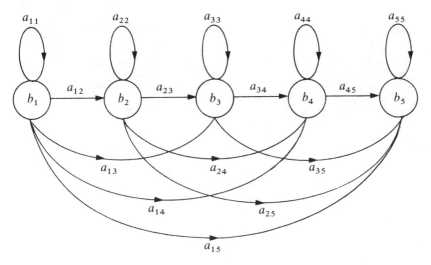

Figure 7.7 Structure of a five-state left–right hidden Markov model

The output probability density functions may either be discrete distributions over a finite set of output symbols, for example the codebook of vectors which might be produced by 'vector quantization', or continuous (parametric) multivariate probability density functions, each defined on an N-dimensional Euclidean space E^N. In the latter case, if v is an arbitrary vector in E^N, then $b_i(v)$ is the probability that v is drawn from state b_i. It is normal to refer to b_i as the ith 'state' of the model. Figure 7.7 illustrates a five-state left–right hidden Markov model. It can therefore be arranged for a hidden Markov model to be a stochastic (probabilistic) generative model of a class of example patterns.

For recognition purposes, it is necessary to be able to find the probability of a particular observation sequence $O = o_1, \ldots, o_T$ being generated by a given model M: $P(O|M)$. Since the hidden Markov model can generate an infinite set of observation sequences, this involves a potentially large search space. However, it turns out that the first-order Markov properties of such models leads to an efficient search algorithm.

The calculation of $P(O|M)$ is achieved for a given model and a given observation sequence by the generation of a two-dimensional data structure or 'lattice'. The lattice specifies all of the admissible state sequences which could generate an output pattern the same length as the observation sequence. The number of nodes in the lattice is a linear function of the number of states in the model and the number of vectors in the observation sequence.

Figure 7.8 shows the lattice data structure defined by a five-state left–

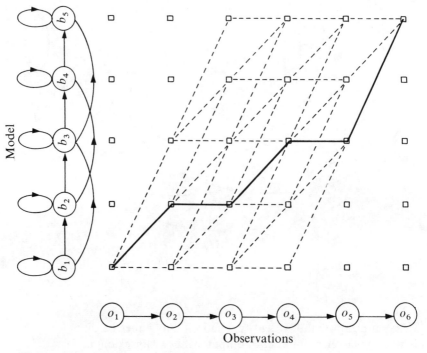

Figure 7.8 Two-dimensional lattice structure containing all of the possible state sequences which map an observation sequence of length six onto a five-state hidden Markov model

right hidden Markov model and an observation sequence of length six. One possible state sequence is highlighted.

Each distinct state sequence defines a path through the lattice and each path has a probability associated with it (the sum of the transition probabilities and the state output probabilities along the path). Hence, for any state sequence $S = s(1), \ldots, s(T)$ of length T, the joint probability $P(O, S|M)$ of O and S conditioned on M is given by

$$P(O, S|M) = b_{s(1)}(o_1) \prod_{t=2}^{T} b_{s(t)}(o_t) a_{s(t-1)s(t)}$$

The probability of the model generating the observation sequence $P(O|M)$ can thus be obtained by summing $P(O, S|M)$ over all possible state sequences.

Let $p_t(i) = P(o_1, \ldots, o_t$ and $s(t) = i|M)$ be the probability, conditioned on M, that the partial observation sequence $o_1, \ldots, o_t$ arises from a state sequence S whose tth state is b_i. Then, setting $p_1(1) = b_1(o_1)$, the following recursive equation holds:

$$p_t(i) = \sum_{j=1}^{I} p_{t-1}(j) a_{j,i} b_i(o_t)$$

Hence

$$P(O|M) = p_T(I)$$

The probability of the model generating the observation sequence $P(O|M)$ can thus be computed quite efficiently by the use of a recursive algorithm which starts at the node in the lattice specified by the initial state of the model and the first observation vector, and then progresses through the lattice calculating at each node the probability of the given state generating the given observation vector, accumulating the results and continuing through the lattice until the node specified by the final state of the model and the last observation vector is reached.

It is also useful to be able to calculate the most probable state sequence S that could have generated an observation sequence O that is the sequence S for which $P(O, S|M)$ is greatest. In this case the summation is replaced by a maximum operation and the recursive computation becomes a *best path* search algorithm or dynamic programming (also known as the 'Viterbi' algorithm).

Let $p'_t(i)$ be the maximum value over all partial state sequences S' of length t terminating in state b_i of the joint probability $P(o_1, \ldots, o_t; S'|M)$. Then $p'_t(i)$ satisfies the recursive equation

$$p'_t(i) = \max_{j=1}^{I} p'_{t-1}(j) a_{j,i} b_i(o_t)$$

subject to the initial condition $p'_1(1) = b_1(o_1)$. In this case, $p'_T(I) = \max P(O, S|M)$, where the maximum is taken over all admissible state sequences S. The probability of this most probable state sequence will be denoted by $P'(O|M)$.

In order to recover the most probable state sequence, it is necessary to maintain a record of the results of each local maximization during the dynamic programming iteration. These decisions are recorded as a matrix $R = [r_t(i): t = 1, \ldots, T; i = 1, \ldots, I]$ such that

$$r_t(i) = j \text{ iff } p'_t(i) = p'_{t-1}(j) a_{j,i} b_i(o_t)$$

The most probable state sequence is then recovered by performing a backward trace through R. If M is a left–right model, then the backtrace starts at $r_T(I)$ and terminates at $r_1(1)$. In general, the backtrace is initiated from the $r_T(i)$ *for which* $p'_T(i)$ *is maximized*.

The resulting path through the lattice is known as a 'time alignment' or 'time registration' path since it defines the most likely correspondence between states in the model and vectors in the observation sequence. (It is interesting to note that this whole process is identical to dynamic time warping in the special case where the output distributions are multivariate normal with diagonal co-variance and unity variance.)

Parameter estimation

Because of the particular probabilities embedded in a given model, some patterns are more likely to be generated than others. An important question is thus concerned with the problem of arranging for these probabilities (the parameters of the model) to reflect a particular class of patterns.

One of the most important features of the hidden Markov modelling approach is that there exist powerful learning algorithms for updating the model parameters and increasing the probability of a model generating a given set of training patterns (Liporace, 1982). This process of parameter estimation is achieved by means of an iterative process which *reestimates* the parameters of a model with respect to a particular set of training data. In practice there are two parameter reestimation algorithms; one is based on the reestimation of $P(O|M)$ and the other is based on the reestimation of $P'(O|M)$.

The maximum likelihood version is called the 'Baum–Welch' or 'forward–backward' algorithm (Baum, 1972) and is guaranteed to improve the model; that is the probability of the model generating all the example data is increased on each iteration. This is obviously a very important property and it is interesting to note that there is a version of the Baum–Welch procedure, known as the 'inside–outside' algorithm, that can be used to reestimate the parameters of a stochastic context-free grammar (Baker, 1979).

It is also possible to use the Viterbi search algorithm to update the parameters of a model. In this case convergence is not guaranteed but, in practice, it has been found to improve the model satisfactorily.

Recognition

The previous sections have been concerned with the process of *modelling* a single class of patterns. This section turns to the use of such models in order to perform *recognition*, that is to identify to which class an unknown pattern belongs. In order to use the foregoing algorithms for pattern classification, it is simply necessary to define a separate model for each class of patterns.

'Maximum likelihood classification' of an unknown observation sequence can be achieved by calculating the probability of the observations given the model $P(O|M)$ for each model in turn. The unknown pattern is then assigned to the class of the model that has the highest probability of generating the observed data; that is for N classes $C = c_1, \ldots, c_N$, where class c_n is represented by model M_n, O is assigned to class c_n if

$$P(O|M_n) = \max_{m=1}^{N} P(O|M_m)$$

Maximum likelihood classification is also termed 'Baum–Welch

RECOGNITION—THE STOCHASTIC MODELLING APPROACH

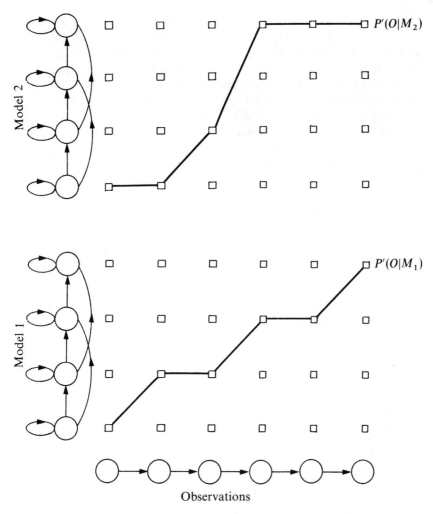

Figure 7.9 Illustration of Viterbi classification; the most likely state sequence is shown for each model

recognition' and mathematically this produces the optimum Bayesian classifier.

An alternative is to use a classification rule based on $P'(O|M)$, in which case the process is termed 'Viterbi classification'. Figure 7.9 illustrates Viterbi classification for an observation sequence of length six with two four-state hidden Markov models. The most likely state sequence for each model is shown. Viterbi classification is particularly important since it facilitates a number of interesting extensions to the basic HMM-based methodology. For example, it is possible to overcome the problems of continuity where patterns for individual classes are joined together in a continuous sequence.

If it is assumed that the observation sequence to be recognized may not be a single pattern class, but is formed from a sequence of pattern classes of unknown length, then the Viterbi recognition technique may be easily extended to calculate the most likely state sequence through the most likely sequence of models: $P'(O|M_1, \ldots, M_N)$. This is achieved by introducing extra transitions which connect all the model 'end' states to all the 'beginning' states, thus allowing a continuous path through any sequence of models.

In this case the most likely state sequence reveals the *optimal* segmentation boundaries in the observation sequence at the same time as revealing the identity of the individual segments. It is thus an example of 'segmentation by recognition' (rather than segmentation *followed* by recognition) which is one of the most important properties of this recognition scheme; the algorithm is able to account for contextual effects by postponing decisions about the precise segment boundaries until all ambiguities have been resolved.

Figure 7.10 shows the lattice resulting from two four-state models and an observation sequence of length twelve. The two models are connected together. A single path shows the most likely state sequence through the lattice, and the path trajectory reveals that the observation sequence is recognized as a pattern from the second class followed by a

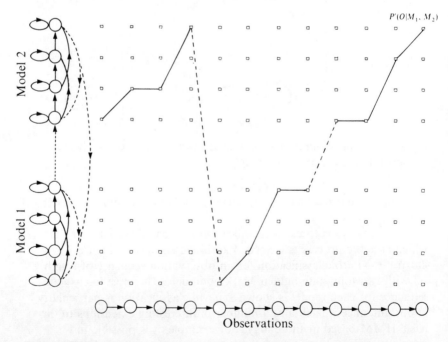

Figure 7.10 Optimal segmentation of an observation sequence using connected models

pattern from the first class followed by another pattern from the second class. The use of the Viterbi algorithm in this context is known as the 'one-pass' algorithm (Bridle and Brown, 1983). Other techniques that achieve essentially the same result are the 'two-level' algorithm (Sakoe, 1979) and the 'level building' algorithm (Myers and Rabiner, 1981).

7.4.2 Hidden Markov models of speech

The acoustic pattern of a speech signal typically consists of a sequence of observation vectors that have been derived from the speech waveform using some form of preprocessing. For example, it is common to perform an analysis based on the Fourier transform in order to produce vectors corresponding to the short-time power spectrum of the signal (see Chapter 2). Hence, in a hidden Markov model of acoustic speech patterns the transition probabilities capture the durational and sequential structure, and the output distributions at the states describe variations in pronunciation. Also, since each state in a hidden Markov model of an acoustic speech signal would describe the statistics of regions of the patterns having an approximately similar spectral content, the number of states in a model is usually smaller than the average number of vectors in the corresponding acoustic patterns. It is usual to employ left–right models.

The importance of hidden Markov modelling for automatic speech recognition is that it is a general modelling paradigm which can be used in a variety of different ways depending upon the level at which speech patterns are being modelled. For example, it is possible to model speech at the phrase level using models to represent word sequences, at the word level using whole-word models or at the phonetic level using models that represent 'subword units'.

Whole-word modelling

The simplest way of using hidden Markov models in automatic speech recognition is to allocate one model for each word in a vocabulary.

Figure 7.11 illustrates an eight-state hidden Markov model for the word 'zero'. The bottom part of the figure shows an example spectrographic pattern of the word as produced by a talker. The pattern immediately above is a corresponding version generated by the model. The parameters of the model were derived from ten utterances of the word 'zero' by the same speaker and ten iterations of the Baum–Welch algorithm. In this figure, the output from the model is not the model's average version of a 'zero', but that which is most like the 'real' example shown. It can be seen that the model approximates the natural pattern by generating a sequence of eight quasi-stationary acoustic segments of different durations. The resulting *synthetic* pattern can be thought of as a 'cartoon' of the word 'zero'.

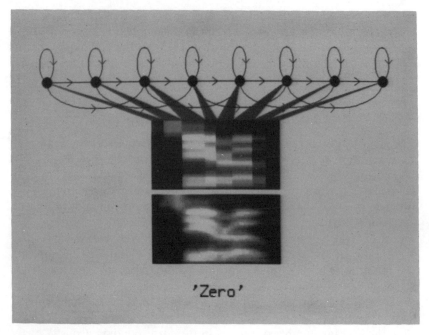

Figure 7.11 Eight-state hidden Markov model for the word 'zero'

Recognition of spoken words can thus be achieved by matching the observed signal against a set of whole-word hidden Markov models. The unknown pattern is then recognized according to the identity of the model (or sequence of models) that has the highest probability of generating that data. For the recognition of 'isolated' words (words with pauses between them) the first and last states in each word model may correspond to the silence before and after the word. For the recognition of connected words it is usual to have a separate model for 'silence' (background noise) in order to explain any pauses between words.

Using this approach, the statistical nature of the models means that, not only is the recognition performance better than that obtainable from simple pattern matching, but the models can be trained on example patterns from a range of different talkers, thereby achieving some degree of 'speaker independence' (Levinson, Rabiner and Sondhi, 1983a, 1983b) or on a range of different speaking conditions thereby achieving a degree of 'robustness' (Paul, 1988).

Subword modelling

The advantage of modelling whole words is that the models are able to capture many of the localized contextual effects that occur in speech. However, the obvious disadvantage is that a large vocabulary system based on whole-word modelling would require a correspondingly large number of models, each of which would have to be trained on a suitable quantity of example speech data. As a consequence, in recent years there has been considerable interest in exploiting the lower level phonetic patterning in speech by modelling phoneme-based 'subword' structures.

In principle it would be possible to model all the words in a given language using one hidden Markov model for each phoneme; this would mean that only about forty models would be required! However, the acoustic realization of any phoneme is highly context dependent (because the temporal and spectral contents of each sound strongly influences the shape and content of its immediate neighbours). Thus the performance of a 'monophone'-based system is very poor indeed due to the extreme variability encountered.

As a consequence, most subword-based speech recognition systems employ context-dependent phone models with one hidden Markov model for each phone triple or 'triphone' (Schwartz et al., 1985). This increases the number of models greatly, but not all combinations of phones occur in a given language; it is also possible to share information between such models, thereby reducing the inventory to a manageable level. In practice good results have been achieved with triphone sets containing several hundred models and, despite the simplistic assumptions involved, the fact that the models are able to be trained on a significant amount of data means that the scheme can be used to transcribe speech phonetically with a respectable degree of accuracy (around 70 per cent).

Word recognition is achieved by linking the appropriate subword models together to make whole-word models. In this situation, the resulting word accuracy is considerably higher than the underlying phonetic accuracy as a direct consequence of the sequential phonetic constraints imposed by the words.

Speech pattern modelling at the subword level means that, in principle, systems are able to become vocabulary 'independent'; that is the amount and type of training material may be chosen to ensure adequate coverage of the phonetic structure of the language rather than complete coverage of the particular set of words associated with any specific application. Of course, there is a price to be paid for this increase in flexibility—a potential loss of performance caused by the reduction in long-range (word-based) contextual modelling.

7.4.3 The state of the art

In contrast with the knowledge-based approach described in Sec. 7.2.2, stochastic modelling was initially seen as just another form of pattern matching and was nicknamed the 'ignorance-based' (!) approach to automatic speech recognition (Makhoul and Schwartz, 1984). However, recent successes in large-vocabulary speaker-independent speech recognition by means of subword and language modelling indicate how stochastic modelling has been able to provide a powerful HARPY-like integration of different types of speech knowledge within a trainable statistical framework.

This has meant that there have been many significant developments during the eighties. For example, several large systems based on this philosophy have demonstrated a very impressive recognition performance. Examples of such systems are the IBM 'TANGORA' system for dictating office memoranda (Averbuch *et al.*, 1987), BBN's 'BYBLOS' (Chow *et al.*, 1987) and Carnegie Mellon University's 'SPHINX' systems (Lee, 1988) for a DARPA-funded 'resource management' task, and RSRE's 'ARMADA' system (Russell *et al.*, 1990) for the recognition of spoken airborne reconnaissance mission reports. The IBM system is an isolated-word speaker-dependent system with a 20 000 word vocabulary and the SPHINX system is a connected-word speaker-independent system with a 1000 word vocabulary. All of these systems have achieved success through the use of stochastic modelling at the subword, word and language levels.

Also, since there is significantly less computation required for recognition than for training, it is possible to produce very low cost implementations of stochastic model-based automatic speech recognition algorithms. As a result, several speech recognizers based on HMMs are currently available commercially either as chip sets or as licensable software. Of particular interest is the 'DRAGON-DICTATE' system which is a 30 000 word, speaker-*adaptive*, isolated word recognition system which has been designed to provide a facility for virtually unlimited vocabulary spoken text entry.

7.5 Future directions

Current concerns in the development of high-performance automatic speech recognition fall into two main categories: those that impact on the research work and the advanced algorithms which are still needed to raise performance to a usable level, and those that relate to the integration of speech recognition within a given application environment.

However, a more important underlying set of issues relates to 'performance characterization', that is the establishment of a suitable methodology for measuring and benchmarking performance (Moore, 1977).

7.5.1 Performance characterization

The effectiveness of any application of automatic speech recognition is crucially dependent on the level of recognition accuracy exhibited by

RECOGNITION—THE STOCHASTIC MODELLING APPROACH

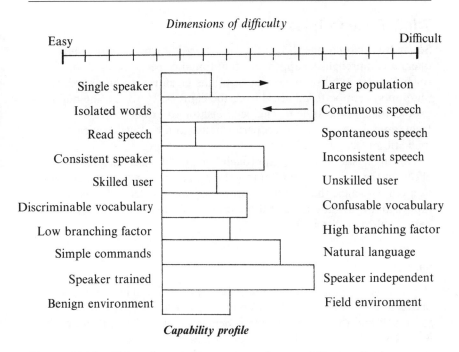

Figure 7.12 Illustration of the concept of a capability profile for characterizing speech recognizer performance

the recognition device. Also, successful exploitation of current and future speech input systems depends on an efficient match between the capabilities of the device and the requirements presented by any particular application. Thus it is important, from both an applications and a research perspective, to be able to characterize device performance in ways that accurately reflect its true capabilities in terms that are relevant to a complete specification of a given automatic speech recognition machine.

Recognition accuracy of course depends on a whole range of factors. Equipment suppliers have a tendency to quote performance as '99 per cent' (or '99 per cent plus'). In practice, a difficult vocabulary, an inconsistent speaker or a noisy environment can lead to very poor field performance.

The performance of a speech recognition device can therefore be usefully characterized in terms of a 'capability profile' across a number of key dimensions (Moore, 1989) (see Fig. 7.12). Capabilities can be traded one against another; not only is there the obvious performance–cost trade-off but, for example, large-vocabulary speaker-dependent recognition can be traded against small-vocabulary speaker-independent recognition for a given level of recognition accuracy.

7.5.2 Research issues

So far this chapter has shown that structured stochastic modelling offers a comprehensive approach to automatic speech recognition (including trainability) which, in turn, has given rise to an impressive jump in system performance during the eighties. However, the main problem facing the systems engineer wishing to exploit current technology is that recognition performance in a real-life application may still not be good enough.

The stochastic modelling approach has already had a significant effect on the performance of automatic speech recognition systems and, as a consequence, the capability profile has been extended in each dimension, providing a general increase in robustness. However, many real problems still remain.

Speaking style

It is currently possible to accommodate speech that has no pauses between the words; however, problems arise when words run together too much and the end of one word begins to affect the start of another. Very few systems are able to model these word-boundary effects in a satisfactory way. Also, considerable effort has been expended on the recognition of speech material that has been prepared by the talkers (that is read material or so-called 'lab-speech'). Part of the reason that field performance is so much lower than laboratory performance is due to the more variable nature of fluent *natural* speech. Therefore, a priority should be placed on modelling paradigms that might be needed to deal with more 'spontaneous' speech material.

Environment

Considerable progress has been made in the area of speech recognition in the presence of interfering signals and noise (Varga and Ponting, 1989). Good recognition results are obtainable at noise levels down to 0 dB signal-to-noise ratio for environments in which the noise itself can also be modelled (including both stationary and impulsive noises) (Varga and Moore, 1990). However, there is still considerable room for improvement since these results are often based on experiments with speech and noises that have been added together electrically. In a 'real' environment the environmental noise can greatly influence the talker's behaviour and this type of interaction also needs to be modelled.

Vocabulary

As has been stated above, current recognition techniques are moving away from whole-word modelling to subword modelling and this means that systems are becoming 'vocabulary independent'. The performance of such systems is dependent on the size of the vocabulary, but systems with vocabularies as high as 30 000 words have been demonstrated already. Such systems operate with a dictionary that contains each word that the system is able to recognize. New words that are not known to the system can only be recognized after the user has provided an explicit indication that the word should be added to the lexicon. A better arrangement might be for systems to be able to handle words that have never been heard before by invoking some form of direct phoneme-to-orthographic spelling conversion process.

User population

Most so-called speaker-independent systems achieve this facility by averaging the speech patterns from many different speakers. This is not adequate for very large speaker populations which cover a wide range of accent groups (or languages). Techniques based on accent (and language) identification are needed to provide improved speaker independence, and this in turn brings a requirement for an ability to adapt rapidly to the voice of each new user. This may mean that the dialogue must be designed to accommodate some preliminary non-task-critical interaction (for example, 'hello, how are you'—see Sec. 7.1.1).

Enrolment protocol

The emergence of subword modelling techniques means that, in principle, users no longer have to provide vocabulary-dependent training material. There is still a significant amount of work to be done in optimizing this process such that the best performance is obtained for the least amount of explicit enrolment.

Dialogue structure

The complexity of the interactive dialogue between a user and an application is a critical area since it is dependent on developments in the field of natural language processing. As the dialogue moves beyond simple menu structures and small artificial languages there are considerable human factors problems associated with the 'habitability' of the interface. It is possible that developments in 'keyword spotting' will facilitate a separation between the capabilities of a recognition system and the interpretation faculties of the task software. In other words, it may not be necessary for a speech recognizer to produce a complete transcription of an utterance. Hence there is a need for a suitably intelligent interface protocol to connect the speech recognizer to the rest of the system which may have to go much further than those that have been proposed so far (Paul, 1989).

7.5.3 Application issues

The successful exploitation of automatic speech recognition depends crucially on the ability of the technology to deliver the goods, in this case, high accuracy (robust) recognition. However, in order to achieve the full benefit of an automated speech input system, it is also necessary to have a better understanding of how to integrate a speech system into the full *multimodal* human–computer interface (Taylor, Neel and Bouwhuis, 1988).

This means that not only is it important to develop tools for characterizing the behaviour of the technology but tools are also needed that can be used to determine the requirements of particular applications and for matching the technology capabilities against the applications requirements. Also, even with a very high accuracy automatic speech recognition system, the use of the technology within any particular application involves a systems integration task which must not be underestimated. Unfortunately it is difficult to specify generic applications requirements; each application has its own special needs which arise from the combination of spoken human–computer dialogue and the overall task that is to be performed by the user. Also, the interactive nature of the dialogue means that not only are the capabilities of the speech-input device important but there are considerable human factors issues involved—many of which are directly dependent on the behaviour of the speech input system. For example, a recognition error by the system may disrupt the dialogue and slow down task completion. These important issues are taken up again in Chapter 11.

References and further reading

References

Alter, R. (1968) 'Utilization of contextual constraints in automatic speech recognition', *IEEE Transactions on Audio and Electroacoustics*, vol. 16, 6–11.
Averbuch, A., L. Bahl, R. Bakis, P. Brown, G. Daggett, S. Das, K. Davies, S. de Gennaro, P. de Souza, E. Epstein, D. Fraleigh, F. Jelinek, B. Lewis, R. Mercer, J. Moorhead, A. Nadas, D. Nahamoo, M. Picheny, G. Schichman, P. Spinelli, D. van Compernolle and H. Wilkens (1987) 'Experiments with the TANGORA 20,000 word speech recogniser', Proceedings of IEEE International Conference on Acoustics, Speech, and Signal Processing, 701–704.
Baker, J. K. (1974) 'The DRAGON system—an overview', Proceedings of IEEE Symposium on Speech Recognition, 22–26.
Baker, J. K. (1975) 'Stochastic modelling for automatic speech recognition' in *Speech Recognition*, D. R. Reddy (ed.), Academic Press, London.
Baker, J. K. (1979) 'Trainable grammars for speech recognition' in *Speech*

Communication Papers for the 97th Meeting of the Acoustical Society of America, D. H. Klatt and J. J. Wolf (eds), 547–550.
Baum, L. E. (1972) 'An inequality and associated maximization technique in statistical estimation for probabilistic functions of a Markov process', *Inequalities*, vol. 3, 1–8.
Bellman, R. E. (1957) *Dynamic Programming*, Princeton University Press, N.J.
Bridle, J. S. and M. D. Brown (1983) 'Continuous connected word recognition using whole word templates', *The Radio and Electronic Engineer*, vol. 53, 167–175.
Bridle, J. S. and M. P. Ralls (1985) 'An approach to speech recognition using synthesis by rule' in *Computer Speech Processing*, F. Fallside and W. Woods (eds), Prentice-Hall, Englewood Cliffs, N.J.
Chow, Y. L., M. O. Dunham, O. A. Kimball, M. A. Krasner, G. F. Kubala, J. Makhoul, P. J. Price, S. Roucos and R. M. Schwartz (1987) 'BYBLOS: the BBN continuous speech recognition system', Proceedings of IEEE Conference on Acoustics, Speech, and Signal Processing, 89–92.
Hopcroft, J. E. and J. D. Ullman (1969) *Formal Languages and Their Relations to Automata*, Addison-Wesley, Reading, Mass.
Jelinek, F. (1979) 'Continuous speech recognition by statistical methods', *Proceedings of the IEEE*, vol. 64, 532–555.
Jelinek, F. (1985) 'Markov source modelling of text generation' in *NATO Advanced Study Institute: Impact of Processing Techniques on Communications*, Martinus Nijhoff, Dordrecht.
Jelinek, F., R. L. Mercer and L. R. Bahl (1982) 'Continuous speech recognition: statistical methods' in *Handbook of Statistics*, P. R. Krishnaiah and L. N. Kanal (eds), vol. 2, 549–573, North Holland.
Klatt, D. H. (1977) 'Review of the ARPA speech understanding project', *Journal of the Acoustical Society of America*, vol. 62, 1345–1366.
Lee, K. F. (1988) 'Large vocabulary speaker-independent continuous speech recognition: the SPHINX system', Ph.D. Thesis, Carnegie-Mellon University, Pittsburgh, Pa.
Levinson, S. E., L. R. Rabiner and M. M. Sondhi (1983a) 'Speaker independent isolated digit recognition using hidden Markov models', Proceedings of IEEE International Conference on Acoustics, Speech, and Signal Processing, 1049–1052.
Levinson, S. E., L. R. Rabiner and M. M. Sondhi (1983b) 'An introduction to the application of the theory of probabilistic functions of a Markov process', *Bell System Technical Journal*, vol. 62, 1035–1074.
Liporace, L. A. (1982) 'Maximum likelihood estimation for multivariate observations of Markov sources', *IEEE Transactions on Information Theory*, vol. 28, 729–734.
Lowerre, B. T. (1976) 'The HARPY speech recognition system', Ph.D. Thesis, Department of Computer Science, Carnegie-Mellon University, Pittsburgh, Pa.
Makhoul, J. and R. Schwartz (1984) 'Ignorance modelling' in *Invariance and Variability in Speech Processes*, J. Perkell and D. H. Klatt (eds), Erlbaum, Hillsdale, N.J.
Mendel, J. M. (1973) *Discrete Techniques of Parameter Estimation*, Marcel Dekker, New York.
Moore, R. K. (1977) 'Evaluating speech recognisers', *IEEE Transactions on Acoustics, Speech, and Signal Processing*, vol. 25, 178–183.
Moore, R. K. (1989) 'Assessment of speech input systems' in *Proceedings of ESCA Workshop on Speech Input/Output Assessment*, Noordwijkerhout, Holland.

Myers, C. and L. R. Rabiner (1981) 'A level building dynamic time warping algorithm for connected word recognition', *IEEE Transactions on Acoustics, Speech, and Signal Processing*, vol. 29, 284–297.

Newell, A., J. Barnett, J. W. Forgie, C. Green, D. Klatt, J. C. R. Licklider, J. Munson, D. R. Reddy and W. A. Woods (1973) *Speech Understanding Systems*, North Holland/American Elsevier.

Paul, D. B. (1988) 'Speaker-stress resistant continuous speech recognition', Proceedings of IEEE International Conference on Acoustics, Speech, and Signal Processing.

Paul, D. B. (1989) 'A CSR-NL interface specification', Proceedings of DARPA Speech and Natural Language Workshop, October.

Pierce, J. R. (1969) 'Whither speech recognition', *Journal of Acoustical Society of America*, vol. 46, 1049–1051.

Rabiner, L. R. and B. H. Juang (1986) 'An introduction to hidden Markov models', *IEEE Acoustics, Speech and Signal Processing Magazine*, vol. 3, 4–16.

Reddy, D. R. (1967) 'Computer recognition of connected speech', *Journal of Acoustical Society of America*, vol. 42, 329–347.

Reddy, D. R. and L. D. Erman (1975) 'Tutorial on system organisation for speech understanding' in *Speech Recognition*, D. R. Reddy (ed.), 457–459, Academic Press, London.

Russell, M. J., K. M. Ponting, S. M. Peeling, S. R. Browning, J. S. Bridle, R. K. Moore, I. Galiano and P. Howell (1990) 'The ARM continuous speech recognition system', Proceedings of IEEE International Conference on Acoustics, Speech, and Signal Processing.

Sakoe, H. (1979) 'Two-level DP-matching—a dynamic programming based pattern matching algorithm for connected word recognition', *IEEE Transactions on Acoustics, Speech, and Signal Processing*, vol. 27, 588–595.

Sakoe, H. and S. Chiba (1978) 'Dynamic programming algorithm optimisation for spoken word recognition', *IEEE Transactions on Acoustics, Speech, and Signal Processing*, vol. 26, 43–49.

Schwartz, R. M., Y. L. Chow, O. A. Kimball, S. Roucos, M. Krasner and J. Makhoul (1985) 'Context-dependent modelling for acoustic-phonetic recognition of continuous speech', Proceedings of IEEE International Conference on Acoustics, Speech, and Signal Processing.

Taylor, M. M., F. Neel and D. G. Bouwhuis (eds) (1988) *The Structure of Multimodal Dialogue*, North Holland, Amsterdam.

Varga, A. P. and R. K. Moore (1990) 'Hidden Markov model decomposition of speech and noise', Proceedings of IEEE International Conference on Acoustics, Speech, and Signal Processing.

Varga, A. P. and K. M. Ponting (1989) 'Control experiments on noise compensation in hidden Markov model based continuous word recognition', Proceedings of ESCA Eurospeech.

Vintsyuk, T. K. (1968) 'Speech discrimination by dynamic programming', *Kibernetica*, vol. 4, 81–88.

Further reading

Ainsworth, W. A. (1988) *Speech Recognition by Machine*, Peter Peregrinus, London.

Allerhand, M. (1987) *Knowledge-based Speech Pattern Recognition*, Kogan Page, London.

Bristow, G. (ed.) (1986) *Electronic Speech Recognition*, Collins, London.

de Mori, R. and C. Y. Suen (eds) (1985) *New Systems and Architectures for Automatic Speech Recognition and Synthesis*, Springer-Verlag, Heidelberg.

Holmes, J. N. (1988) *Speech Synthesis and Recognition*, Van Nostrand Reinhold, Wokingham, Berks.

House, A. S. (ed.) (1988) *The Recognition of Speech by Machine—A Bibliography*, Academic Press, London.

Lea, W. A. (ed.) (1980) *Trends in Speech Recognition*, Prentice-Hall, Englewood Cliffs, N.J.

Mariani, J. J. (1989) 'Recent advances in speech processing', Proceedings of IEEE International Conference on Acoustics, Speech, and Signal Processing, 429–440.

8 Pattern recognition and its application to speech

GRAHAM LEEDHAM

8.1 Introduction

Automatic pattern recognition (APR) is a wide multidisciplinary subject that has been under investigation for more than thirty years. During this time there have been various attempts to define what is meant by pattern recognition (Verhagen, 1975) but these definitions are frequently specific to a particular class of patterns. The statement by Guiliano (1967) summarizes the problem of definition as:

> For a human it is probably not possible to give a better definition than: a pattern is something which one recognizes as a pattern.

The potential applications of APR are, therefore, only limited by the requirements and ingenuity of people working in many different environments and professions. For example, APR systems are currently used for such diverse applications as visual inspection of components, character recognition, medical data analysis and speech recognition. Many other applications have been investigated and others are constantly being proposed.

While computers are far superior to humans when carrying out well-defined tasks such as database searching and numeric calculations they have proved inferior at solving the reasoning and pattern recognition problems which occur so often in everyday human activities. The main reason for this is, as Sayre (1965) observed:

> We simply do not understand what recognition is. And if we do not understand the behaviour we cannot reasonably hold high hopes of being successful in our attempts to simulate it.

A number of general pattern recognition techniques have, however, been developed over the past few decades and applied successfully to a wide range of patterns. In the previous chapter a particular approach to APR, using stochastic modelling, was described in detail as it applies to the speech recognition problem. In this chapter some of the more general aspects of automatic pattern recognition are discussed and related to how they have been applied to automatic speech recognition (ASR). Examples of some ASR systems exhibiting these general pattern recognition techniques are described.

8.2 Pattern recognition techniques

Traditional APR techniques fall into three categories: template matching methods, statistical methods and structural methods. These three techniques are described briefly below. While none of these techniques can achieve the recognition performance of humans they are effective in numerous constrained recognition environments with any class of patterns whether they are visual or non-visual signals.

8.2.1 Template matching

This is the simplest APR technique as patterns are identified by comparing the input pattern to each member of a list of stored pattern representations (template patterns). This principle is illustrated in Fig. 8.1. An input pattern is isolated, digitized and passed in some suitable form to a comparator which calculates a similarity measure between the input pattern and each pattern in a set of prestored template patterns. The pattern must be represented in an appropriate form to enable the template comparison to operate efficiently and effectively. For a visual signal this may involve passing the individual pixel intensities of the image (binary, grey scale or colour) or the result of some transform (e.g. discrete cosine transform (DCT) or Hadamard transform) of the image (Gonzalez and Wintz, 1987). For a non-visual

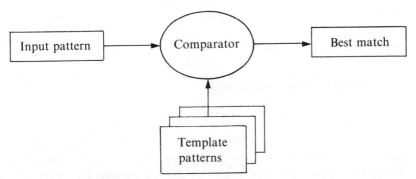

Figure 8.1 Pattern recognition using template matching

signal, such as speech, the passed representation may be the sampled time signal itself or parameters extracted from the frequency spectra of the signal. The most suitable representation of the input and template patterns depends to a large extent on the type of patterns that are under investigation. The requirement is to choose a representation that produces the best discrimination between different pattern classes but is relatively insensitive to variations of patterns in the same pattern class, thus minimizing the chance of misrecognition and maximizing the chance of correct recognition.

Template matching produces a list of best-to-worst matches between the input pattern and the templates. In a simple system the comparison that produces the best match is deemed to be the recognized pattern if the match exceeds a predefined threshold. In more sophisticated systems the final pattern classification is made after contextual postprocessing has determined the most likely interpretation. There are various methods of performing the pattern comparison and some are discussed in Sec. 8.4. The template patterns have to be obtained and stored prior to running the recognizer during a training session and the quality of these template patterns has a significant effect on the performance of the recognizer.

Simple template matching provides satisfactory performance when:

1 The number of different template patterns is small.
2 There is little variation in each pattern class from time to time.
3 There is little similarity between different pattern classes.

These criteria are interrelated as template matching can operate successfully with a very large number of different template patterns if an input pattern always matches exactly with its associated stored template pattern; that is if there is no variation in patterns of the same class. This is rarely the case in real pattern recognition problems as the number of different template patterns that can be successfully accommodated in a template matching scheme is determined by the amount of between-class (intraclass) variation (e.g. in a speech recognition system phonetically similar words such as 'that', 'hat' and 'hot' may be easily confused), the within class (interclass) variation, the acceptable error rate for the particular application, the robustness of the particular pattern representation and the effectiveness of the comparison technique employed.

8.2.2 Statistical methods

Statistical pattern recognition techniques (also known as decision-theoretic techniques) determine which class a given input sample pattern belongs to, based on selected measures or features extracted from the pattern. Figure 8.2 shows the main components of a statistical pattern recognizer. It essentially consists of two parts: a feature

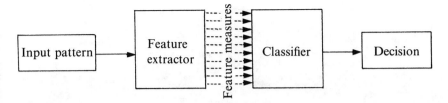

Figure 8.2 Pattern recognition using a statistical pattern recognizer

extractor and a pattern classifier. Each isolated input pattern is passed to the feature extractor which extracts significant features from the pattern and passes this feature vector to the classifier, which, in turn, makes a decision as to which class the pattern belongs. In many ways this is similar to the template matching technique described above but the statistical method can be less prone to noise and pattern variation as it does not match the whole pattern with a reference pattern but uses a set of features that are resilient to pattern variation. The simple template matching described above, in Sec. 8.2.1, is a special case of statistical pattern matching.

The selection of features to form the feature vector is a difficult problem but, in general, they should be chosen such that they are less sensitive to interclass pattern variation. The features may not necessarily have any physical meaning but are nevertheless good measures for the pattern classification. There are at present no real, tangible rules for the selection of features. Selection is usually based upon experimentation with a training set of known patterns. In Chapter 3 of Young and Fu (1986) and Chapter 5 of Devijver and Kittler (1982) the issue of feature selection is discussed further.

In statistical pattern recognition a feature vector containing N features maps each pattern as a point in an N-dimensional feature space, as shown in Fig. 8.3. Statistical information obtained from observations on a known set of representative patterns (the training set) is used to determine suitable features and the boundaries between the classes to maximize the recognition performance for each pattern class. If a given pattern is identical to a reference pattern then they will both be mapped to the same point in the feature space. If this was the case then the definition of the boundaries between patterns would be trivial. This is, however, not the case with real patterns. The requirement is to choose features such that patterns of the same class are tightly clustered in the N-dimensional space and patterns of different classes are in other tightly clustered regions which are well separated from the clusters produced by other pattern classes. If this is achieved the task of the pattern classifier is simplified and improves the classification performance.

There are two methods of determining whether a pattern belongs to

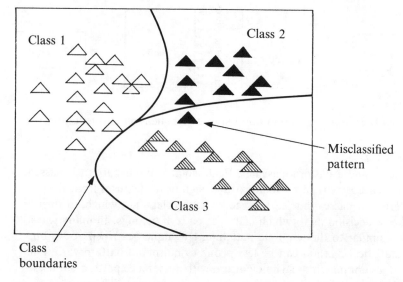

Figure 8.3 Subdividing the feature space

a certain class. These are the non-parametric and the parametric decision methods. In the non-parametric method a discriminant function $\delta_j(X)$ is defined, based on F extracted features for each feature vector X, where $X = \{x_1, x_2, x_3, \ldots, x_F\}$. If $\delta_j(X)$ is a maximum then the input pattern with the feature vector X belongs to the particular class W_j. This is expressed in the following equation where the boundary between classes is where $\delta_i(x) - \delta_j(x) = 0$:

$$\delta_i(x) > \delta_j(x) \qquad i,j = 1, \ldots, C; \; i \neq j \qquad (8.1)$$

where C is the number of distinct classes.

The commonly used discriminant functions are the linear discriminant function, the minimum distance classifier and the nearest neighbour classifier (these are all described in detail in Chapter 4 of Devijver and Kittler, 1982). For example, the linear discrimination function is a weighted linear combination of the feature measurements:

$$\delta_j(x) = \sum_{k=1}^{F} w_{jk} x_k + w_{j,N+1} \qquad i = 1, \ldots, C \qquad (8.2)$$

In the parametric classifier the decision rule involves the class conditional densities ($p(X|W_j)$) and the *a priori* probabilities of occurrence of the pattern classes ($P(W_j)$). The objective is to design a statistical decision classifier to minimize the risk of misclassification. The decision rule to determine the class boundaries is given by

$$P(W_i)p(X|W_i) \geq P(W_j)p(X|W_j) \qquad \text{for all } j = 1, \ldots, C \qquad (8.3)$$

This is generally referred to as the Bayes classifier. Further discussion

PATTERN RECOGNITION AND ITS APPLICATION TO SPEECH

of the parametric decision rule can be found in Chapter 2 of Devijver and Kittler (1982) and Chapter 1 of Young and Fu (1986).

Statistical classifiers need to be trained on a large database of known patterns if they are to be effective. This large database is needed to accurately define the feature set and the classifier decision boundaries and involves repeating the following five operations until the result is acceptable:

1. Extract the feature vectors and map them into the feature space.
2. Find the boundaries that separate the classes.
3. Test the classifier with a set of known samples (the training set).
4. Estimate the error rate by estimating the overlap between regions.
5. If the error rate is greater than an acceptable threshold then find an alternative measurement vector or, alternatively, modify the classifier.

If a large database is available but the number of distinct pattern classes is unknown the pattern identification can be on the basis of looking for clustering or grouping together of certain samples. This enables automatic pattern grouping to be achieved and is a form of automatic learning.

8.2.3 Structural or syntactic methods

The principle behind structural pattern recognition is the observation that many patterns contain structure and can be expressed as an ordered composition of simple subpatterns or pattern primitives; that is patterns can be formed from a sentence or grammar of simple pattern primitives in a pattern description language. For example, many visual images can be represented by a small set of simple straight and curved lines. This is illustrated in Fig. 8.4, where two simple patterns, a square and a brick, are expressed, using a tree representation, as a

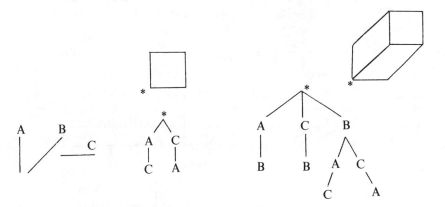

Figure 8.4 Examples of structural pattern representation

combination of three basic straight line pattern primitives A, B and C. Pattern representation is made on the basis of the identification and connectivity or syntax of the subpatterns; that is the square can be represented as a string of four pattern primitives and a brick can be represented as a string of nine pattern primitives. The connection of the pattern primitives to create the pattern is expressed in a pattern grammar.

The main components of a structural pattern recognizer are shown in block diagram form in Fig. 8.5. Each incoming pattern enters a segmenter which splits the pattern into its component pattern primitives. This list of pattern primitives is then passed to a recognizer which classifies the pattern primitives and the relationship between them. This list of pattern primitives and their connectivity is then passed to a syntax analyser which makes the final classification of the pattern based on the observed syntax of the incoming pattern primitives and on its own set of pattern syntax rules which describe patterns known to the recognizer.

While the overall process of recognition appears as a simple logical progression there are several difficulties that must be overcome. The selection of pattern primitives is a difficult primary task. There are no general rules available for this task and, in essence, it depends on heuristics and the experience or preference of the user. On the subject of selecting pattern primitives, Fu (1982) has only given the guidelines that:

1 The primitives should serve as basic pattern elements to provide a compact but adequate description of the data in terms of specific structural relations and
2 The primitives should be easily extracted.

It is important that the pattern primitives are easily extracted as it is difficult, if not impossible, to correct gross errors at this stage.

As a general rule selecting simple pattern primitives results in complex syntax grammars while selecting complex pattern primitives

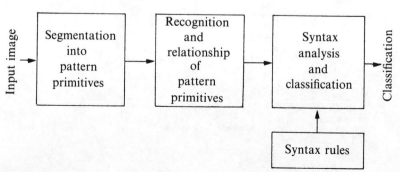

Figure 8.5 Pattern recognition using a structural recognizer

results in simpler syntax grammars. It is the designer who must decide upon the trade-off between these two criteria after a careful study of the patterns that are to be recognized.

Structural recognition techniques would appear to suit the speech recognition problem very well as spoken language has well-defined structure at both the phonetic level and the word level. In this case the pattern primitives could be phonemes or words and the syntax rules would represent the pronunciation of individual words or the concatenation of words to form syntactically correct sentences. We shall see later in Sec. 8.6 how such structural techniques have been used to develop connected word speech recognizers.

8.2.4 Incorporating knowledge

One of the most important aspects of pattern recognition is the use of *a priori* knowledge in the pattern matching process. It is this knowledge that relates past and present experiences enabling people to recognize patterns. For example, if we study the picture of an object we have never seen before we are subsequently able to pick this object out in other scenes, even if it appears in different orientations and sizes. Similarly, if we are listening to a radio programme and hear the two sentences 'One of the major challenges in speech research is to recognize speech' and 'Today, vandals were observed to wreck a nice beach' the utterances for 'recognize speech' and 'wreck a nice beach' may be indistinguishable when heard in isolation but world knowledge and knowledge of the language on the part of the listener will usually enable them to distinguish the different words.

While it is a valid exercise to analyse the speech signal and classify it into basic segments of phonemes, allophones, words, formant frequencies and their trajectories, etc., such analysis does not consider the complex higher level mechanisms by which humans use the speech communication channel to convey information from person to person. It is this higher level knowledge that resolves such difficulties. It has already been discussed (Chapter 2) how prosodic variation in stress, pitch and timing can be used to modify meaning in a spoken sentence, but it is obviously the choice of the words in the sentence that conveys much of the meaning. Rules of grammar determine this aspect of the language.

Speech can be thought of as hierarchical levels of units and structures which are concatenated together according to a set of rules to form the language. At the lowest level we have the speech signal itself and knowledge about its acoustic construction (see Chapter 1). At a higher level we know that speech is formed from a small number of basic speech sounds (phonemes or allophones) which are contatenated to form a large number of words or morphemes (a morpheme is the smallest unit of grammar which conveys meaning as part of a word or a complete word in its own right, e.g. the words 'copy-ing', 'copy-ist', 'cop-ier' and 'cop-ied' are all composed of two morphemes which have separate meanings). At a higher level syntax rules are required to concatenate words and morphemes into a virtually infinite number of phrases and sentences.

The grammar of a language is defined by rules that govern the concatenation of verbs, nouns, adverbs, articles, conjunctions, etc. (see Chapter 2). For example, the sentence 'Do you watch television?' is composed of (auxiliary verb + pronoun + verb + noun) and is a question enquiring whether the listener watches television. If the words are ordered differently as 'Watch television do you?' or 'You do watch television?', they still ask a question but the meaning of each sentence is subtly different. They now also convey information about the prior knowledge of the questioner concerning the television viewing habits of the person being questioned and consequently affects the subsequent reply to the question. It must also be noted that each language has its own grammar. For example, the question 'Do you watch television?' in French is 'Regardez-vous la télévision?', which literally translated is 'Watch you the television?'. While the grammar is correct in French it is incorrect in English. We also have knowledge about the subject or application of the speech that helps the listener to classify correctly ambiguous words or phrases. Prosodic knowledge is a still higher level of information which modifies the words and sentences to subtly change their meaning.

It is encoding and interaction with these different levels of knowledge (termed acoustic, phonetic, syntactic, semantic and pragmatic knowledge in Sec. 7.2.2) that will ultimately lead to high performance speech recognition.

There are numerous methods of encoding knowledge including nets (Ballard and Brown, 1982), production rules (Leith, 1983) and neural networks (Rumelhart and McClelland, 1986). The difficulty is in defining and obtaining the information, encoding it and interacting with it to improve the performance of an APR system. The structural pattern recognition approach described above in Sec. 8.2.3 incorporates low-level structural knowledge into the recognition process in the form of a pattern grammar. Statistical APR systems often also contain low-level knowledge about patterns in the form of class conditional densities and the probability of occurrence of each pattern class.

8.3 Approaches to speech recognition

The current performance of speech recognizers is not sufficient to tackle the problem of recognizing the unconstrained, normal speech that humans use in their everyday environment. There are several simpler

PATTERN RECOGNITION AND ITS APPLICATION TO SPEECH 265

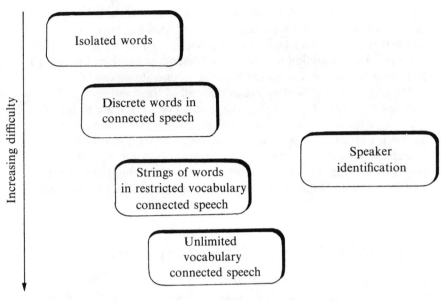

Figure 8.6 Increasingly difficult problems in speech recognition

levels of complexity at which speech recognition can be approached (see Fig. 8.6).

The simplest level is to attempt to recognize single isolated words chosen from a highly restricted group of words spoken by the same speaker in a controlled noise-free environment. In this case the problems of word separation are avoided, as is the background noise, and the recognizer need only classify each sound segment into one of a small number of predefined classes using a template-matching or statistical technique.

A higher level of complexity is to attempt to detect the utterance of one of a small number of keywords which may be spoken in normal connected speech. For example, it may be required to detect and perform some task each time the words 'London', 'Birmingham', 'Bristol' or 'Manchester' occur in unconstrained spoken text where these words may be connected to any other word. This is a simplified case of the more complex problem of recognizing a string of words in connected speech and could be approached using a form of template matching.

When recognizing short strings of connected speech a restricted vocabulary of words may be connected together as normal connected speech to form a sensible sentence. This is a considerably more complex problem as word separation needs to be performed. As syntax is an important aspect in this type of recognition a syntactic pattern-matching algorithm is most appropriate.

As observed earlier, by far the most difficult task is the recognition of all words in unlimited vocabulary connected speech. This problem is one that requires a very high level of artificial intelligence in the form of knowledge interaction for speech understanding combined with high-performance pattern recognition in order to operate successfully.

The recognition difficulties escalate for all of these levels of complexity as soon as the recognizer is required to operate with any speaker rather than one specific speaker to whose voice the recognizer has been trained. This leads to the additional recognition style of speaker identification where the prime interest is not in recognizing the words that are spoken but determining who is speaking the words by spotting features about the changing voice patterns that characterize a particular voice.

8.4 Isolated word recognition

This section considers the problem of recognizing a relatively small number of single isolated words, that is words that are spoken singly, in clear isolation from each other. This problem is usually tackled using a template-matching technique and is appropriate for voice-operated command systems where only a small number of simple commands are required.

8.4.1 Word-start and word-end detection

The first task to be accomplished is the detection of each spoken word and accurate determination of where it starts and where it ends. Approximate start and end points are relatively easily located when the word starts or ends with strong or plosive consonants like /T/, /K/ or /G/ (e.g. 'tick', 'gate'), as a simple thresholding technique on the amplitude of the speech signal will usually suffice; that is if the amplitude exceeds a threshold for more than a short period of time a word start is assumed to have been detected, but if the duration of the sound is short it is assumed to be noise. Difficulties are encountered when the first or last phoneme of the word has a weak sound such as /F/, /S/ or /H/ (e.g. 'fish', 'hash'). When this happens simple level thresholding is often inadequate, especially when background noise is present. Some, or all, of the weak sound will not be detected and therefore accurate recognition is less likely to occur. For example, consider the difficulties in distinguishing the difference between the words 'sit', 'fit', 'lit', 'hit' and 'it' if the start of the word is poorly located. Even if there is no background noise or it is reduced to a minimum using a head-mounted directional microphone there are often short bursts of noise present just prior to the beginning of the word and just after it has finished due to lips opening and closing and slight inhalation and exhalation of air. In addition, it is quite common for words with strong plosive or stop consonants within them to have short silent breaks between the word start and end points. Typically silences of less than 150 ms are considered to be silent pauses within words and silent periods of > 150 ms are considered to be word breaks (Wilpon, Rabiner and Martin, 1984).

Background noise not only increases the problem of locating the start and end points of isolated words but also causes the signal level in quiet parts of speech to be significantly distorted, especially when the signal level is less than the background noise. If the templates were trained in a quiet background then, even if word start and end points are accurately located, the high noise level in the quiet parts of the input word will not match well with the corresponding quiet parts in the template word, thus degrading the match. If some knowledge of the background noise is known it is quite possible to formulate a template match that detects all parts of the input word that fall below a certain level which is close to the background noise and avoid matching those particular frames. The template match is then only made with the parts of the input segment that are well above the background noise level.

In addition to locating start and end points of a word it is usual to normalize the maximum amplitude of the word to a standard level to take account of louder or quieter versions of a word due to such factors as the speaker's distance from the microphone and different recording levels. This normalization retains the relative spectral components but normalizes their energy for matching purposes.

8.4.2 Recognition

The recognition of single isolated words is usually approached by the method of template matching. This is illustrated in Fig. 8.7. The incoming speech is preprocessed (start and end points detected and amplitude normalized) and, if the recognizer is in the training mode, a suitable representation of the word is extracted and stored as a template for the selected word. If the recognizer is operating in the

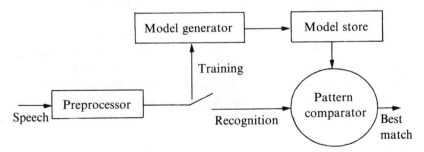

Figure 8.7 A typical template-matching technique for isolated word recognition

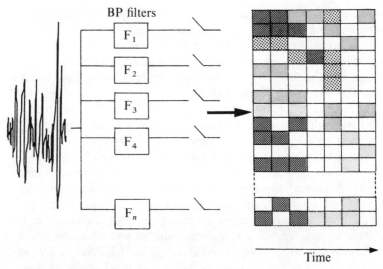

The shaded boxes represent the average energy output from each filter during a fixed sample period: the darker the box the higher the energy.

Figure 8.8 Filter bank analysis of speech

recognition mode the input word will be compared to each of the stored templates using a suitable distance metric to determine the best match. If the best match exceeds a decision threshold the match is considered to have been found, but if the match is less than the decision threshold the input word is not considered to be any of the stored templates and consequently not recognized.

The features that are commonly extracted from the speech waveform for recognition purposes are the spectral components and their energies during time-frame intervals of typically 20 to 50 ms. These are frequently LPC (linear predictive coding) coefficients (see Chapter 5), FFT (fast Fourier transform) coefficients (Gonzalez and Wintz, 1987), the output of a bank of filters as illustrated in Fig. 8.8 or mel-based cepstral coefficients (Furui, 1981). The bank of bandpass filters usually covers the whole range of significant frequency components for speech (100 to 8000 Hz) and typically up to 20 filters are used. The output of these filters are sampled over the frame interval and the energy of each output averaged over the frame interval to produce a single energy intensity value for each filter output. Templates are often stored in this format as they occupy less storage than the sampled time waveform and pattern comparison is faster.

8.4.3 Distance metrics

The comparison of input pattern and stored template pattern is by means of a suitable distance metric $D(x, y)$ which measures the similarity between pattern x and pattern y. There are a number of simple distance measures that are commonly used in speech recognition.

The Euclindian distance is one of the simplest metrics and is the square root of the sum of the squares of the differences between spectral energies in each channel of the input signal and the corresponding channel in the template signal. As a simple example consider the two patterns $x = a, b, c, d$ and $y = p, q, r, s$ in Fig. 8.9. The Euclidian distance between these two patterns is

$$D(x,y) = \sqrt{(a-p)^2 + (b-q)^2 + (c-r)^2 + (d-s)^2} \quad (8.4)$$

In the case of patterns represented as sets of spectral intensities measured at 20 to 50 ms frame intervals the distance measure between each frame will be the square root of the sum of the difference squared of all the corresponding spectral components.

To reduce computation the square root and the squaring of the differences are often omitted and the sum of the magnitude of the differences is used (this is sometimes referred to as the 'city block' distance); that is for the illustrative example in Fig. 8.9 the distance metric becomes

$$D(x, y) = |a-p| + |b-q| + |c-r| + |d-s| \quad (8.5)$$

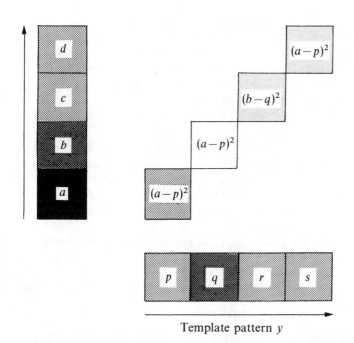

Euclidean distance $D(x,y) = \sqrt{(a-p)^2 + (b-q)^2 + (c-r)^2 + (d-s)^2}$

Figure 8.9 Illustration of the Euclidean distance metric

Other, slightly more complex, distance metrics are also used in simple whole-word pattern matching. These include the Mahalanobis distance, the Itakura distance, the Itakuro–Saito distance and the log-likelihood ratio. See Chapter 10 of O'Shaughnessy (1987), Chapter 7 of Holmes (1988) and Chapter 2 of this volume for more information on these metrics.

8.4.4 Linear and non-linear time alignment

In the previous discussion of distance metrics the assumption was that the two patterns to be compared were of exactly the same length. This is rarely the case in speech as the same word spoken by different speakers or the same speaker at different times is often of different time duration. For example, the word 'catch' could be spoken quickly with a very short /A/ vowel sound or it could be spoken slowly and the /A/ vowel drawn out and so lengthening the word duration. Variation in word length for the same word is illustrated in Fig. 7.2. In simple distance measures the words must either be normalized in time duration or the longer word truncated to the length of the shorter word (as illustrated in Fig. 8.10) so that the distance can be calculated. Truncation can occur either at the beginning or end of the word or at both by stepping the shorter word along the longer word one frame at a time, performing the template match at each step until the end of the word is encountered. The position that produces the best match is selected. However, this method is computationally expensive. Neither of these normalization techniques are entirely satisfactory as the time alignment in spoken words is not a linear process. Vowels and other voiced sections are more variable than stop or plosive consonants. In addition, the errors that are likely to occur in word start and end point detection are also non-linear.

What is needed to overcome this problem is a technique that allows parts of words to be stretched or normalized differently than other parts of the word; that is a non-linear time alignment technique, as illustrated in Fig. 8.11, is required. The favoured solution of this form of normalization is *dynamic time warping* or *elastic template matching* as it is variously known. The technique is based upon dynamic programming which is a widely used technique for solving minimum cost or shortest path problems.

8.4.5 Dynamic programming

Dynamic programming was first proposed by Bellman in 1957. Essentially the idea behind dynamic programming is the principle of optimality which states:

PATTERN RECOGNITION AND ITS APPLICATION TO SPEECH

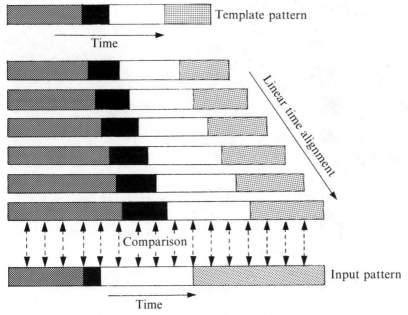

(a) Linear time alignment to match template and input patterns

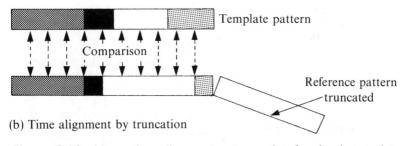

(b) Time alignment by truncation

Figure 8.10 Linear time alignment or truncation for simple template matching

If the optimum path from state K_s to state K_e goes through a state K then the optimal path includes as a portion of it the optimum partial path to the state K.

Relating this to the state transition diagram in Fig. 8.12 shows that there are a number of paths from state 1 to state 5 and each subpath is subject to a state transition cost $c(i,j)$ of going from state i to state j. The objective is to find a path through the network such that the total accumulated cost of going from state K_s to state K_e is a minimum.

Let $C(n, K)$ be the minimum cost of going from state K_s to state K passing through n (or less) path segments and $c(0, K_s) = 0$ and

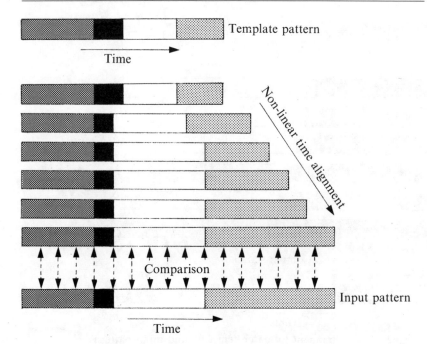

Figure 8.11 Non-linear time alignment for simple template matching

$c(0, K) = \infty$ for all $K \neq K_s$. Then, since the costs are additive and a path to state K must pass through a preceding state K^-,

$$C(n, K) = \min[C(n-1, K^-) + c(K^-, K)] \quad \text{for all } K^- = 1, \ldots, K$$

(8.6)

This is the general form of the dynamic programming template match and is a recursive function.

In speech recognition this recurrence relates to having a distance measure between pairs of frames in input and template patterns. The overall distance is obtained as the path which accumulates the minimum overall cost in traversing from the start of the input and reference words to the end of both input and reference words subject to some local constraints. This minimum path concept can be expressed as

$$s(i, j) = d(i, j) + \min \begin{bmatrix} s(i-1), j) \\ s(i-1), j-1) \\ s(i, j-1) \end{bmatrix}$$

(8.7)

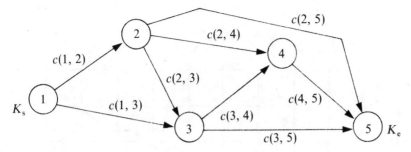

Figure 8.12 Illustration of the principle of optimality

with initial conditons
$$s(1,1) = d(1,1)$$
$$s(1,j) = d(1,j) + s(1,j-1)$$
$$s(i,1) = d(i,1) + s(i-1,j)$$

where $s(i,j)$ is the cumulative distance from the beginning of the word $(d(1,1))$ to the current position in the word $(d(i,j))$ and $d(i,j)$ is a distance measure between component i of an input word and component j of a template word.

This recurrence relation essentially computes the minimum path through a $d(i,j)$ matrix from $d(1,1)$ at the beginning of the word to $d(I,J)$ at the end of the word. This is shown in Fig. 8.13 where two words are compared for correct spelling. Note that in this case the minimum path results in warping by repetition of letters when the two cases of repeated letters are encountered in the input word. The path is allowed to progress through the matrix subject to the local constraints contained within the 'min' term of Eq. (8.7).

This algorithm differs from the simple template match described earlier in that non-linear time-scale warping is allowed in order to obtain the best overall match between the two word patterns. Obviously, if the two words are identical the optimum path is the diagonal path through a square matrix. If the two patterns are not identical in duration the optimum path will stray from this diagonal and a path through a rectangular matrix is sought. The more deviation there is from the diagonal path the more the warping comes into effect.

A problem with the recurrence relation in Eq. (8.7) is that the length of the path in terms of the number of $d(i,j)$ components in it will vary from pattern to pattern and consequently the overall cumulative distance will differ depending on the path taken, thus making it impossible to compare the total path cost between different patterns. This can be compensated for by adding twice the weighting of $d(i,j)$ for diagonal steps than those for horizontal or vertical steps:

$$s(i,j) = \min \begin{bmatrix} s(i-1,j) + d(i,j) + P_h \\ s(i-1,j-1) + 2d(i,j) \\ s(i,j-1) + d(i,j) + P_v \end{bmatrix} \quad (8.8)$$

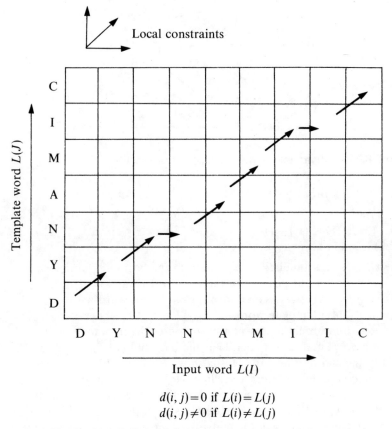

$$d(i,j) = 0 \text{ if } L(i) = L(j)$$
$$d(i,j) \neq 0 \text{ if } L(i) \neq L(j)$$

Figure 8.13 Illustration of the dynamic programming template match technique

The horizontal and vertical penalty functions, P_h and P_v respectively, are included to penalize movement away from the diagonal of a square matrix. The values of these penalties can be zero if unrestricted warping is allowed or non-zero if some restriction is required. The actual value of the penalty values is determined by experimentation with typical examples of patterns that are to be recognized.

The value of $s(I,J)$ (the overall word match) will still depend on the length of the words to be compared and will favour matches with short words where the number of $d(i,j)$ contributing to the overall path is less than for long words, even though the actual matches are good.

To overcome this the value of $s(I,J)$ has to be normalized to the length of the words being compared. Typically this is achieved by dividing $s(I,J)$ by the sum of the length of the template and input words; that is $D(I,J)$ the distance metric is obtained as

$$D(I,J) = \frac{s(I,J)}{M+N} \qquad (8.9)$$

PATTERN RECOGNITION AND ITS APPLICATION TO SPEECH 275

where N and M are the numbers of frames in the template and input words respectively.

In the above analysis we have assumed that the local constraints are that each frame can be repeated in either template of input word (corresponding to horizontal and vertical steps in the $d(i,j)$ matrix) or the frame in both words compared (corresponding to the diagonal step in the $d(i,j)$ matrix) as shown in Fig. 8.14(a). The recurrence relation can be reformulated to allow different forms of local constraint where a frame in a word can be stepped over or omitted from the analysis producing local constraints as shown in Fig. 8.14(b), (c) and (d). Other forms of local constraint can be formulated as long as compensation is made to ensure that the overall path distance is independent of the number of steps in the path.

In the foregoing, penalty functions were introduced to favour a diagonal path through the $d(i,j)$ matrix. When words differ from one another the optimum path will deviate from this diagonal. The more the path deviates from the diagonal the more warping is being performed. There is a point at which the warping is such that the final match is unlikely to be good and can be *pruned* or terminated to reduce computation. This is, in effect, applying a global constraint which dictates that the path must lie within an allowed region around

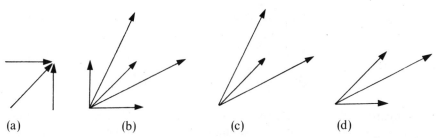

Figure 8.14 Examples of different forms of local constraint used in dynamic time warping template matching

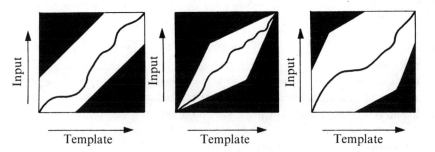

The path through the matrix is not allowed to enter the shaded regions.

Figure 8.15 Different types of global constraint used in dynamic time warping template matching

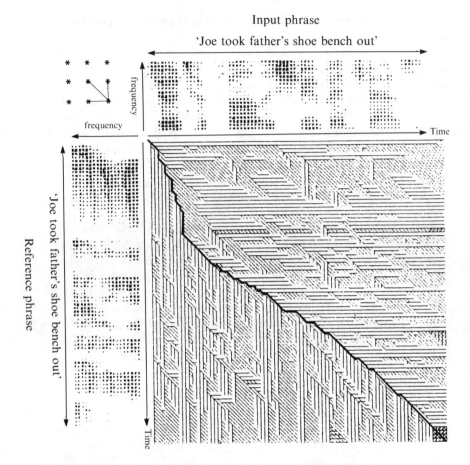

Figure 8.16 Comparison of two phrases using dynamic time warping with unrestricted path consideration. The optimum path is indicated by the dark line. (© Controller HMSO, London, 1984, reproduced with permission)

the diagonal of the matrix. Examples of the type of fixed global constraints used are shown in Fig. 8.15. Alternatively, a dynamic global constraint can be applied using what is termed a *beam search* technique. This technique only considers a range of possible paths which lie within a given range of the current best match, thus narrowing the possible path region when the match is good and widening it (and thus allowing more time warping) when the match is poor. In doing this we are, however, removing the guarantee that the optimum match is found.

PATTERN RECOGNITION AND ITS APPLICATION TO SPEECH 277

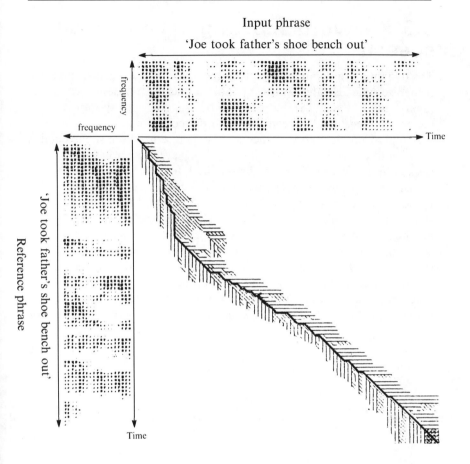

Figure 8.17 The effect of applying a beam search global constraint on the dynamic time warping pattern match between the same two phrases shown in Fig. 8.16. The optimum path is indicated by the dark line. (© Controller HMSO, London, 1984, reproduced with permission)

Figure 8.16 shows the effect of comparing two versions of the phrase 'Joe took father's shoe bench out' using dynamic programming with no global path constraint imposed (Moore, 1985). The two versions of the phrase are shown in spectrogram form as the sampled output of a bank of 19 bandpass filters: the darker the blob the greater the energy in that spectral band. The optimum path is shown as the darker line starting at the beginning of the phrases (top left corner) and proceeding to the end of the phrases (bottom right corner). Figure 8.17 shows the same two versions of the phrase used in Fig. 8.16 but with the path constrained using the beam search.

8.5 Keyword spotting in connected speech

The dynamic programming template match technique was first used to recognize single isolated words, or short connected phrases of words (Sakoe and Chiba, 1978), but can also be used for word spotting in connected speech and for connected word recognition.

A typical problem occurs when a short phrase has been spoken and we wish to determine whether or not one of several keywords was spoken in the phrase. This may be required, for example, in an application such as a simple environmental controller which operates by detecting command words like 'on', 'off', 'light', 'television', etc., in simple sentences like 'I would like the television turned on' and 'time for the light to be turned on'. The word separation problem is not trivial because silent breaks rarely occur at word boundaries. Instead, speech naturally connects many words together and co-articulation effects (the changing of a phoneme sound due to preceding and following articulatory positions as, for example, in the /E/ sound between the two words 'we ate') can distort the start and end sounds of

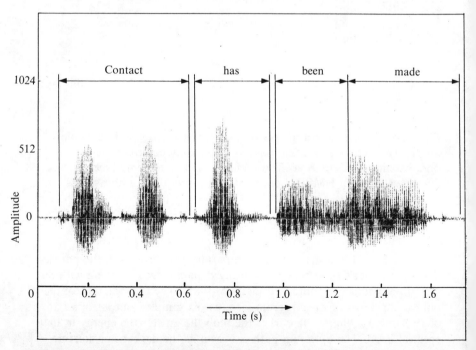

Figure 8.18 Illustration of the problem of word isolation in connected speech. Note that the word 'contact' is composed of two distinct pulses of speech, the end of 'contact' and the start of 'has' is not obvious, and the words 'been' and 'made' form a continuous single pulse.

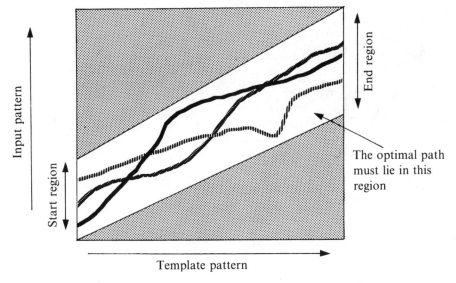

Figure 8.19 The use of ambiguous word-start and word-end regions in word spotting using a dynamic programming template match

the words. The problem of word separation is illustrated in Figs 8.18 and 7.1.

Because of the difficulty in locating word-start and word-end points in connected speech the basic approach to connected word recognition is to search for the central parts of a word that is relatively unaffected by connectivity and co-articulation. Therefore, a segmentation algorithm searches for start and end *regions* rather than frames (Myers, Rabiner and Rosenberg, 1981). Alternatively, the whole phrase is considered as the word and the start region is assumed to exist from the start of the phrase and the end region is assumed to continue backwards from the end of the phrase. The word-spotting algorithm searches for the best match for each template within that region, as illustrated in Fig. 8.19. The amount of computation required to solve this problem can be excessive, but it has been shown that by suitable selection of local constraints and weighting functions it can be significantly reduced (Myers, Rabiner and Rosenberg, 1981).

8.6 Connected word recognition

Connected word recognition differs from word spotting in that rather than detecting a small quantity of words in unlimited vocabulary speech we are attempting to recognize all words that are spoken in connected speech of limited vocabulary. This makes the problem somewhat more complicated.

280 SPEECH PROCESSING

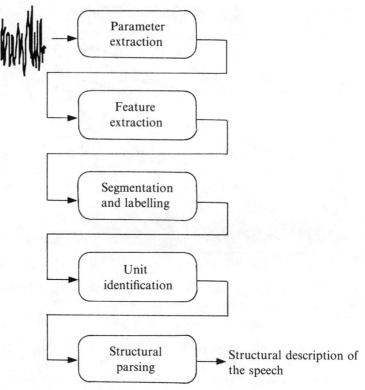

Figure 8.20 Elements in connected word recognition of large vocabularies

One approach to the connected word recognition problem is to consider it as an extension of the word-spotting problem by searching for template matches of single whole words but to allow some relaxation in the position of start and end points of the word. This approach, however, severely restricts the vocabulary as template matching can only cope accurately with a few tens of words.

The alternative is to attempt to perform the recognition of a larger vocabulary of words by locating and recognizing subword sound segments and concatenating them together to form words using structural information as illustrated in Fig. 8.20. In this case parameters are extracted from the speech (e.g. by filter bank or mel-based cepstral analysis) and encoded with other features such as voiced speech, silences and loudness. These features are segmented into likely speech segments such as phonemes or allophones and identified (or, more usually, a list of likely interpretations for each speech segment is produced) by a suitable distance metric. The sequence of likely speech segments is then compared to a structural description of words in the

PATTERN RECOGNITION AND ITS APPLICATION TO SPEECH

allowed vocabulary and an overall word level interpretation of the phrase is arrived at based on likely word pronunciations and word contexts.

This approach was attempted in the HARPY system at Carnegie-Mellon University (Lowerre and Reddy, 1980) and is an extension, to two levels, of the dynamic programming principles described in Sec. 8.4.5. In the HARPY system speech was segmented into discrete components using such features as silence, voicing, frication and peak-and-dip detection. After segmentation each speech component was matched using a distance metric.

The training of the HARPY system involved the speaker repeating 20 sentences from which all allophones were extracted and averaged to form templates for the distance metric. HARPY allowed sentences of connected speech to be spoken and recognized. The only constraint was that the words must be from a 1011 word dictionary and all sentences must make linguistic sense.

The recognition was based on a knowledge network based on the 1011 allowable words. Syntactic knowledge was encoded in the form of a word network to provide all acceptable sentences, as illustrated in Fig. 8.21. In addition, lexical knowledge was also stored in terms of all the acceptable pronunciations of the allowable words, as shown in Fig. 8.22.

Recognition of the connected speech was based on a variation of dynamic programming, but the optimum path was searched for through a network of speech segment template matches and interacted

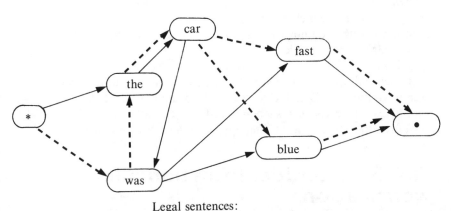

Legal sentences:
Was the car fast, Was the car blue
The car was fast, The car was blue

Figure 8.21 Illustration of syntactic knowledge of all allowable word combinations

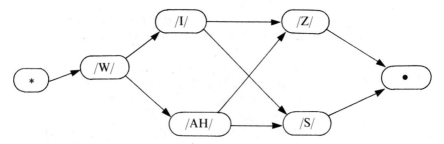

Four pronunciations for the word 'was': /W/I/Z/, /W/I/S/, /W/AH/Z/, /W/AH/S/

Figure 8.22 Illustration of lexical knowledge of all word pronunciations

with the knowledge source. A *beam search* technique was used and involved:

- Analysing a speech waveform from beginning to end.
- Segmenting it into phones and performing a template match on each phone.
- Keeping all likely phone matches.
- Generating a recognition tree based on good phone matches.
- Assigning an overall path probability at each node of the tree (*based on previous and present match and interaction with the knowledge source*).
- Pruning all paths that fall below a threshold.

When complete the beam search finds an interpretation of a sequence of speech sounds which consists of high match segments and is consistent with the knowledge source. Because of the high level of knowledge stored in this system only limited vocabulary recognition is possible.

Other major continuous speech systems that incorporate knowledge representation are DRAGON (Baker, 1975) and HEARSAY (Lesser *et al.*, 1975). Later systems, such as BYBLOS (Chow *et al.*, 1987), have favoured the Markov modelling approach described in Chapter 7.

8.7 Speaker identification and verification

In this task we are attempting to recognize the speaker and not the words that are spoken. In many ways this is analogous to handwritten signature verification and has a number of potential applications in personnel identification and authorization applications. A detailed description of this subject can be found in Chapter 11 of O'Shaughnessy (1987).

The problem can be approached in two different ways. It can be either *speaker verification* where the user is asked to confirm his or her identity by voice print or *speaker identification* where the system attempts to determine whether the identity of the speaker is one of a set of voices with which the system has been trained. Speaker verification involves analysing input speech and comparing it to parameters of the stored speech for a single user. The decision required is a binary decision between whether the speaker is who he or she claims to be or an imposter. Speaker identification involves comparing input speech to the stored parameters of a number of different speakers and deciding whether the speech does indeed belong to one of the previously stored speakers or none of them.

The techniques employed usually make use of statistical features of the speech. A short passage of speech is required to train the system and statistical parameters are extracted from this passage of speech. These parameters take into account such features as formant trajectories, speech rates and vowel durations (O'Shaughnessy, 1986).

8.8 Summary

In this chapter we have introduced three general pattern recognition techniques: template matching, statistical pattern recognition and structural pattern recognition. These three basic techniques can be applied to any type of pattern. Template matching is appropriate when patterns are easily isolated from other patterns, there is little variation in patterns of the same class and the number of different pattern classes is small. If the pattern classes are noisy or exhibit more within-class variation and statistical information is available about the class conditional densities and/or the probability of occurrence of each pattern class a statistical pattern classifier is appropriate. Alternatively, if the pattern has a well-defined structure which can be expressed as a combination of a small number of subpatterns (pattern primitives) then the structural pattern recognition approach is appropriate. In practice it is not always clear which pattern recognition technique to choose for a particular class of patterns and it is not unusual for classifiers to combine all three techniques.

Automatic speech recognition is a difficult problem which current technology has only partially solved. The difficulties are associated with the continuous nature of natural speech, coping with large vocabularies and complex grammars, isolating individual voices from background noises and resolving the large variation in speech from person to person.

The speech recognition problem can be constrained to some extent by requiring the user to wear a head-mounted directional microphone to reduce the problems caused by background noise, training the recognizer to each individual user and restricting the vocabulary and style of speech. These restrictions can take the form of requiring the user to only speak single isolated words from a restricted vocabulary. Systems of this type are available and can perform with fairly high accuracy when the vocabulary is restricted. These recognition systems frequently use a template-matching technique, usually based on dynamic programming principles.

More sophisticated systems attempt to recognize short strings of words in connected speech. These systems also have restricted vocabularies but the words need not be spoken in isolation but may be joined together as in normal speech. There are a number of recognizers available which attempt this form of recognition. To operate successfully these systems require a detailed knowledge of the syntax of the vocabulary and usually involve a structural knowledge-based recognition strategy.

All current speech recognition systems require a co-operative user and their performance is severely constrained by a limited vocabulary and the accuracy with which they are trained.

The recognition of normal unlimited vocabulary continuous speech remains, at the present time, unresolved and a solution is not likely to be reached for some considerable time. The general pattern recognition techniques discussed in this chapter have not proved capable of coping with unlimited vocabulary continuous speech. The problem is that they do not provide a sufficiently accurate model of the recognition process. Future work in all areas of pattern recognition will need to concentrate on improving the artificial intelligence aspects of encoding and interacting with knowledge about the patterns and gaining a better understanding of the fundamental issues of the patterns under investigation. For speech recognition, a better understanding of the speech signal itself and how it is affected by such factors as stress and speed of speech is required. Until an accurate model is available the success of ASR systems with normal continuous speech will remain poor.

There are, however, a number of applications where currently available recognizers using the techniques describes in this chapter can, even with their limitations, still be used effectively in selected applications. This aspect of speech recognizers is discussed in Chapter 11.

References and further reading

References

Baker, J. K. (1975) 'The DRAGON system—an overview', *IEEE Transactions on Acoustics, Speech and Signal Processing*, vol. 23, no. 1, 24–29.
Ballard, D. H. and M. B. Brown (1982) *Computer Vision*, Prentice-Hall, Englewood Cliffs, N.J.

Bellman, R. (1957) *Dynamic Programming*, Princeton University Press, Princeton, N.J.
Chow, Y. L., M. O. Dunham, O. A. Kimball, M. A. Krasner, G. F. Kubala, J. Mahoul, S. Roucos and R. M. Schwartz (1987) 'BYBLOS: the BBN continuous speech recognition system', *IEEE Transactions of the International Conference on Acoustics, Speech, and Signal Processing*, vol. 1, 89–92.
Devijver, P. A. and J. Kittler (1982) *Pattern Recognition: A Statistical Approach*, Prentice-Hall, Englewood Cliffs, N.J.
Fu, K. S. (1982) *Syntactic Pattern Recognition and Applications*, Prentice-Hall, Englewood Cliffs, N.J.
Furui, S. (1981) 'Cepstral analysis technique for automatic speech verification', *IEEE Transactions on Acoustics, Speech, and Signal Processing*, vol. 29, 254–272.
Gonzalez, R. C. and P. Wintz (1987) *Digital Image Processing*, 2nd edn, Addison-Wesley, Reading, Mass.
Guiliano, V. E. (1967) 'How we find patterns', *International Science and Technology*, 42, February.
Holmes, J. N. (1988) *Speech Synthesis and Recognition*, Van Nostrand Reinhold, London.
Leith, P. (1983) 'Hierarchically structured production rules', *The Computer Journal*, vol. 26, no. 1, 1–5.
Lesser, V. R., R. D. Fennell, L. D. Erman and R. D. Reddy (1975) 'The Hearsay II speech understanding system', *IEEE Transactions on Accoustics, Speech, and Signal Processing*, vol. 23, no. 1, 11–24.
Lowerre, B. and R. Reddy (1980) 'The Harpy speech understanding system' in *Trends in Speech Recognition*, W. A. Lea (ed.), 340–360, Prentice-Hall, Englewood Cliffs, N.J.
Moore, R. K. (1985) 'Systems for isolated and connected word recognition' in *New Systems and Architectures for Automatic Speech Recognition and Synthesis*, R. de Mori and C. Y. Suen (eds), 73–143, Springer-Verlag, Berlin.
Myers, C. S., L. R. Rabiner and A. E. Rosenberg (1981) 'On the use of dynamic time warping for word spotting and connected word recognition', *Bell Systems Technical Journal*, vol. 60, no. 3, 303–325, March.
O'Shaughnessy, D. (1986) 'Speaker recognition', *IEEE Acoustics, Speech, and Signal Processing Magazine*, vol. 3, 4–17, October.
O'Shaughnessy, D. (1987) *Speech Communication—Human and Machine*, Addison-Wesley, Reading, Mass.
Rumelhart, D. E. and J. L. McClelland (1986) *Parallel Distributed Processing*, vols 1 and 2, The Massachusetts Institute of Technology, Cambridge, Mass.
Sakoe, H. and S. Chiba (1978) 'Dynamic programming algorithm optimization for spoken word recognition', *IEEE Transactions on Acoustics, Speech, and Signal Processing*, vol. 26, no. 1, 43–49.
Sayre, K. M. (1965) *Recognition, A Study in the Philosophy of Artificial Intelligence*, University of Notre Dame Press.
Verhagen, C. J. D. M. (1975) 'Some general remarks about pattern recognition; its definition; its relation to other disciplines; a literature survey', *Pattern Recognition*, vol. 7, no. 3, 109–116.
Wilpon, J., L. Rabiner and T. Martin (1984) 'An improved word-detection algorithm for telephone quality speech incorporating both syntactic and semantic constraints', *Bell System Technical Journal*, vol. 63, 479–497.
Young, T. Y. and K. S. Fu (1986) *Handbook of Pattern Recognition and Image Processing*, Academic Press, Orlando, Florida.

Further reading

Bristow, G. (1986) *Electronic Speech Recognition: Techniques, Technology and Applications*, Collins, London.

Lea, W. A. (1980) *Trends in Speech Recognition*, Prentice Hall, Englewood Cliffs, N.J.

Peckham, J. (1988) 'Talking to machines', *IEE Review*, vol. 34, no. 10, 395–388, November.

Schwab, Eileen, C. and Howard C. Nusbaum (eds) (1986) *Pattern Recognition by Humans and Machines*, vol. 1, *Speech Perception*, Academic Press, Orlando, Florida.

Tou, J. T. and R. C. Gonzalez (1974) *Pattern Recognition Principles*, Addison-Wesley, Reading, Mass.

9 Speech on packet networks

GILL WATERS

9.1 Introduction

In this chapter, we shall be looking at flexible techniques for switching encoded speech signals in digital communication systems. By carrying speech, data and other digitized information on a single network the best use can be made of available bandwidth, and new services that combine different media such as speech, data and images can be accommodated.

In this section, packet networks are introduced and new integrated services are discussed. Techniques for the packet switching of data are described in Secs 9.2 and 9.3; readers familiar with these techniques may wish to skip these sections. We then look at some approaches to integrated networks, at the characteristics of packetized speech and at speech coding implications. Finally, we look in more detail at integrated packet network techniques and give examples.

9.1.1 What is a packet network?

A packet network carries information in blocks of a maximum length. Each packet contains user information and other information including the destination address and possibly a sequence number, a checksum and details of how the user information should be interpreted. By dividing information into packets, each of which is individually switched, it is possible to make economical use of the network by sharing trunk links and even user access links between many conversations or calls.

Packet networking techniques are now widely accepted for data transmission and have the following advantages:

1 Computers can deal easily with blocks of information. (It is easy to split memory into fixed size buffers for processing, and direct memory access encourages block working.)

2. Bit errors which occur when a packet is transmitted can be detected by means of a checksum. Errors can also be corrected by asking for retransmission of faulty packets.
3. Packet interleaving offers economical use of a communications network.
4. A wide variety of data transmission rates can be supported, and variable bit-rate sources can also be accommodated.
5. Most packet networks will transmit any bit pattern transparently, allowing many different forms of information representation to be carried.
6. Unlike pieces of equipment can communicate (e.g. a small intelligent terminal communicating at 1200 bit/s could communicate with a large mainframe that had a 48 kbit/s link to the network).
7. It is possible for several applications in the user's system to communicate simultaneously with different remote systems over the same physical network connection.

9.1.2 Requirements of different types of traffic

The introduction of packet networks among others has meant that a number of information networks are now running side by side, for example the existing telephony network, telex network and public packet switched network. Many different types of communication services can be coded and transmitted digitally, and the possibility of using a single network for the transmission of all services is being actively pursued. The integrated services digital network (ISDN) offers a standard interface for combined integrated services, based on circuit switching with gateways to existing networks. Packet networks may offer a flexible alternative to circuit switching. In order to consider their suitability for integrated service networks it is necessary to consider the service requirements. A comparison of the transmission rate, nature, toleration of errors and toleration to delay for a number of services is given in Table 9.1. Note that data services generally occur in short bursts and are sensitive to bit errors. Speech requires regular bandwidth and low consistent delays but can tolerate some loss. We shall be looking in more detail at speech service requirements later in the chapter.

There are three main reasons why we might want to packetize speech and carry it on the same network as data. Firstly, integration with computing systems enables applications such as conferencing, voice messaging and the retrieval of multimedia documents to be developed. Secondly, packetization offers an economic way of dealing with variable bit rates which may be

Table 9.1 Typical service requirements

Type	Rate	Nature	Errors tolerated	Delays tolerated
Data (e.g. file transfer remote login electronic mail)	0–10 Mbit/s	Bursty	None	Yes
Speech	64 or 32 kbit/s	Periodic	Some	No
Facsimile	0–64 kbit/s	Bursty	None	Yes
Fixed-rate video	140 Mbit/s	Periodic	Some	No
Variable bit-rate video	0.5–30 Mbit/s	Bursty	Some	No

useful, for example, in adapting quality of service to network availability, or by implementing silence detection and suppression. Thirdly, packet networks offer a more flexible multiplexing technique in terms of the user's access to the network and the ability to handle several calls to different destinations at the same time.

It is already possible to make speech calls on some packet networks, but for a number of reasons this is not a very widely available facility. Good medium access protocols are required and these need VLSI interfaces to reduce costs. (Several have been proposed and implemented as we shall see later.) Other factors are that circuit switching technology is well understood and works, and there is a vast existing investment in telephony equipment.

9.2 An overview of existing packet networks

9.2.1 LANs, MANs and WANs

Packet networks can be categorized into local, wide and metropolitan area networks. The names are derived from the area covered, but the three categories also have different characteristics. (For a good comprehensive introduction to computer networks see Tanenbaum, 1988.)

A *local area network* (LAN) is generally contained within a building or a small site and is controlled by one organization. It is

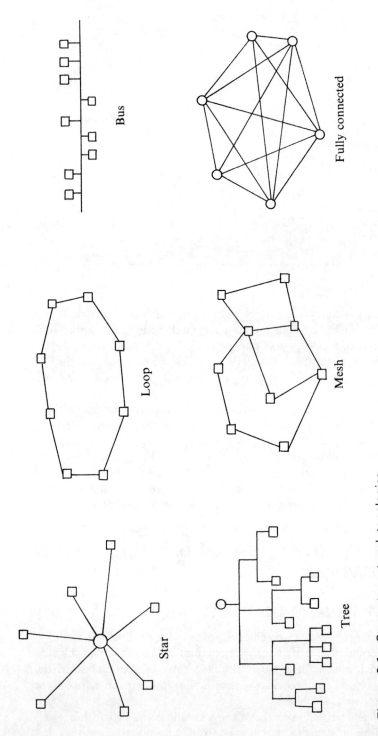

Figure 9.1 Computer network topologies

characterized by high speed (typically 10 Mbit/s), a low bit error rate (almost negligible) and low latency. A LAN offers single routing, that is a single path from any one station to another station. LAN protocols are simple and amenable to VLSI implementation, and are generally low cost. It is possible to use LANs without invoking retransmission strategies for error recovery; full reliability can be provided if required by a higher level procedure than that used to access the medium. LAN characteristics make them suitable for consideration for integrated traffic but conventional LANs have some drawbacks, as we shall see in Sec. 9.4.

A *wide area network* (WAN) may cover any area; typically they are contained within a single country, but some cross national boundaries. This results in longer and more variable delays. Because of the availability and cost of the transmission media WANs have to cope with higher bit error rates than LANs (e.g. 1 in 10^6) and use slower transmission links (typically 9.6 or 48 kbit/s). WANs usually carry traffic by packet switching, a technique where each link in the network can be used to transmit packets belonging to many 'conversations' and the network exchanges use addressing information within each packet to route it to its destination. This is a very efficient way to handle data applications where information tends to occur in bursts of variable length with large periods of silence.

A *metropolitan area network* (MAN) typically covers the area of a city. A number of MANs has now been installed, but research is continuing, the ideal being a high-speed network with simple access techniques which will offer facilities for integrated traffic. They will have more connections than a typical LAN and will therefore require more complex topologies, although delay will have to be minimized.

9.2.2 Network topologies

Computers can be connected together in many different ways, and the layout or topology of the network will influence how reliable the network is and how easy it is to access. Figure 9.1 illustrates a number of possible topologies.

In a *star* topology, the central exchange is very important, since if this fails so does the rest of the network. It is easy to add new systems and easy to police centrally, but may need a large amount of cabling.

A *loop* topology is simple, needs a minimal amount of cabling and can be controlled in a distributed manner. Failure of any individual system or link between systems will cause the whole network to fail unless extra measures are taken to avoid this.

A *bus* topology needs the minimum amount of cabling and can also be controlled in a distributed manner. Failure of any part of the bus results in a network that is split into two segments. Systems on bus networks generally only put signals onto the network when they have data to transmit, so that a single system breakdown does not usually affect the rest of the network.

In a *tree* topology, it is easy to add or remove systems without affecting the rest of the network, and this arrangement can make good use of lines by sharing capacity where appropriate. However, link or individual system failures can result in a disjoint network.

A *mesh* topology contains redundant links. A packet may have to pass through several exchanges to get to its final destination. It is reasonably easy to add systems, but there are design problems in deciding where to add exchanges or links. This topology is generally used for packet-switched wide area networks.

In a *fully connected* network topology there is always likely to be an alternative path if one link or exchange fails. This arrangement requires a good deal of cabling and each system or exchange will have to control many links. This topology is only necessary if a high degree of reliability is required.

In networks with a *broadcast* topology, such as those based on radio or satellites, a single packet generally reaches all the participating systems. Packets are usually labelled with a particular destination, but it is also possible to make use of this broadcast capability to achieve economical use of the channel where the same data is to be sent to many systems.

9.2.3 Physical media

A wide variety of communication media can be used to carry information in packets. These include a wide variety of guided systems such as twisted pairs, coaxial cable and fibre optic links. Some networks also include satellite or radio segments. Use can be made of telephony networks, and this is becoming easier with the introduction of digital techniques.

9.2.4 Protocols for multiple access

In order to carry information across a network in an orderly manner and to preserve its meaning, it is necessary to follow rules of procedure and to insert signalling information within the packets. These rules of procedure are called protocols. Where a single channel is shared between a number of users as on a LAN or in a packet radio or packet satellite network, a protocol is required to determine which user is allowed access to the channel, to use the channel efficiently and to prevent conflicts. Techniques fall into a number of categories:

1 *Dedicated access* Each user is allocated a fixed share of the channel. Examples are time division multiplexing (TDM) and frequency division multiplexing (FDM), both commonly used in telephony.

2. *Reservation access* Users are allocated a share of the channel on request, the reservations being determined by a central allocator.

3. *Polling* Each user is allowed to use the channel in some predetermined order. (This scheme has an implicit ordering.)

4. *Token passing* Permission to use the channel is passed from user to user in the form of a token. The order in which this is done may be implicit (e.g. when a token is passed around a network with ring topology) or explicit (e.g. a predetermined order assigned to systems connected to a bus).

5. *Random access* In these techniques users access the channel when they have information to send. This incurs the possibility of collision. In a completely random scheme this can drastically reduce performance, so many schemes rely on detection of ongoing activity ('listen before talk') before a decision is made to use the channel. Early detection of collision can be made by a system checking if its own transmissions are being corrupted ('listen while talking'). Random access techniques fall into three subcategories:
   (a) No 'listen before talk' and no 'listen while talking' (Aloha).
   (b) 'Listen before talk' (carrier sense multiple access, CSMA).
   (c) 'Listen before talk' and 'listen while talking' (carrier sense multiple access with collision detection, CSMA/CD).

9.3 Local area networks

Packet networks which cater for the transmission and switching of data, voice and other real-time services are generally based on LAN protocols. This is because LANs cater for higher speeds, lower delay and simpler protocols than wide area networks. However, most LAN protocols have one significant drawback: they treat all connected systems fairly. This is an admirable aim for bursty traffic, but real-time traffic such as speech needs to use network capacity regularly and with minimum delay, as we shall discuss more fully in Sec. 9.5.

In order to discuss integrated networks in detail we must first look at the conventional LAN techniques on which they are based. Three LAN techniques are described here: CSMA/CD, token passing and the slotted ring. We shall use the term 'station' to mean a system (software and hardware) directly connected to a LAN.

9.3.1 Carrier sense multiple access with collision detection (CSMA/CD) (IEEE Standard 802.3)

This method of accessing a bus network was first described for the Ethernet by Metcalfe and Boggs (1976). It is the technique used in the majority of installed LANs. It uses both 'listen before talk' and 'listen while talking'.

The procedure followed by a system wishing to transmit onto the network is as follows:

1. Enclose the packet with synchronization bits to form a frame and pad it to a minimum length.

2. Wait until there is no carrier on the bus and then start to transmit the packet.

3. Detect whether the packet collides with a packet being transmitted from another station, by listening to the signals on the bus. This must be done throughout a maximum possible 'transmission window' which includes the time required for the packet to propagate across the network.

4. If a collison occurs, send sufficient jamming information so that all stations can detect the collision. Abandon the transmission. Wait a random time and they try again, starting from step 2 up to a maximum number of times. (For each retransmission attempt the random time taken is taken from a larger range including longer periods.)

5. If no collison is detected, continue to transmit the packet.

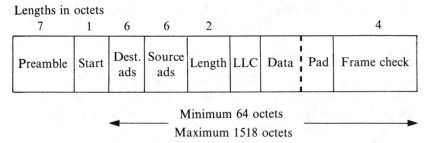

Figure 9.2 CSMA/CD (IEEE 802.3) frame format

Frames are of the format shown in Fig. 9.2.

The topology can consist of a number of interconnected buses as shown in the example in Fig. 9.3 with the following restrictions. When using baseband coaxial cable at 10 Mbit/s, this is subject to a maximum span of 500 m without repeaters, a maximum 2.5 km between any two stations and there must only be one possible path between any two stations.

9.3.2 Token passing (IEEE Standard 802.4 and IEEE Standard 802.5)

The basis for token passing is very simple: a unique token is circulated among the stations connected to the network in a logical ring. When a

SPEECH ON PACKET NETWORKS

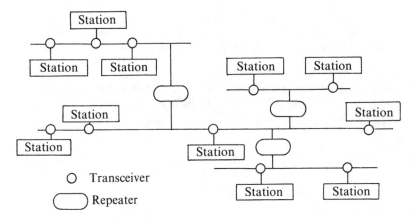

Figure 9.3 Example of a CSMA/CD configuration

station holds the token, it may transmit a packet or packets and must then regenerate the token. This access scheme is applicable to several topologies. The IEEE Standard 802.5 describes the token ring.

Figure 9.4 shows the example of the token bus (IEEE Standard 802.4). All stations wishing to communicate are admitted to a logical ring of stations. The token is normally passed to the station with the next lower address. The figure shows a logical ring of six stations, with stations H and F not participating.

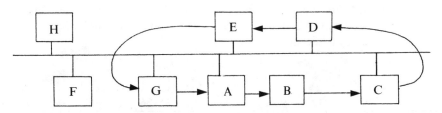

Figure 9.4 Token-passing bus

The protocol is augmented to deal with any problems that may arise, as follows. A number of special frame formats are used.

1. *Lost token*   After transmitting the token, the sender checks that its successor transmits valid information (either an information frame or another token). If not, it retries once; if this also fails the sender sends a 'who follows' frame so that the next station but one can respond and take the token.

2. *Duplication of token*   The sending station does not relay any tokens it may receive while in possession of the token.

296 SPEECH PROCESSING

3 *Station entering logical ring* Following transmission, the sender may send a 'solicit successor' frame and wait for a reply. A valid response from a previously inactive station allows it to enter the logical ring.

4 *Station leaving the logical ring* The station wishing to leave the ring does not transmit the token when it has received it. This will then be detected as indicated for a lost token.

Because the token is a bit string, its value can indicate extra information, for example a priority indication which only some stations have permission to use. The IBM token ring can periodically switch to synchronous mode suitable for traffic such as packetized speech, the switch of mode being indicated by the type of token.

9.3.3 Slotted rings

This technique is used for the Cambridge ring (Wilkes and Wheeler, 1979). A slotted ring consists of a closed ring of cable and active repeaters which co-operate to circulate a fixed number of constantly circulating bits. The bits are divided into a number of fixed-size slots called minipackets, each 40 bits long. The number of slots depends on the size of the ring and the number of stations. The minipacket format is shown in Fig. 9.5.

A slotted ring interface comprises a repeater and a station for each connected system. The repeaters receive clock and data from the ring, demodulate the incoming bit stream, modulate the required output and transmit it to the ring. They take d.c. power from the ring cables. The stations provide serial-to-parallel conversion, maintain minipacket synchronization and detect errors.

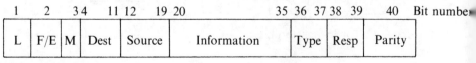

Key: L = leader bit
F/E = full/empty bit (0 empty, 1 full)
M = monitor pass bit
Type = type bits
Resp = response bits
  11 ignored
  01 accepted
  00 busy
  10 not selected

Figure 9.5 Cambridge ring minipacket format

Access procedure

A station wishing to *transmit* a packet must first wait for any empty slot, indicated by a full/empty bit value of 0. It then sets the full/empty bit to 1 and the monitor pass bit to 1, inserts the destination address, data and type required and supplies its own source address. The response bits are set to zero and the parity bit is set to maintain even parity over the entire minipacket.

If the receiving station is disabled, there will be no response. Otherwise, a *receiving station* will accept only full minipackets addressed to self and from the currently selected source which may be either none or all or one specific source station. If the station is capable of processing the packet, the data field is copied to memory and the response bits are set to indicate accepted. Otherwise the response bits may indicate either that the station is busy or that the source is not selected.

The minipacket is not removed from the ring by the destination; this is done by the source station setting the full/empty bit to 0 when the packet returns. The source station then checks the response bits; if the packet was not accepted it may attempt to send it later. The source station is not allowed to use another empty slot until the one it used has returned and one more slot has passed, ensuring fair access to all stations on the ring.

For maintenance procedures and in order to recover from errors, one station on the ring acts as the monitor station. This station resets the monitor pass bit of full slots to 0. A slot passing the monitor station marked full and with the monitor pass bit set to 0 is on its second way round the ring and is discarded by the monitor station.

9.3.4 LAN Standards

Computer network standards are defined within the terms of the open systems interconnection (OSI) reference model. The model provides for seven layers of protocol. Each layer has a particular function, and offers services to the next higher layer.

Standardization of LANs has taken place within the IEEE 802 committee, which has defined four layers which correspond to the lower two layers of the OSI model as shown in Fig. 9.6. All LAN types share a common logical link control (LLC) protocol, but each type has its own medium access control, the three first definitions being:

IEEE 802.3 CSMA/CD
IEEE 802.4 Token-passing bus
IEEE 802.5 Token-passing ring

Other definitions will follow. LLC can provide reliable communication above the unreliable medium access control protocol.

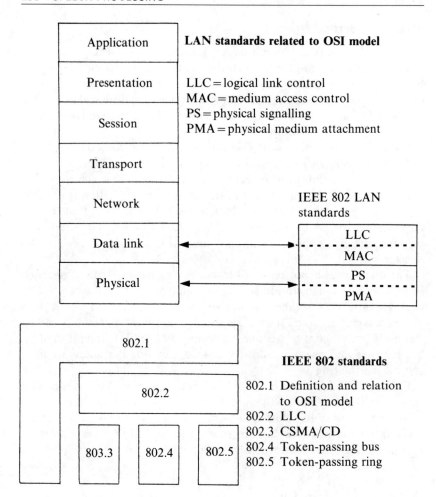

Figure 9.6 IEEE 802 LAN standards

9.3.5 The performance of LAN protocols

Let us define the efficiency, E, of a protocol to be

$$E = \frac{\text{transmission time}}{\text{transmission time} + \text{acquisition time}}$$

For CSMA/CD, it can be shown that the loss of efficiency which occurs as the acquisition time approaches the transmission time is given by

$$\frac{P}{C} = 5.4L$$

where

P = packet length (in bits)
C = channel transmission rate (bits/s)
L = end-to-end propagation delay

This becomes significant for high-speed LANs and for LANs with a long cable length.

For the *token ring* when N nodes are transmitting

$$E = \frac{NP}{NP+R}$$

$$= \frac{P}{P+R/N}$$

where R = length of the ring in bits. In this case the acquisition time reduces with increasing node activity. Both CSMA/CD and the token ring improve in efficiency for longer packet length, P.

Graphs of the delay versus throughput of the token ring at both 4 and 16 Mbit/s are shown in Figs 9.7 and 9.8 respectively.

Performance summary

CSMA/CD is very fast for low traffic load, less efficient with heavy load. Larger packets give better throughput. Delays are highly variable due to back-off.

In the *slotted ring*, time is wasted due to source release when considering point-to-point throughput. The overheads are high because of the small minipacket size, and delays are higher than CSMA/CD but less variable.

The *token ring* performs well under heavy loads and performs better with larger packets. Synchronous working is possible.

9.4 Approaches to integrated networks

9.4.1 Integrated services digital network (ISDN)

In order to meet some of the needs of new services and provide a common network on which to carry them, the PTTs have developed the concept of an integrated service digital network (ISDN), which uses digital technology and allows many kinds of traffic to be carried. ISDN is now coming into service in a number of countries. ISDN is described by the CCITT I series recommendations. For a more approachable introduction, see, for example, Ronayne (1987).

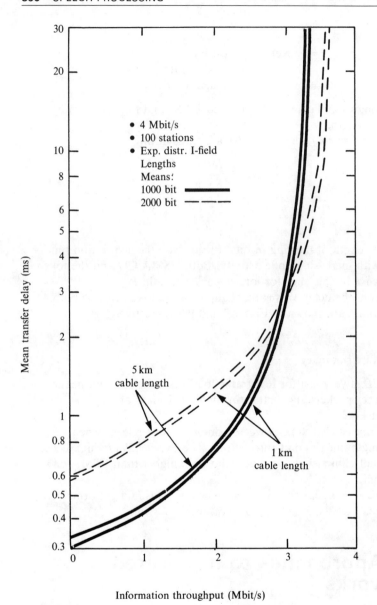

Figure 9.7 Delay versus throughput of 4 Mbit/s token ring (from Bux, 1985; reprinted by permission of Springer-Verlag and the author)

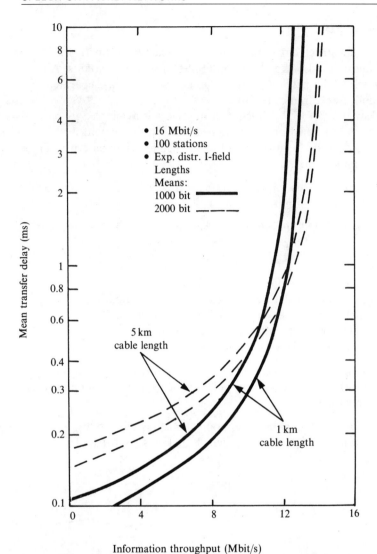

Figure 9.8 Delay versus throughput of 16 Mbit/s token ring (from Bux, 1985; reprinted by permission of Springer-Verlag and the author)

What does an ISDN offer?

An ISDN is constructed around a circuit-switched digital network based on multiplexed 64 kbit/s and multiples of 64 kbit/s circuits. All signalling within the network and to the user is carried as separate channels and uses a frame structure.

A small number of well-defined user interfaces is available, giving access either to switched or non-switched circuits. Equipment is connected to an ISDN using a set of recommended channel types. The two main channel types are:

1 *The B channel* This offers 64 k/bits for a wide variety of voice or non-voice user information streams.

2 *The D channel* (16 kbit/s basic, 64 kbit/s primary) This channel is primarily intended for signalling, but additionally may carry packet switched data.

Additionally, *H channels* offer multiples of 64 kbit/s.

The *basic* interface structure at 144 kbit/s as illustrated in Fig. 9.9 is two B channels plus one D channel. The *primary* interface at 2048 kbit/s offers 30 B channels plus one D channel.

The services provided by an ISDN cover two broad categories:

1 *Bearer services* The means of transmitting digital information transparently between two end points which the user can employ as he or she wishes.

2 *Teleservices* Full capability equipment such as telephones, teletex, videotex and message handling, which make use of the underlying network.

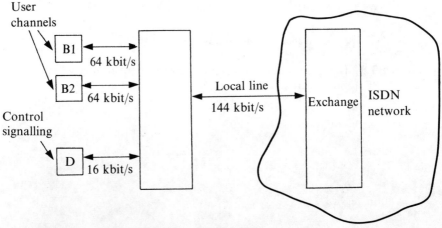

Figure 9.9 Basic interface to ISDN

Packet mode access to an ISDN

Because an ISDN is basically a circuit-switched network the full flexibility of packet switching does not fit in very easily. However, since an ISDN is aimed at all service users including those who use public packet-switched data networks, it is necessary to include packet mode access. This is being introduced in four stages as follows (Cooper, 1988):

Stage 1 Access to external packet-switched services through the user's D channel (CCITT Recommendation X.31 (I.462)).

Stage 2 Access to a packet-switching service provided within an ISDN using either the B or the D channels using Recommendation X.31 (I.462).

Stage 3 Provision of an additional packet mode bearer service. This would allow variable length blocks of data to be switched through either the B or D channels using the following techniques. (Note that the virtual circuit set-up is always performed through the D channel.) The techniques are:
  (a) *Frame relaying* (two versions) Information is transmitted across the network in HDLC-like frames but no explicit acknowledgement is provided by the network.
  Multiplexing takes place at the data link layer. Frame sequence is maintained and errors are detected. This mode is particularly suitable for the interconnection of local area networks or high-speed data applications where flow control is best left to the end users.
  (b) *Frame switching* Acknowledged exchange of frames across the network.

Stage 4 Broadband ISDN which is now the subject of a considerable amount of research will be required to cater for higher speeds to include real-time video transmission of, say, 140 Mbit/s and to offer a more flexible approach to integrated services. Recently it was agreed that broadband ISDN would be based on switching techniques for small fixed-length packets for all services—a technique called asynchronous transfer mode (ATM) (Littlewood, Gallagher and Adams, 1987). ATM can be seen as an extension of the techniques we shall be discussing later in this chapter for integrated voice/data packet networks, and is characterized by a small fixed packet size.

The provision of ISDNs is still in its infancy, but prototype frame relay networks are being developed. See, for example, Lamont, Doak and Hui (1989).

9.4.2 The impact of optical fibres

Optical fibres currently offer capacities of multigigabits per second and higher speeds are envisaged. Much of the UK's trunk network now uses optical fibres which have much better performance in terms of loss and the need for repeaters than the copper cables they replace. They also make the speeds required to integrate all services possible. Optical fibre therefore offers a readily available means of transmission for integrated services. Switching is not so advanced and ATM will require extremely fast switching, perhaps optical switching.

The maximum flexibility for new services, especially those involving a mix of traffic types such as interpersonal communication including simultaneous voice, video and data, can only be provided by packet switching. However, it is not clear that this will always be the best solution for all services. Techniques such as wave division multiplexing and subcarrier modulation enable many high-speed channels to be carried simultaneously on a single optical fibre, some of which could be packet switched and others circuit switched. It would also be possible to carry both analogue and digital signals; with low attentuation, analogue signals may be the best choice for some services. Another technique that could be considered is fast circuit switching—which may be particularly appropriate to voice conversations where silence detection is used.

9.5 Characteristics of packetized speech

If speech is transmitted in packets it has rather different requirements to those of data. Loss of some packets is acceptable, so packets can be dropped occasionally, but a speech circuit is very sensitive to delay, and in particular variable delay, which can be a problem in packet networks. Gruber and Le studied the requirements of speech in integrated networks; we now summarize the most important characteristics for speech on digital networks (Gruber and Le, 1983). Variable delay and speech clipping are dealt with in Gruber and Strawczynski (1985).

9.5.1 Delay components

When speech is packetized and transmitted over a network it is subject to several different forms of delay. Figure 9.10 shows where these delays arise. The components are as follows:

1 *Transmission time* for each packet.
2 *Propagation delay* which is fixed, and should be kept to less than 24 ms for national connections (which has no subjective effect). For

SPEECH ON PACKET NETWORKS

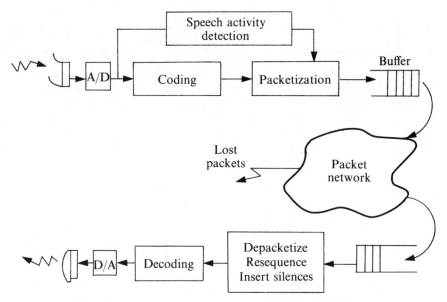

Figure 9.10 Delays incurred in sending packetized speech over a network

paths including a one-way satellite link, delays should be kept to less than 300 ms.

3 *Processing delay* due to coders and processors (including filters and speech interpolators which provide bandwidth economy by, for example, silence detection).

4 *Packet-switching delay* which is variable, and is due to the time a packet spends getting access to the network and being routed or queued by the network.

5 *Packetization delay* which is the time taken to fill a packet with information and will depend on the size of the packet.

6 *Resynchronization delay* at the receiver which compensates for the variable delay experienced by the packets.

For a LAN a typical requirement would be to keep the total delay caused by packetization, propagation and access delay to within, say, 5 ms mean one-way.

The effect of variable delay

The variable delay that can occur on packet networks can seriously affect the perceived quality of speech. The effect of variable delay on a speech burst depends on the setting of the speech detector, both the threshold at which a decision is made between speech and silence and

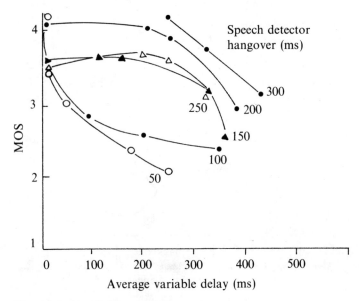

Figure 9.11 Subjective effect of variable speech burst delay (Gruber and Le, 1983; © IEEE 1983)

the hangover or amount of time before the speech to silence transition is implemented.

Figure 9.11 shows the effect on the mean opinion score when speech bursts are delayed by a certain amount with a threshold hangover. (A score of 5 is good; lower scores indicate poorer subjective quality.) From the figure, it can be concluded that speech detector hangovers of 100 ms or less are not recommended; 200 ms is a good design value.

9.5.2 Transmission error effects

A bit error may result in a lost packet. This will depend on the error detection characteristics of the network. In some networks packets with errors may be presented to the user flagged to indicate that an error has occurred.

Where packets are lost or discarded in the middle of a speech burst, this results in mid-speech burst clipping. Gruber's studies showed little or no impairment results from less than 1 per cent clipping with a maximum clipping duration of 5 ms. Other studies suggest that from 2 to 5 per cent mid-speech burst clipping is acceptable with speech interpolation, although for very low bit rate coding, low percentage clipping is needed. Intelligibility is not affected by much larger amounts of clipping.

We can conclude from this that packet dropping could be a viable

SPEECH ON PACKET NETWORKS

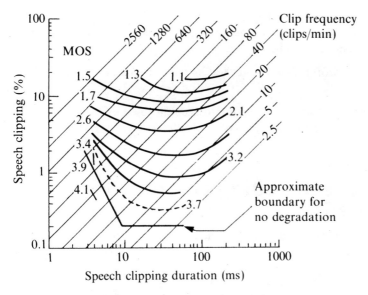

Figure 9.12 Subjective effect of mid-burst speech clipping (Gruber and Le, 1983; © IEEE 1983).

way to handle transient overload conditions. Figure 9.12 shows some mean opinion score curves for speech clipping probability against speech clipping duration.

9.5.3 Bandwidth

The current bandwidth for telephone conversations is 300 to 3300 Hz or 3000 Hz and can be coded using PCM techniques into 64 kbit/s. Digital techniques allow better coding strategies while preserving speech quality thus it is now possible to consider either lower bit rates with a similar quality or the same bit-rate requirement with a higher quality wider bandwidth. Packet networks offer the flexibility to use these new techniques as appropriate.

9.5.4 Implications for the transmission of speech over packet networks

We can summarize some of the constraints for carrying speech on packet networks as follows:

1. A packet size of up to about 30 ms for digital networks or networks employing echo cancellers should be tolerable.

2. Variable delay on the network can be absorbed into variable delay of a whole speech spurt.

3 Packet loss should be kept between 0.5 and 1.0 per cent unless sophisticated packet interpolation strategies are used, in which case up to 5 per cent packet loss may be tolerable.

4 Packet networks should cater for encoded speech at a variety of bit rates.

An integrated network must take into account these different characteristics and provide all types of traffic with the service they require.

9.6 Speech coding and protocol implications

When speech is to be transported across a packet network, the effects of the network can be catered for or to some extent overcome by appropriate speech coding techniques, and by a network voice protocol above the medium access protocol (Bradlow and Hall, 1988), as shown in Fig. 9.13. The network voice protocol techniques that may be considered are packetization, prioritization and congestion control for transmission; and depacketization, synchronization and fill-in for reception.

9.6.1 Packetization

A number of factors must be considered. The packet length must be kept small to minimize delays and to minimize the perception of packet

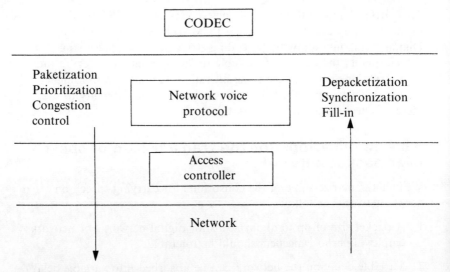

Figure 9.13 A packet voice terminal

loss, but small packets impose higher overheads on the network. Block coding imposes higher delays than sequential coding, but is less sensitive to packet loss. It may be desirable to vary the coding rate resulting in a variable packet length or a variable packet rate—this could improve response to congestion. Variation in coding rate can be produced by, for example, varying the sample rate or the number of bits per code word. However, variable-length packets can impose serious synchronization problems. Typically, fixed-size packets of 10 to 20 ms are used.

9.6.2 Prioritization

A number of factors affect the rate at which we might want to send packetized speech onto the network, and thus produce changing requirements on the network. The most obvious factor is silence detection where packets are transmitted only in talkspurts. The coded packet pattern may also be varied in response to the signal-to-noise ratio achieved by the coder, the short- or long-term transmission history or the activity of received signals resulting in double talk. It may also be necessary for speech traffic to ensure that it is carried at an appropriate priority with respect to other traffic such as data or packet video.

9.6.3 Congestion control

Queuing of packets is undesirable because it causes delay, but it can help to minimize loss. The conventional way of responding to excess telephony traffic has been call blocking or, in the case of analogue speech interpolation, freeze-out. In packet networks it should be possible to cater for reduced transmission capacity by adapting the coded signal and varying the packet rate, possibly with some reduction in signal quality. 'Embedded' coding (where both coding parameters and coded information are carried in each packet) is particularly useful in this context, because it allows the bit rate to be reduced by discarding packets at any point in the network. This requires an ongoing knowledge of the load on the network, which is offered by some network protocols.

9.6.4 Receiver synchronization and fill-in

Packetized speech carried on networks is subject to variable delay which must be compensated for at the receiver. This can be done by noting an average delay for packets, identifying missing packets, discarding late packets and filling in gaps or varying silent periods slightly. Filling can be achieved by silence, low-level noise, repeating the previous packet or by pitch-related interpolation.

9.7 Integrated packet networks

9.7.1 Techniques

When handling speech or other real-time services on a packet network, there are two extremes and a number of options between. The first extreme is to treat voice as having completely synchronous requirements and to allocate a fixed amount of bandwidth to the voice circuits (blocking any calls which would exceed that). The other extreme is not to differentiate between voice and data packets at all.

In practice, a flexible integrated packet network will fall between these extremes and a boundary is provided. Other important choices in integrated network design are between distributed and centralized control, and the method of handling circuits or packets related to a single logical connection.

Boundary position

There are three possibilities in deciding how to implement the boundary between synchronous and asynchronous services: fixed, movable or none. A *fixed boundary* leads to inefficient use of capacity but is simple. With a *movable boundary* a communication mechanism is needed to inform of changes in boundary position. If there is *no boundary* it may not be possible to guarantee the required capacity for synchronous services. In general, a movable boundary is preferred.

Centralized or distributed control

For *centralized* control, a request for a channel access is made to a central controller, which allocates capacity, releases it after use and blocks if no capacity is available.

For *distributed* control, each station monitors the channel and uses the same algorithm to take over its share of the bandwidth, relinquishing capacity when it has finished without explicit communication to other stations.

The disadvantages of centralized schemes are that they are prone to failure of the central station, a processing bottleneck can occur round the central station and additional protocol elements are required to make and grant requests. The disadvantages of distributed control are that they are more difficult to implement and maintain, and billing and authorization are more complex. The advantages of both can be combined by using centralized control for new speech calls and distributed access to the channel on a per-packet basis.

Circuit, virtual circuit and packet working

Where circuits are used, a bit field of a certain size is allocated guaranteeing bandwidth, and there is no addressing within the packet. In virtual circuit working, slots are reserved periodically for packets. The packets will include a channel number allocated for a particular conversation. In packet working, traffic of all types takes it chances with the rest of the traffic on the network and there is no specific reservation.

For circuits bandwidth is preallocated and minimum signalling is required, so delay can be kept small. For virtual circuits the signalling overhead in each packet is reduced, resulting in smaller packets and shorter delays. Where all packets are treated in the same way, no call set-up protocol is required, so there is no delay in setting up a call. The most flexible approach is to provide virtual circuit working, reducing signalling overhead in the packets, guaranteeing bandwidth for real-time traffic such as speech where required, but offering the possibility of reuse if not required, for example in silent periods.

9.7.2 Some approaches to integrated packet networks

The final sections of this chapter will discuss three specific examples of integrated packet networks. These have been chosen to indicate a variety of approaches and represent a very small selection from the many designs that have been reported.

Research into integrated packet networks has concentrated particularly on LANs which offer characteristics that are suitable to both voice and data, having simple access protocols to a fast shared medium. In general, LANs support data traffic well, and data transport on LANs is well understood. The work on integrated service LANs has therefore mainly been concerned with producing acceptable delay and bandwidth-sharing characteristics for voice.

Most of the integrated LANs that have been produced can be described as 'hybrid' in the sense that packets are divided into two classes, with some amount of bandwidth allocated to each class. Many have movable boundaries between the classes. Proposals for metropolitan area networks show a common ancestry with those for integrated LANs. Notable examples of integrated LANs and MANs are BID (Ulug, White and Adams, 1981), Expressnet (Tobagi, Borgonovo and Fratta, 1983), Fasnet (Limb and Flores, 1982), FDDI-2 (Ross, 1986), FXNet (Casey, Dittburner and Gammage, 1986), Magnet (Lazar et al., 1985), Orwell (Adams and Falconer, 1984), Prelude (Thomas, Coudreuse and Servel, 1984), Philan (Brandsma, Breuhers and Kessels, 1986), QPSX (Newman et al., 1988) and SCPS (Takeuchi et al., 1987). These represent a spectrum of designs from those that look like a traditional private automatic branch exchange (PABX) to those that look like computer communication LANs.

Starting at the computing end of the spectrum, Fasnet is a pair of unidirectional data buses. The buses carry fixed-size packets and the end stations co-operate to provide cycles during which traffic of different classes can be transmitted onto the bus. If cycles for delay-sensitive traffic such as voice are not needed, the end stations can allow data traffic to use the unwanted portions of the cycles. Voice slots are a minimum of 10 ms long, longer if system efficiency is a major criterion. Expressnet and BID resemble Fasnet to a large extent—in all three cases the protocol for integrated services support is implemented in hardware of considerable complexity.

Magnet is a slotted ring with two classes of slot and a variable boundary between voice and data traffic. The voice slots offer guaranteed delay characteristics, but if not needed then more data slots can be allocated, in a similar manner to Fasnet. This change is accomplished centrally by a monitor station. Magnet is discussed more fully in Sec. 9.8.

The Orwell slotted ring adopts a rather different approach—that of 'counters' and 'epochs'. Each node is allocated a share of the ring's bandwidth represented as a counter, which is decremented each time a packet is sent. When a node's counter reaches zero it is not allowed to transmit until the next epoch is declared and all counters are refreshed. The Orwell protocol is the subject of Sec. 9.9.

At the PABX end of the spectrum, Prelude has a hardware architecture very much like a digital circuit switch, although it is based on packets and supports a very wide range of traffic data rates. SCPS uses a synchronous ring structure linking conventional concentrator switches to handle both circuit switched traffic and HDLC packets with guaranteed fixed delay. It, too, can support a very wide range of data traffic rates. Philan and FXNet both use a 125 μs slot scheme, with some parts of the slot allocated for circuit-switched traffic and some parts available for intermittent transmission of data packets by any node.

FDDI-2 and QPSX have both been proposed as metropolitan area networks. Both combine slot reservation for isochronous traffic with access to individual unreserved slots for data. The IEEE 802.6 committee, which has the task of producing a MAN standard, has adopted the QPSX MAN and this is the subject of Sec. 9.10—our third example of an integrated packet network.

There are many other LANs that support some type of voice/data traffic mix; in fact it is rare for a new type of LAN to be designed without some ability to support voice being claimed. MSTDM (Maxemchuk, 1982) is an example of a modified Ethernet that supports voice. The IBM token ring (Bux *et al.*, 1983) also incorporates voice/data provision, but despite efforts to base a commercial PABX upon it, its voice-handling facilities are very limited.

High-speed LANs which do not have this hybrid approach have also been used to carry voice traffic. One particular example is the Cambridge fast ring (Hopper, Temple and Williamson, 1986) which supports a mix of voice and data well but has very simple interfaces reflecting simple protocols. This network has been used as the basis of the ISLAND project which is discussed in the following chapter by Stephen Ades.

9.8 Magnet

(Lazar et al., 1985; Patir et al., 1985)

The Magnet testbed was intended to provide a practical system to investigate the performance of integrated service LANs and to address other issues such as reliability and ease of expansion. A fibre-based LAN was built using two 100 Mbit/s channels (the channels being provided by wavelength division multiplexing) which rotate in opposite directions. The arrangement is shown in Fig. 9.14. The head station provides clocking for each of the rings. Each of the network stations is connected to both rings, allowing them to be used if all components are working, or a single ring to be configured in the case of a station or single link failure.

The LAN is based on a slotted ring technique with slots of 256 bits, of which 224 bits are available for information (see Fig. 9.15). Source and destination fields of 8 bits allow up to 256 stations on a single network. Empty slots are filled by the transmitting station under the medium access control protocol and the information is deleted from the ring by the destination station, allowing the slot to be used by other

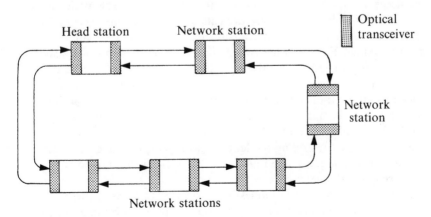

Figure 9.14 Configuration of Magnet testbed

314 SPEECH PROCESSING

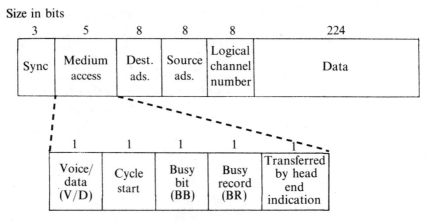

Figure 9.15 Magnet frame format and medium access control field

stations. When both rings are operating the ring offering the shorter route can be selected, thus enabling the expected throughput to reach twice the nominal link capacity. The protocols allow user applications to use the network either in 'circuit-switched mode', in which bandwidth is guaranteed, or in 'packet-switching mode', where access depends on the statistical pattern of traffic from other stations. In circuit-switched mode the head end ensures that new real-time traffic (voice and video) calls are blocked if the capacity of the network is already in use for existing calls. In packet-switching mode the threshold for the number of real-time traffic calls is set higher assuming, for example, that silence detection is employed for speech calls. This enables the network to handle more calls but some packets may be lost.

The medium access control is implemented in hardware to achieve the necessary speed. Three protocols are possible to enable performance comparisons to be made. These are called zero cycle, one cycle and two cycle. Zero cycle allows all stations to use empty slots in a random access manner. For the other two modes voice and video traffic (called V traffic) are treated differently from data (D traffic), and the head end imposes a cycle structure on the slots. In the one-cycle mode V-type packets may be transmitted into empty slots without restriction, but a station may only transmit one D packet in any cycle. In the two-cycle mode slots can only be filled by V traffic in the V cycle and D traffic in the D cycle. By controlling the length and repetition rate of the V cycle the head end can guarantee that delays on V traffic are bounded.

To see how the network can be used efficiently we describe the two-cycle mode (the other two modes being simpler versions of these). During the V cycle, the head end station sets the V/D bit in every slot to V, and during the data cycle to D. If a slot arrives back at the head end station with the 'busy record' (BR) bit still 0 this indicates that the

SPEECH ON PACKET NETWORKS

slot has travelled round the ring without being used. The head end will then generate a new cycle. Alternatively, a new cycle is generated when all the slots of the previous cycle have been used.

Magnet thus offers several solutions to the integrated LAN problem based on a decentralized scheme for slot access and a centralized scheme for the number of voice and video calls carried at any time.

9.9 Orwell

The Orwell protocol is named after the River Orwell in Suffolk, England, and was devised by J. Adams of British Telecom (Falconer and Adams, 1985). Orwell is based on a slotted ring but with destination release of slots. This results in short delays and high bandwidth utilization. As can be seen in Fig. 9.16, it is possible to use different sectors of the ring for traffic belonging to two different conversations at the same time. However, if stations were allowed unlimited access to free slots, they would be able to 'hog' the network, making the network unusable for other stations, so some extra control is needed. The way in which this control is managed is rather unusual and is described below.

9.9.1 Control in Orwell

Stations in an Orwell ring must behave as follows:

1 Stations may transmit up to a locally agreed maximum Di of packets; then must PAUSE.

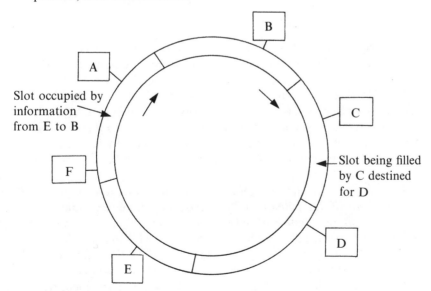

Figure 9.16 The Orwell ring

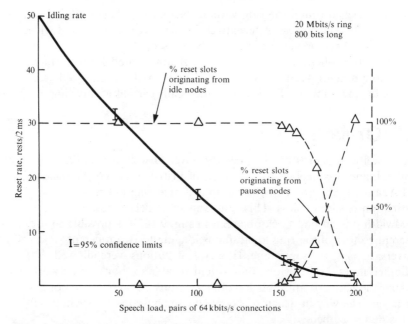

Figure 9.17 Orwell—the effect of load on resets (Falconer and Adams, 1985; reprinted by permission of British Telecom Research Laboratories)

2 This continues until all stations are either PAUSED or IDLE.

3 A RESET is issued. (This refreshes Di for all stations.)

4 Activity starts again as in 1.

Guaranteed bandwidth is offered by allocating a suitable value of Di and by keeping the reset interval less than some maximum value (say 2 ms).

RESETS themselves are issued in a distributed manner; that is any station may decide to issue a RESET as follows. IDLE or PAUSED nodes convert empty slots to TRIAL slots (addressed to themselves). TRIAL slots may be filled, and thus converted to data packets. If a TRIAL slot arrives back at its source station without being filled, it is converted to a RESET.

The value of each station's Di is determined by a central station. New requests for an increase in Di succeed only if the reset interval can be kept within the predetermined maximum value. Requests are made when a new speech call is made, but while the call is in progress, access to the ring continues in a distributed manner. All stations are allocated a minimum value of Di, so that they can send data if required.

As can be seen in Fig. 9.17, the reset interval is more frequent when the load is light, thus allowing more data to be handled.

9.9.2 Asynchronous transfer mode (ATM)

A great deal of interest is being shown in the possibility of networks such as Orwell which use fixed-size packets called cells and allow both flexible and guaranteed access to capacity. This type of network has been described variously as asynchronous time division (ATD) and currently in the international community as asynchronous transfer mode (ATM). ATM has been adopted as the means of transmission and switching for broadband ISDNs. In order to make the right decisions about the correct ATM protocol for use in broadband ISDNs it is important to gain practical experience with these integrated networks.

9.10 Metropolitan area networks

In order to provide flexible and fast networking to a city or region of similar size the techniques used for integrated local networks need to be applied to this larger area. IEEE's project 802, having standardized a number of LAN protocols, is now working on project IEEE 802.6 which addresses a standard for metropolitan area networks (MANs) (Mollenauer, 1988).

After considering a number of alternatives, the technique which has now been chosen was originally known as QPSX (the queued packet and synchronous switch) and was a proposal from Telecom Australia. This is now known more generally as the distributed queuing dual bus (DQDB), and because of its importance in the current standards arena it is described here.

The DQDB is based on two contradirectional buses as shown in Fig. 9.18 (Newman, Budrikis and Hulett, 1988). Each station is attached by both read and write taps and a slot structure is imposed by a frame

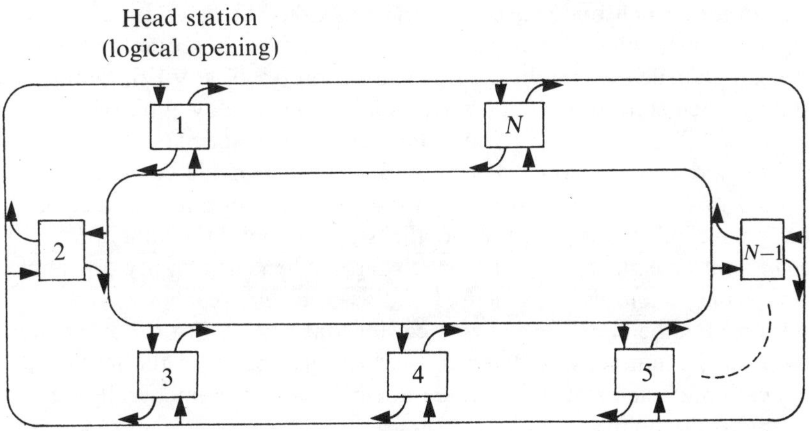

Figure 9.18 Dual bus arrangement of DQDB

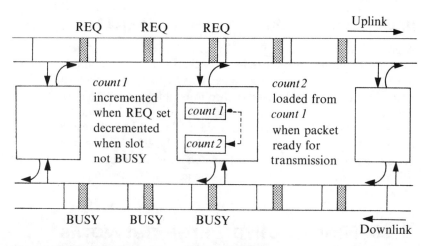

Figure 9.19 Distributed queuing on DQDB

generation at the leading edge of each bus. (Similar arrangements have been used, for example, in Fasnet; see Limb and Flores, 1982.) By looping the bus into an incomplete ring topology it is claimed that the ring will be reconfigurable if a break occurs elsewhere by healing the ring at the original opening.

Apart from reliability, attention has also been given to the interconnection of DQDB networks and their interaction with existing switches and existing wide area networks. The DQDB network is another example of ATM suitable for use with emerging broadband ISDNs.

One of the most interesting features of the DQDB is its medium access protocol called the distributed queuing protocol. This makes full use of the dual bus arrangement to provide a prioritized queuing system for asynchronous access to the network. The basis for joining a queue in one direction is dependent on the number of requests to join the same queue arriving in the other direction (see Fig. 9.19). Quite simply, each station wishing to access a slot on the downlink of the network will examine the REQ indication in each slot passing on the uplink and keep a count (*count1*) of such requests made before it wished to access the network (i.e. it knows its position in the queue), and then send its own request. At this stage it copies the value of *count1* into *count2* and sets *count1* to zero. When each empty downlink slot passes, *count2* is decremented and when *count2* reaches zero the station may then access the first free slot that passes. Priority is given by having a number of different bits which may be set in the REQ indication of each slot. Distributed queuing works symmetrically for access to the uplink.

Slots on the DQDB are grouped into frames which occur at intervals of 125 μs, so that access to one octet in each frame gives a 64 kbit/s channel. Some of these are used by asynchronous and some by synchronous traffic. Higher or lower speed channels can be allocated by assigning more or less than one octet per frame respectively. Thus, again this system requires centralized control on a per call basis to ensure that sufficient bandwidth is available, while access to each slot is determined by a distributed procedure. The ability of a station to detect activity on the network may be very useful if applied to variable bit-rate sources such as some forms of encoded speech and video where the encoding rate could be adjusted depending on the perceived load on the network as a whole.

9.11 Conclusion

In this chapter we have looked at some of the problems and possibilities of using a packet network approach to give access to a wide variety of communication services. The seriousness with which those investigating broadband ISDN are considering the methods described in the later sections of this chapter indicate that these techniques will indeed be employed in communication networks which will carry not only traditional speech conversations but also add speech to other types of traffic to form totally new services.

References

Adams, J. L. and R. M. Falconer (1984), 'Orwell: a protocol for carrying integrated services on a digital communications ring', *Electronics Letters*, vol. 20, no. 23, 970–971.

Bradlow, H. S. and S. C. Hall (1988) 'The design of an integrated voice/data terminal and voice protocols', 3rd Fast Packet Switching Workshop, Melbourne.

Brandsma, J. R., A. A. M. L. Breuhers, and J. L. W. Kessels (1986). 'PHILAN: a fibre optic ring for voice and data', *IEEE Communications Magazine*, vol. 24, no. 12, 16–22.

Bux, W. (1985) 'Performance issues' in *Lecture Notes in Computer Science 184: Local Area Networks—An Advanced Course*, Springer-Verlag, Heidelberg.

Bux, W., F. H. Kloss, K. Kummerle, H. J. Kieller and H. R. Mueller (1983), 'Architecture and design of a reliable token ring network', *IEEE Journal on Selected Areas in Communications*, vol. 1, no. 5, 756–765.

Casey, L. M., R. C. Dittburner and N. D. Gammage (1986), 'FXNet: a backbone ring for voice and data', *IEEE Communications Magazine*, vol. 24, no. 12, 23–28.

Cooper, N. J. (1988) 'Packet mode services for ISDN', *British Telecom Technology Journal*, vol. 6, no. 1, January.

Falconer, R. M. and J. L. Adams (1985) 'Orwell: a protocol for integrated services on a local network', *British Telecom Technology Journal*, vol. 3, no. 4, October.

Gruber, J. G. and N. H. Le (1983) 'Performance requirements for integrated voice/data networks', *IEEE Journal on Selected Areas in Communications*, vol. SAC-1, no. 6, 981–1005, December.

Gruber, J. G. and L. Strawczynski (1985) 'Subjective effects of variable delay and speech clipping in dynamically managed voice systems', *IEEE Transactions on Communications*, vol. COM-33, no. 8, 801–808, August.

Hopper, A. S. Temple and R. Williamson (1986) *Local Area Network Design*, Chapter 8, Addison-Wesley, Wokingham, Berks.

IEEE Standard 802.2 (1985) 'Logical Link Control', IEEE.

IEEE Standard 802.3 (1985) 'CSMA/CD', IEEE.

IEEE Standard 802.4 (1985) 'Token passing bus', IEEE.

IEEE Standard 802.5 (1985) 'Token ring', IEEE.

Lamont, J., J. Doak and M. Hui (1989) 'LAN interconnection via frame relay', Proceedings of IEEE Globecom, 686–690.

Lazar, A. A., A. Patir, T. Takashi and M. El Zarki (1985) 'MAGNET: Columbia's integrated network testbed', *IEEE Journal on Selected Areas in Communications*, vol. SAC-3, no. 6, 859–871, November.

Limb, J. O. and C. Flores (1982) 'Description of FASNET, a unidirectional local area communications network', *Bell Systems Technical Journal*, vol. 61, no. 7, pt 1, 1413–1440, September.

Littlewood, M., I. D. Gallagher and J. L. Adams (1987) 'Evolution toward an ATD multi-service network', *British Telecom Technology Journal*, vol. 5, no. 2, 52–62, April.

Maxemchuk, N. F. (1982) 'A variation on CSMA/CD that yields movable TDM slots in integrated voice/data local area networks', *Bell Systems Technical Journal*, vol. 61, 1527–1550, September.

Metcalfe, R. M. and D. R. Boggs (1976) 'Ethernet: distributed packet switching for local computer networks', *Communications of ACM*, vol. 19, no. 7, 395–404, July.

Mollenauer, J. F. (1988) 'Standards for metropolitan area networks', *IEEE Communications Magazine*, vol. 26, no. 4, 15–19, April.

Newman, R. M., Z. L. Budrikis and J. L. Hulett (1988) 'The QPSX MAN', *IEEE Communications Magazine*, vol. 26, no. 4, 20–28, April.

Patir, A., T. Takashi, Y. Tamura, M. El Zarkl and A. A. Lazar (1985) 'An optical fiber-based integrated LAN for MAGNET's testbed environment', *IEEE Journal on Selected Areas of Communication*, vol. SAC-3, no. 6, 872–881, November.

Ronayne, J. (1987) *The Integrated Digital Services Network—From Concept to Application*, Pitman, London.

Ross, F. E. (1986) 'FDDI—a tutorial', *IEEE Communications Magazine*, vol. 24, no. 5, 10–17, May.

Takeuchi, T. with T. Yamaguchi, H. Niwa, H. Suzuki and S. Hayano (1987), 'Synchronous composite packet switching, a switching architecture for broadband ISDN', *IEEE Journal on Selected Areas in Communications*, vol. SAC-5, no. 8, 1365–1376.

Tanenbaum, A. S. (1988) *Computer Networks*, 2nd edn, Prentice-Hall, Englewood Cliffs, N.J.

Thomas, A., J. P. Coudreuse and M. Servel (1984) 'Asynchronous time division techniques: an experimental packet network integrating videocommunication', International Switching Symposium, Florence, May.

Tobagi, F. A., F. Borgonovo and L. Fratta (1983) 'Expressnet: a high performance integrated services local area network', *IEEE Journal on Selected Areas in Communications*, vol. 1, no. 5, November.

Ulug, M. E., G. M. White and W. J. Adams (1981) 'Bidirectional token flow system', Proceedings of 7th Data Communications Symposium, Mexico City, 149–155, October.

Wilkes, M. V. and D. J. Wheeler (1979) 'The Cambridge digital communication ring', Proceedings of Local Communications Network Symposium, Boston.

10 Integrated services on local networks

STEPHEN ADES

10.1 The office of the future

This chapter discusses a vision of the future, which is not now so far from becoming a reality. In the fairly near future we can expect to see offices equipped with just a single communications system, which provides for all of the office's needs. It will provide:

- Telephony, comparable with the service offered by a modern PABX;
- 'Personal computer' facilities, with shared file storage and shared peripherals such as printers, all accessible from the desk-top personal computer or workstation;
- Voice recording, storage and editing;
- Communications access from the personal computer to other machines, including job submission and database queries to remote mainframes, file transfer and electronic mail;
- 'Shared screenfuls', where, for example, a group of people at remote workstations can work together on editing a document visible to them all;
- Various other data communications, such as facsimile and telex;
- A wide variety of image transfer services, from workstation-to-workstation single image transfers to videophones and full-rate video conferencing, plus image editing and various derived services such as the 'distributed blackboard' (where two people sit at remote workstations and can sketch with their mice on the screen, each seeing both sets of sketches, distinguished by different colours).

Each of these services is useful in itself, but the most important thing that will become available when all of these types of service are handled in the office by a single network is what one might term the 'integrated workstation', a personal computer that combines the individual services listed above in ways that provide new facilities. Imagine, for example, a truly integrated diary and answerphone system. Each office worker keeps an electronic diary, into which he or she can enter appointments and into which other personnel can 'pencil things in'. When the worker leaves the office (the departure recorded via an integrated clock in/clock out system) he or she may leave various messages or an indication as to whether callers are to be told of his or her whereabouts. When a caller phones the office, a voice synthesizer will give information as to where the worker is and may ask the caller to leave a message. There may also be a voice recognizer in the system, so that specific callers can be recognized and any messages left for them (or only for the ears of other company employees, or so on) can be played or synthesized.

If this type of facility sounds a long way from reality, it is not so very far-fetched. The basic components required (networks, video CODECs, voice recognizers and synthesizers) have been demonstrated and are being refined and improved by researchers at present. Many of these components are discussed in preceding chapters of this book. The basic components are, however, only a part of the story: design of the electronic office of the future requires a lot of high-level systems design work. Integrating the various communications services is as much of a challenge as is providing each component.

This chapter is concerned with the systems design aspects of integrated services and will discuss work done towards the 'vision' above, both on design of the networks which can support the services and on integration of those services within the workstation. The chapter will draw on experiences from a wide variety of research projects, with much of the discussion centred around the 'ISLAND' project at the University of Cambridge Computer Laboratory.

10.1.1 The ISDN versus the integrated office

It is worth contrasting the aims in our vision with those of the integrated services digital network (ISDN). The ISDN is essentially a set of standards being agreed by the various telecommunications network providers (PTTs), so that equipment offering one of a range of services can communicate with other equipment offering the same service at a different site. A computer in Italy will be able to send data to a computer in England, and so on. The principal motivation of the PTTs in setting up the ISDN is that their networks (mainly consisting of the circuit-switched phone networks) will then be able to offer more types of service and hence earn more revenue.

To the end user, wide-area ISDN offers less benefits than it does to the PTTs. The end user does benefit when equipment in different countries can communicate, due to agreement of common standards. (We have seen this benefit in the huge growth of facsimile transmission.) Eventually domestic users will also benefit from the fact that existing telephone lines will be able to carry voice and data to and from the user's premises, although this is some way off. However, the ISDN does not actually set out to offer service integration to the user. Although traffic is integrated for carriage on the PTT's network, different traffic types are then separated out and directed to different types of terminal equipment on the user's premises—facsimile going to the fax machine, telephony to the phones, data files to host computers and so on.

This is quite different from our vision. For integrated services to be of significant interest to the customer, they must be integrated within the customer's premises. The customer then gains two major benefits.

The first is reduction in cost and complexity of wiring in the premises. This may seem mundane, but it is likely to be a sufficient financial justification in the short term for companies to buy integrated communications products, as they start to come on to the market.

Practically every company has a PABX or telephone key system, with star wiring to the phones within the company's premises. Increasingly companies also have data networks. In previous decades these used to consist exclusively of star connections to front-end processors of mainframes, but they typically now consist of LAN-networked personal computers. The growth rate of networked PC installations in Europe is currently around 30 per cent per year and it has been predicted (Anderson, 1986) that over the next three years the balance of data to voice traffic carried in Europe will alter from its present 17:83 per cent to 43:57 per cent. (The growth rate in telephony is currently very small.) It is clear that using current technology a large number of companies will go from having a single communications system (the PABX) to having two communications systems, side by side.

About half of the cost of installing any communications system into the office is spent in wiring up the building. Therefore there are substantial savings to be made by using the same network, and hence the same wiring, for all communications services.

The second benefit of an integrated system lies in high-level service integration, the vision with which this chapter started. Once all communications are carried over a single network, they can be delivered to a single workstation and combined to offer new integrated services, which save time, save money and make the working environment of the office more pleasant.

In practice, communications services do not have to be carried over a single network in order to be combined in the same workstation. Indeed, this chapter gives various examples of commercially available products where the facilities of a PABX and separately wired LAN-based PCs are combined to offer new possibilities. However, from the point of view of systems design, the fact that all of the traffic is carried on a single network makes organization and implementation of integrated services much easier. As will be discussed below, the cost of implementing the services usually dwarfs the cost of the network and associated hardware: hence techniques that make implementation easier are generally to be welcomed.

We have concentrated on offices in this discussion, for it is in the office that an integrated services network is most likely to be installed—most likely in the size of office that has a PABX rather than a smaller key system. However, one can envisage use of an integrated services system in much smaller businesses. Each small shop or office in a High Street may well have use for the various integrated facilities, although it obviously will not have its own network. It may become attractive to a PTT to install a network within the High Street, rather than providing integrated facilities via star connections to the local exchange.

10.2 The ISLAND project

As noted above, much of the technical discussion within this chapter will be based upon the ISLAND project, which was begun in 1982 at the University of Cambridge Computer Laboratory. With the financial support of British Telecom Research Laboratories, the project set out to investigate the provision of integrated services on a local area network, that is to investigate the various topics which required work in order to realize the vision with which we started the chapter. When work on ISLAND was first begun, it was clear that there were four broad areas which deserved attention.

10.2.1 Infrastructure

Under this heading came the network itself, the terminal equipment by means of which users access integrated facilities and the interface between the network and workstations.

The network itself must be capable of carrying voice, data and video. Preferably it should carry the different types of traffic in their own 'natural modes'. Allocation of circuit-switched channels is suitable for voice connections, for example, but is not well suited to datagrams or other bursty data traffic.

The terminal equipment needed to support the various integrated services needs to incorporate a framestore able to handle images and graphics. It should also encompass or be coupled closely to the telephone, since many of the services that ISLAND was designed to support would require the resources of the two to be coupled together.

Special work was called for on the interface from network to workstation, because of the combination of services that need to be handled simultaneously. The problem is that the characteristics of voice, video and data interfaces have traditionally been very different. For voice and video, circuit-switched interfaces have been used: these are very efficient, because only a small amount of processing is required per byte. In contrast, data protocols require flexibility and hence much more processing per byte, with the result that the network interface has for a long time been the performance limitation in LAN systems. (For example, LANs able to switch around 10 Mbit/s have been available for more than a decade (Metcalfe and Boggs, 1976; Wilkes and Wheeler, 1979), but LAN interfaces have rarely been able to switch more than a tenth of this.) It is clear that an interface that has to be able to handle voice, data and video traffic simultaneously in a flexible manner needs careful design.

10.2.2 A new approach to PABX functions

One of the two main motivations for starting the ISLAND project was a belief that the experience gained from the 'Cambridge Distributed Computing System' (Needham and Herbert, 1982) could usefully be applied to the design of a PABX. The Cambridge Distributed Computing System was an attempt to build a large operating system (of the same complexity as that running on a mainframe computer) using a series of small computers linked by a LAN. Each of the functions within the operating system was given to a single machine, with individual machines serving as access/privilege controller, fileserver, file cache and locking manager, boot server, resource manager, terminal access concentrator, printer driver, spooler and so on. Users' processing requirements were provided for by 'processing servers' in a 'processor bank'. Each user, on request to the resource manager, would be granted a single processing server, accessed through the terminal access concentrator. This processor would behave as the user's personal computer for as long as the user required it. As soon as the usage was finished, the processor would be returned to the bank, ready for allocation to somebody else.

The idea of using a processor bank was shown to be very attractive, and is becoming increasingly so as the number of personal computers installed continues to grow rapidly. The reasons for this have been discussed in other texts (e.g. Ades, 1987).

An equally important lesson that the Cambridge Distributed Computing System showed was that the idea of splitting up a large and complex operating system into a series of modules, each in a separate machine, eased the task of producing and then upgrading the system.

It seemed likely at the inception of the ISLAND work that the same ideas would also make the implementation of a modern PABX rather easier. The separate functions within the PABX, such as conference bridge, voice recording and playback, tone and message generation and so forth, could be split into separate hardware. If this were to ease the tasks of both implementating and upgrading a PABX then considerable savings could be made, for 80 per cent of the cost of developing a typical PABX currently lies in the software. Also the life expectancy of a PABX is now typically 5 years (from figures given in British Telecom internal studies), as against 20 years in the days of mechanical PABXs. While some of the change-out can be attributed to changing sizes of installations, most replacements are due to the desire by a company for additional PABX features and the fact that additional features are very hard to add to an existing design.

10.2.3 Reliability

If a combined voice and data system is to be built out of computer-like objects, as proposed above, then the issue of reliability needs to be addressed. The reason is that user expectations of reliability are very different in voice and data systems. In the case of a PABX, the facilities available are comparatively unsophisticated, but must be highly reliable. Typically a PABX will have some type of 'drop-out' mode, so that in the case of a failure, such as power failure, the very basic telephony functions (outgoing calls, plus incoming calls routed to a fixed set of extensions) are still available. Drop-out is important to provide access, for example, to emergency services.

Expectations of computer systems are generally rather different. The flexibility of services available from a computer installation will be very great, but users have a general expectation that computers can crash and also 'go down' (except in the case of, for example, large tandem machines used for financial services, where a considerable price is paid for the extra reliability). As the ISDN approaches, the problem of providing drop-out will hit most PABXs, which are becoming more and more complex and which make it less easy to incorporate simple drop-out mechanisms.

The reliability of voice services within an integrated system can be addressed by using more than one technique:

1 The basic control functions for voice can be made redundant.
2 Self-monitoring can be used to detect and handle faults.
3 Key components which provide the basic voice service can be made very simple, since the simpler the hardware and software comprising an object the more reliable it will be.
4 Simplifying these components is also likely to imply that the amount of stand-by power required during a power failure will be minimized.

10.2.4 User-level services

The last area of research within the ISLAND programme was aimed at demonstrating various integrated services which could be provided once a multimedia integrated workstation was available. Various examples of services which have been demonstrated within both the ISLAND and other projects will be discussed later in this chapter: as noted above they are in general aimed at making life in the electronic office more efficient, productive and pleasant.

10.3 Design of a network

In order to produce a local area network suitable for both voice and data, able to handle each in its natural mode, the first thing that the designer needs to understand is the very different natures of voice and data.

10.3.1 Characteristics of voice and data

Table 10.1 lists the key factors that distinguish voice and data. The amount of delay in a voice connection is very important. Considering a connection between two people, in the absence of echo, there is a length of 'round trip' delay below which the parties will not be bothered. This is given by various authorities as between 100 and 125 ms (Ades, 1987). Above this level the parties may find themselves repeatedly talking simultaneously and then backing off. The amounts of delay tolerated in the presence of echoes are less straightforward. Echoes with very short delays will be confused with sidetone and certain delay/echo combinations can induce speech impediments.

Table 10.1 Characteristics of voice and data

Voice	Data
Low delay across connection vital	Delay usually unimportant (except for control applications)
High error rates are acceptable (10^{-5} random error rate, bursts up to 2 ms long)	Any undetected errors unacceptable
Requires constant bandwidth allocation (during activity only, if silence detection used)	Bursty traffic, requiring unpredictable short periods of high bandwidth
Only feasible to set up as many connections as system can offer adequate bandwidth	Little restriction on number of connections allowed simultaneously
Each connection must be given the bandwidth it requires until the user closes the connection	During periods of high offered load, each connection runs slowly

All of the above applies to a connection between two speakers, the connection considered as a whole. This connection may typically consist of several components: trunk lines, satellite link, mobile phone connections, PABXs and LANs. The 'delay budget' for the whole connection must be shared between them all and naturally the PABX or LAN receives a very small amount of the quota. For example, at present in the United Kingdom (UK Office of Telecommunications, 1986) a delay budget of 10 ms round trip is allowed for 'private networks', that is any system of LANs, PABXs etc., entirely owned by a customer.

Although voice connections set exacting delay requirements, they can tolerate a high error rate. An oft-quoted figure is 0.1 per cent (Bullington and Fraser, 1959), which applies to burst errors due to TASI 'freeze-out'. A more generally applicable figure, given by two more recent studies (CCITT, 1974; CCIR, 1976), is 10^{-5} random errors under low sound level conditions. However, in the case of burst errors, it has been found that periods of speech of up to 2 ms length can be omitted for up to 1 per cent of the time without noticeable effect (Gruber and Strawczynski, 1983). (This assumes that the bursts are random. The ear is very sensitive to noise that forms a coherent pattern. In the course of ISLAND experiments it was observed that the ear would very definitely detect a 2 ms burst of lost samples which occurred once every two seconds.)

Voice connections in general call for a fixed amount of bandwidth. In the simplest of systems, a voice connection is allocated bandwidth all the time, even though there may not be any information flowing. On a connection where the cost of bandwidth is very high, such as a satellite link or undersea cable, some use is usually made of 'silence detection'; that is when the signal energy level of a call drops below a certain level then the channel bandwidth for that call is deallocated. Typically, in a fully duplex connection carrying a voice call only 47 per cent of the bandwidth is actually used (Bullington and Fraser, 1959). If a number of voice calls have to be multiplexed onto a set number of channels, then as that number rises the ratio of calls that can be carried, without impairment, to the number of channels available (the so-called TASI advantage) rises.

The TASI advantage is only useful where a significant number of calls are being carried, that is where the probability of bandwidth being available as soon as the call becomes active is sufficiently high. In the case where only one or two calls are being carried, for example on a mixed voice/data network, then the bandwidth available when a call becomes inactive can only be used by other traffic if that traffic is immediately preemptible as soon as the voice call becomes active again. Otherwise voice quality will be severely impaired.

The final characteristic of voice, which is of interest in this section, is that once a call is allowed to start by the network, it must be given full resources until it completes. In other words, it is perfectly normal (although irritating) to be told that 'all lines to London are engaged' and to 'please try again later', but once a call has been set up the caller expects it to be allocated sufficient bandwidth until the caller (*not* the network) decides to terminate the call.

If we now turn to consider data traffic, it will become clear that all of the properties of voice set out above are exactly reversed for data.

Firstly, data is not critically affected by delivery delays. Suppose that you are logged on to a remote computer via a network and are using a screen editor to edit some text. When scrolling to a new page, you may be annoyed if the screen refresh happens more slowly at some times than others, but it does not affect comprehension of the data. (Many types of data traffic—file transfers, electronic mail deliveries—are even less affected by delays.) In contrast, if some voice samples in a voice stream are delivered later than others then this can severely affect the intelligibility of the conversation, producing pitch variations, clicks or other artefacts.

Although data can tolerate delay, it cannot tolerate errors: data link protocols are designed to make the probability of undetected errors getting through negligible. Clearly a single bit error introduced into a piece of software as it is transported across a network may have disastrous consequences and nobody would buy a text editor that offered a probability of 10^{-5} that each text character would be written to disk wrongly.

The traffic pattern of a data connection will generally be bursty. Considering a link between a terminal and a mainframe, or from a personal computer sending database queries to a remote machine, or a fileserver supporting file access across a LAN for a number of personal computers, then for most of the time that a connection is set up it will be idle. Occasionally there will be a burst of activity. At that time one would like to offer all of the available network bandwidth to that connection. Supposing several personal computers are all executing commands from a LAN-based fileserver, then we would like to load one command into the first as fast as possible, then make the bandwidth available for the second, and so on. In this way each user waits the smallest amount of time for his command to start executing.

It may of course happen that all of the data connections set up on the network suddenly become active simultaneously. In that case they will obtain only a small share of the bandwidth, and the system will appear to be 'going slowly'. This is unavoidable. It is clearly sensible to allow many connections to be set up on the basis that normally most of them will be idle and those that are active will see a very good network response: it is not too serious that occasionally a high proportion become active together and see a much poorer response.

Again this is quite the opposite of voice, where we may restrict calls in progress to those that we can handle all together; once set up, a

INTEGRATED SERVICES ON LOCAL NETWORKS

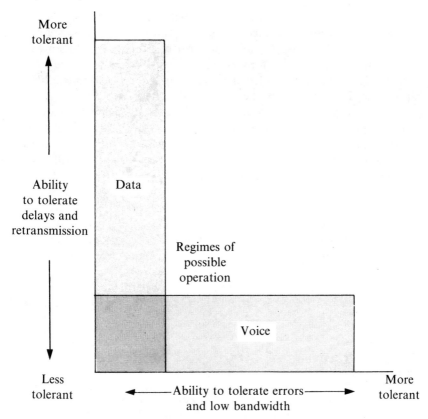

Figure 10.1 Voice and data tolerance of network characteristics

voice call requires only a steady fixed level of bandwidth but that bandwidth must be available on demand.

The above discussion on voice and data characteristics is summed up in Fig. 10.1, which plots out on one axis the ability of a traffic type to tolerate network delays and information retransmission (which of course itself introduces delays) and on the other the ability to tolerate errors and to work with low available bandwidth (low here should be taken in the context of a local area network). On each axis, as we move towards the origin it becomes harder for the network provider to offer the required service. It should be clear from the figure and from the above discussion that voice and data impose exactly opposing requirements on the network.

Before moving on to look at networks which are able to handle both data and voice, it is appropriate to comment on video traffic, for video on the network is as much a part of our vision as traditional voice or data traffic. The area of Fig. 10.1 that the video will occupy varies very much according to the way in which it is handled. The first digital video CODECs simply sampled analogue video signals using a linear convertor and generated a data rate of around 140 Mbit/s for broadcast-quality colour signals. The error rate that can be tolerated on a channel carrying this form of video is very high, since there is so much redundancy in the video signal, and much degradation can be applied to the signal before the eye is able to notice.

Because of the high data rate and redundancy of such a simple video signal, a great deal of effort has been put into video compression (see, for example, papers on image compression published monthly in *IEEE Transactions on Communications*). The compression achieved will depend on many things: whether lossless compression is required or whether degradation of the images is acceptable up to some subjective level, whether interframe coding or motion detection is employed, whether a transform coding is used and so on. Regardless of the method used, one comment normally remains true: as the amount of compression used is increased and the bandwidth required falls, the error rate becomes more critical. For example, where predictive coding of some kind is used, within a frame and more particularly between frames, a single error can propagate and affect the reconstructed image for a considerable time.

10.3.2 Examples of integrated services local networks

Most of the integrated local networks that have been developed to date have been packet networks, and it is likely that this will continue. This is in contrast to the development of the ISDN for wide area communications, where voice and data are both offered mainly circuit-switched bandwidths. Bearing in mind the bursty nature of data traffic, it can be seen that circuit switching is hardly ideal. The reasons for the choice are not technical. The ISDN is the offering of the PTTs and the PTTs already have very large circuit-switching systems from which they wish to generate revenue from data traffic as well as voice traffic. Also, since data traffic is still in the minority, the network needs to be designed primarily for voice. (This assumption has always been asserted since the beginning of the ISDN, but given the three-year prediction already cited in Anderson (1986) it may not be such a good assumption by the time that the ISDN is seen in widespread service.)

In the case of the office environment, the voice/data balance is rather different. As far back as 1983 a study by Sincoskie and others at Bell Communications Research, Murray Hill, N.J., on LAN and conventional PABX traffic in a business premises showed that in the 'electronic office' there is more data than voice traffic. Such offices will become more and more commonplace as the number of networked PC installations grows.

Since data is so poorly suited to circuit switching, research has

INTEGRATED SERVICES ON LOCAL NETWORKS

always been concentrated on packet networks and in particular on LANs which could offer characteristics suitable to both voice and data. LANs in general support data traffic well, and data transport on LANs is well understood. The work on integrated services LANs has therefore mainly been concerned with producing acceptable delay and bandwidth sharing characteristics for voice.

Chapter 9 has already described several notable examples of integrated LANs (Sec. 9.7.2), each of which can be described as 'hybrid'; that is the packets are divided into classes, with some amount of bandwidth allocated for each class. These hybrids represent a spectrum of designs, from those that look like a traditional PABX to those that look like a computing LAN, and are shown grouped accordingly in Table 10.2.

Table 10.2 Spectrum of network designs

LAN-like	Mixed	Switch-like
Magnet (slotted ring)	Philan, FXnet (fixed allocation and slotted ring)	Prelude (circuit switch)
Fasnet, Expressnet, BID (pair of unidirectional buses)	QPSX/DQDB (reserved and contention bus pair)	SCPS (fixed allocation ring)
Orwell (destination-delete ring)	FDDI-2 (fixed allocation and token passing)	
Other examples: MSTDM (CSMA/CD variant with collision preemption by voice)	IBM token ring (can generate periodic bursts of isochronous packets)	

Magnet, Fasnet, Expressnet and BID all achieve variable allocation of bandwidth between voice and data. They guarantee enough voice slots in a fixed time to offer guaranteed delay characteristics. However, if not all voice connections actually require slots (e.g. because a call is silent) then these LANs can change the balance and allocate more data slots. This can be done on a demand basis in each fixed time period without violating the critical time requirements for voice traffic.

The approach used in the Orwell slotted ring, that of 'counters and epochs', is rather different, reflecting the fact that Orwell was designed for very large voice loads which are heavily unbalanced between nodes. The Orwell node logic is one of the more complex in existence.

At the opposite end of the spectrum from Magnet and Fasnet, Prelude and SCPS have hardware architectures very much reminiscent of a digital circuit switch, although based on packets. SCPS uses a synchronous ring structure as a data highway between conventional concentrator switches.

In between the extremes, QPSX looks in outline like Fasnet, but at a lower level it is a circuit-switching/packet-switching hybrid, with slot reservation for isochronous traffic and Orwell-like sharing of unreserved slots for data. QPSX (known by the standards committees as DQDB) is as yet not fully developed—of the many implementation options, none has yet been finalized. Also, some aspects of the specification (such as the circuitry required for resilience in the face of line breaks) seem not yet to have been thought through fully. The same comments can also be applied to FDDI-2, which is jostling with DQDB to be adopted as the high-speed mixed traffic LAN/MAN standard. FDDI-1 is a high-speed LAN based on token passing—similar to the IBM token ring but with a low-level protocol modified for higher throughput. FDDI-2 attempts to combine the FDDI-1 structure with isochronous slots, to produce a mixed packet/circuit-switched system.

Further details on these and other hybrid LANs are to be found in Chapter 9. The next section will describe an integrated services LAN which has no formal hybrid structure.

10.3.3 The Cambridge fast ring

The LAN used for the ISLAND project was the Cambridge fast ring (Hopper and Needham, 1986). The fact that the Cambridge fast ring appeared to be very attractive for integrated services was the second major reason why the ISLAND project came into being.

The Cambridge fast ring is derived for the 10 Mbit/s Cambridge ring (Wilkes and Wheeler, 1979), the original slotted ring LAN. While the fast ring supports voice/data mixed traffic well, it has no hybrid bandwidth reservation scheme or any other mechanism specifically provided for voice/data support. For this reason, it is much simpler than the networks discussed above.

The fast ring exists as a VLSI implementation, which at the time of writing runs reliably at up to 75 Mbit/s. Each slot accommodates 32 bytes of information and each node on a ring has a 16 bit address. Most of the complexity of the ring is contained within a single CMOS IC, which can function as a station, a bridge (using two ICs back to back on different rings) or as the 'monitor station' which sets up and maintains the slot structure of the ring. The bridge mode is designed to support a mesh of linked rings, with a flat 16 bit address space across the whole mesh.

The principle of a slotted ring is fairly simple. A fixed-format train of slots circulates indefinitely and transmission is achieved by waiting for an empty slot, inserting into it a destination address and some information and awaiting its return with a marker indicating whether the data has been accepted by the destination. On return the slot is marked empty and passed on.

The properties of slotted rings have been discussed extensively in the literature (Temple, 1984). Those that make the fast ring suitable as an integrated services carrier all stem from its bandwidth sharing mechanism. In particular, since each station must release a slot to the next node in the ring after using it, fair bandwidth sharing is guaranteed among requesters and occurs at a fine granularity (32 bytes).

For mixed traffic on the fast ring we can split the features that make it attractive and unusual into three parts:

1   Where n stations all request maximum available bandwidth they each obtain an nth share of the total. While this is not unusual for LANs, there are many LANs, such as CSMA types, whose total throughput falls as the offered load increases. This is not the case for the fast ring.

2   The fine granularity of bandwidth sharing, bandwidth being allocated in 32 byte slots. On a Cambridge ring, it is perfectly normal for some stations to be transmitting very large blocks and some very small. It will be seen in the next section that this is of great use in the ISLAND design. In contrast, on many LANs, including CSMA and some token-passing types, the use of very small data packets will drastically reduce the overall available bandwidth. This is simply because the 'guard bands' between transmissions will dominate the transmissions themselves. Worse still, the actual available bandwidth will remain constant even if the LAN signalling rate is increased, so long as small packets are used. This is because the guard band is a function of the physical length of the network and not of the signalling rate.

3   Where p stations request small amounts of bandwidth and q stations request the maximum available, then the p stations' requests will be satisfied in a very short time and the q stations will then share the total bandwidth remaining after deduction of that required by the p stations. This clearly suits a voice/data traffic mix very well. The p stations represent the telephony traffic, which requires a very short network access delay but whose bandwidth requirements are very small indeed compared with the total

available. On a 75 Mbit/s fast ring of length 1 km and with 10 stations the access delay to obtain a single packet cannot exceed 70 μs. (For a 20 or 50 station ring the figures will be typically 70 and 210 μs respectively.) The q stations represent bursty data traffic. At any time only a few of them will be requesting bandwidth and the total remaining bandwidth will be divided between them.

All of these properties can also be achieved using hybrid networks, but as remarked above the various hybrid networks discussed are rather more complex than a fast ring. The fast ring does have limitations compared with some other networks; for example Magnet and Orwell are able to guarantee individual stations very high levels of isochronous bandwidth. The Cambridge ring can only guarantee each node its fair share. However, for a 50 node 75 Mbit/s ring, this gives 1.2 Mbit/s of data throughput per station. As will be seen below, most of the nodes in the ISLAND architecture are only required to handle a single voice stream. For those that need to handle several at once (translators and conference servers) a 50 node ring sets the limit at 21 streams, adequate for even large PABXs. 1.2 Mbit/s places some limit on the types of video traffic that can be sent. However, discrete cosine transform-based full frame-rate CODECs operating at around 1.2 Mbit/s can be expected to become available in silicon within the next two years, whereupon the fast ring will be able to act as a satisfactory bearer of most video-based services.

10.4 Providing office voice facilities

Having discussed the network adopted for ISLAND, we now move on to outline how the basic functions of a PABX, as well as those functions to support voice recording and replay, were provided using a distributed computing style of design.

10.4.1 Telephones

The basic ISLAND telephone was deliberately designed as a very cheap and simple object. The phone is capable of notifying an external agent when the handset is picked up, of relaying keystrokes, and then receiving instructions to 'talk to x and listen to y', where x and y are network nodes; x and y may be identical or one may be null. This is the limit of the phone's capabilities. The functionality is not much different from that of a plain analogue phone.

The ISLAND design rejected the type of approach where the phone is able to:

1 Look up the name or number of a desired station to find its network address;

INTEGRATED SERVICES ON LOCAL NETWORKS 337

2 Form speech samples into different lengths of block and run different types of protocol, according to whether the samples are going to another local phone, a fileserver or a bridge to another type of network;

3 Perform the compounding operations associated with conference calls.

This is the approach taken by many researchers whose backgrounds are in distributed computing. It is very much along the lines of, for example, the Etherphone (Swinehart, Stewart and Ornstein, 1983), built for voice experimentation over Ethernet, at Xerox PARC. The ISLAND design also rejected the ISDN phone approach, where the human interface to many of the network facilities is built into phone hardware and the phone understands and negotiates about a variety of network facilities.

These are approaches that were rejected deliberately. The reason is that it is not easy to predict how the functions required within the phone will alter during its life and it is hence undesirable to fix the features available from the system by the design of the phone firmware. Many find it attractive to design PABXs with some of their features implemented within the actual phone instruments, and ISDN interface specifications to an extent promote this. Not only does this mean that full interworking with standard sets is impossible, but it also means that 'new' features of the PABX may require that the existing phones be discarded. The complexity of phones normally supplied by PTTs to small customers to provide particular functions (manager/secretary, etc.) is considerable. These phones are entirely inextensible and therefore either bind to the user to a particular set of facilities or have a very short installed life.

From the above description, there must be more to ISLAND voice facilities than just phones. We will discuss in Sec. 10.6 how more advanced telephony features are provided within the ISLAND design. First, in the next section, we will describe how the phone communicates with other nodes.

10.4.2 A packet voice protocol

Once a phone has received an instruction to send voice to some other node on the LAN, it starts up the ISLAND voice protocol. This protocol is very lightweight. It consists of sending a sequence of 2 ms voice packets, each occupying a single slot on the fast ring. The packets each contain a sequence number and a silence detection flag. At the destination, supposing it to be another telephone, the packets are put into a circular buffer, from which they are then played via the earpiece.

The protocol has no associated flow control and there is no retransmission of packets lost or received with bad checksums. This is for two reasons: firstly, it makes the protocol very cheap and simple to implement and, secondly, such protocol techniques are inappropriate for voice within a LAN environment. They are the characteristic features of data transfer protocols, which have very different aims. The contrast is illustrated in Table 10.3.

Data stream protocols (e.g. X.25, TCP) aim to transfer a fixed quantity of data exactly once without errors on top of a physical channel that has some noise and hence error rate associated with it, through a network and to a destination both of which have a limit to the amount of data they can accept. Our voice protocol has a very different aim. The most important thing is that if voice packets are delivered at all then they must be delivered within a critical time. If they arrive later than that time they are of no use, since they are too late to be played back. In other words, packets that arrive late are exactly as useful as packets that do not arrive at all.

Flow control is not of interest, for the destination phone's CODEC can clearly accept any amount of data sent to it at its stream rate. The network must also be able to handle the voice traffic, otherwise it would not be acceptable as an integrated service network.

Retransmission of packets in a standard data protocol occurs due either to errors or to a packet being completely lost. It introduces a step function delay, since either a retransmission request has to pass back to the transmitter or the transmitter has to time out the awaited acknowledgement and retransmit. Thus additional delay is introduced.

The retransmission would achieve comparatively little on a voice stream, even if it could arrive in time. The retransmission will have occurred either due to a checksum error or due to a packet loss. It was noted above that the ear does not notice random errors in speech provided that the rate is better than 10^{-5}. The typical error rate of a Cambridge ring is better than 10^{-10} (Dallas, 1980), so that it is better to ignore checksum errors and play back voice samples even in the presence of errors.

If a single packet goes completely missing, this too is of little consequence. The ISLAND 2 ms length for voice packets was chosen for two reasons. Clearly the longer the packet the longer the transmission delay becomes (due to packetization delay) and 2 ms is an acceptably low figure. Equally significantly, it was noted above that the ear will notice up to 1 per cent of random burst errors up to 2 ms in length. This means that the 2 ms packet is effectively of a length such that we hardly care whether all packets reach their destination or not.

This is a very convenient state of affairs, for it is very much easier to design a voice protocol for real-time delivery of packets, in which it does not particularly matter whether all packets get delivered, than it is to design a real-time packet network which guarantees that they all do get there. As discussed above, the Cambridge fast ring is well suited to delivering small packets and therefore this lightweight protocol can

INTEGRATED SERVICES ON LOCAL NETWORKS 339

Table 10.3

Attribute	Packet data protocol	Packet voice protocol
Aim	To transfer a fixed quantity of data accurately and exactly once	To deliver a bit stream with low delay and acceptable bit error/loss rate
Flow control	Protects destination and network against overload	Not useful: destination and network must be able to handle the stream
Error detection	Vital for data integrity	Pointless unless sufficient errors for the ear to detect
Retransmission	Used to resend data following packet loss or corruption	Adds delay and causes packets to be delivered too late to be of use
Sequencing	Vital for data integrity	Useful to maintain correct relative timing of bit stream upon packet losses

easily be implemented without problems of diminishing the available network bandwidth.

There is, of course, a small element of variable delay in the delivery of packets across the ring, since there is no prereserved bandwidth in the LAN and the time to obtain an empty slot varies. This variation in access time was given above and is very small. It can be accommodated by delaying playback of the first voice sample sent along a connection by a small amount; this builds up a small reserve such that packets which from time to time are delayed more than the first packet will arrive in time to be played back.

The purpose of including a sequence number in the protocol is to ensure smooth running in the event of a lost packet. If a packet goes astray, when the next packet arrives at the receiver the sequence number will enable the receiver to register the omission and hence leave an appropriate gap in the buffer, to maintain correct timing.

Practical implementation of the ISLAND voice protocol uses a driver known as the 'ISLAND voice provider'. This has a circular buffer at both transmitting and receiving stations, as discussed above. It is designed to be as lightweight as possible. In a normal data protocol, the driver would make a call to the application which is consuming the data, whenever a new packet has arrived. In the ISLAND voice protocol, the application and the voice make no explicit calls to each other but communicate using pointers into the circular buffer.

In the context of a telephone, this makes for a very simple protocol, which can be implemented in an IC of comparatively little complexity. In the context of an operating system it makes it possible for a normal processor to handle the voice stream. Suppose voice is being sent from a phone to a computer, to be stored on disk. Then it matters little if the packets are delayed before being stored, since this is not a real-time connection. What does matter is that the average operating system would be totally unable to cope with a stream of voice packets which arrive every 2 ms. (If it is considered that the task switching time of a typical UNIX implementation is around 10 ms, the problem is clear.) Therefore what is required is that at a low level in the network protocol, voice and data packets are separated. Data packets are handled by a standard link layer protocol. Voice packets are handled by the voice provider, which places them in the correct position within a circular buffer, according to the sequence number. Since the connection is not a real-time one, the circular buffer can be large. Every so often a high-level task within the computer looks at the position of the pointers and when appropriate transfers a large block of samples to disc.

10.4.3 Building up a PABX

The above description covered how phones interact with other phones and discussed the protocol interaction between phones and other objects. In this section we will discuss how the ISLAND design provides for some of the services expected of an integrated PABX.

The first important design principle has already been discussed, that of removing all intelligence from the phone. Change in status of the phone handset or keystrokes on the phone's keypad are notified to some external object, the call controller. After a meaningful number has been dialled, the controller instructs the phone to 'talk to x, listen to y', as discussed above.

This set-up is sufficient for simple phone-to-phone calls. From the phone's viewpoint it is sufficient also for holding conference calls, for conference calls are handled in the ISLAND system in a way that avoids the phones needing to know what they are talking to. A 'conference server' is provided, to which all of the parties in a conference send their voice streams. The conference server then combines all of the streams and sends back to each party the sum of the inputs on all of the other streams. In the ISLAND conference server implementation, 'summing' is performed by a dedicated Motorola 68000 processor. Summing does not use simple addition instructions, for the voice samples sent around the fast ring are in the A-law PCM format, which is a logarithmic representation of amplitude. The fastest way to perform the 'summing' is therefore by table look-up, with an addition table 64-kbytes in size. The ring interface of the conference server runs the voice provider, its 'client' being the dedicated 68000. This arrangement illustrates another feature of the voice provider: because the voice provider and its client are very loosely coupled, it is possible for them to run in different processors which share memory, with a minimum communication overhead. (A single IC is now commercially available that replaces the client 68000 and its look-up tables; due to the separation of functions from the voice provider, it is very easy to use in this IC as a 'slot-in' replacement.)

An important facility provided by the ISLAND system, as the basis for many integrated applications, is a voice recording/playback service. This calls for a fileserver. The voice provider enables voice and data packets to be handled within a common computer operating system, but this is not in practice sufficient for the fileserver. The types of integrated applications that we wish to support using voice recording/playback entail combining voice with other media, and so it is convenient to use a conventional fileserver. The network interface of a fileserver is generally fine-tuned for the task of transferring large blocks of data to clients that request them. This interface is likely to perform very much less well if combined with a heavy load of voice packets to be transmitted and received within critical times. There are several ways around this problem.

One could design the phone to send short packets to other phones and longer packets to fileservers. This is undesirable for two reasons. It increases the complexity of the phones and it means that the phone needs to know what it is talking to, rather than just knowing where voice samples are to be sent. The latter is important: we will attempt to show that there is much to be gained by avoiding the individual elements of the system knowing much about each other.

One could implement a special voice fileserver, which stores voice in a different environment from data. This approach has been adopted within, for example, the Universe (Adams and Ades, 1985) and Etherphone (Swinehart, Stewart and Ornstein, 1983) projects. However, when applications such as multimedia documents are considered, much implementation complexity is saved by holding the whole document on one fileserver—the whole document including its voice and data components.

The solution adopted in the ISLAND system is a machine called the 'translator'. This is a machine that runs a voice provider for each connection to a telephone, plus a data protocol for interaction with the fileserver. When voice is to be recorded, the samples build up in the provider's buffer, until sufficient of them have been collected to be sent to the fileserver in a large block. When voice is to be played back, the reverse happens, with large blocks being retrieved from the fileserver and placed in the buffer, to be sent out 2 ms at a time by the provider. Provision of the translator realizes the two aims of storing voice on a general purpose fileserver and keeping the phone simple.

10.4.4 Universal connections

What has been built up above, shown in Fig. 10.2, is a system of components in which only the very minimum number of interfaces exists between the different ISLAND components. All real-time voice connections are identical, regardless of whether between a phone and a

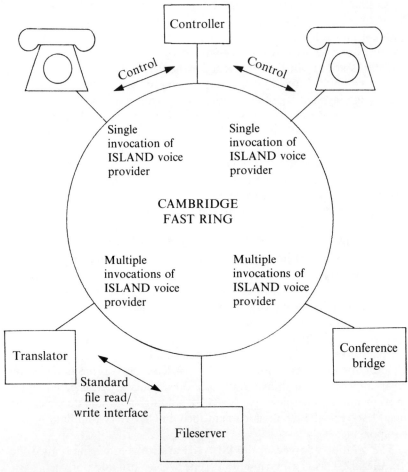

Figure 10.2 System components in ISLAND

phone, a phone and the conference server, a phone and the translator, or the conference server and the translator. The interface between the translator and the fileserver is the standard fileserver interface (read/write n bytes of file f starting from byte b).

Keeping the interfaces as general as possible and avoiding any knowledge within each component of what it is connected to has very significant benefits when it comes to adding new features to the system, as well as when different system configurations are required at different installations. Suppose, for example, that some installations make only very moderate use of voice filing, but at particular sites very large quantities of voice are required to be held on line. At these sites, it will be cost effective to provide a voice compressor between the phones and the file server, since this will cost much less than the additional disk drives that would otherwise be required.

Since none of the ISLAND components knows what it is connected to, it is easy to place a voice compressor between, say, the translator and the fileserver. (It is best placed there rather than between phone and translator because nearly all high compression ratio algorithms, for example the various predictive schemes, operate on blocks of speech rather longer than 2 ms.) The voice compressor receives long blocks of speech from the translator at some frequency and outputs equal-sized blocks to the fileserver rather less often.

As remarked above, most of the cost of developing a PABX lies in the software. By making the PABX out of a series of standard connections we reduce the complexity of adding new features and, most significantly, of getting the features to interact correctly. It is the interactions between features that typically cause the development of a PABX to overrun time-scales and budgets; it is also the difficulty of redesigning the interactions that makes an existing PABX so costly to upgrade.

10.4.5 Wide area connections

The last area of connectivity that will be considered in this section is connection to external and wide area systems.

Clearly a gateway is required to connect an ISLAND PABX to the PTTs' circuit-switched networks, to convert between ISLAND packet streams and either analogue circuits or 64 kbit/s digital channels. However, in the future there will be various packet-switched MANs and WANs, to which an ISLAND-style system will interface. This raises the question of whether to install a gateway or whether to use a single end-to-end protocol between two phones in separate ISLAND installations, when connected together across a wide area network. Some researchers have tried the latter approach (see Adams and Ades, 1985), but the design philosophy of ISLAND calls for the use of gateways. Just as in the case of the fileserver–phone connection, there should not be different protocols running on two connected phones according to whether they are within the same installation or not. It is desirable that each phone treats a connection to a WAN as identical to a connection to another local phone. Experiments have shown (Ades, 1987) that the protocols required for running over the whole of a LAN–WAN–LAN link need to be rather more heavyweight than those for a use within a LAN only—there will be a need for error control and other mechanisms that make the wide area protocol more complex to implement than the simple ISLAND protocol. Also the delays will be greater. Therefore the very simple protocol is used exclusively within the boundaries of ISLAND, with protocol translation at a gateway to interface to more harsh network environments.

There are other benefits to using the gateway approach. Over the wide area connection, the cost of bandwidth will be important. Therefore suppression of silence and use of TASI statistics will bring a significant benefit. This is particularly true of an undersea cable or a satellite link. In contrast, within the local area network bandwidth is comparatively cheap; what is much more important is the complexity of the phones and their interface to the LAN. Silence detection is fairly complex to implement satisfactorily: it requires a number of parameters to be measured in order to provide reliable detection. There are also problems of intermittent background noise, such as how to stop the speech from being masked, and the listener irritated, by background noise suddenly cutting in at the end of a silent period. (Silence detection has been used for a long time on submarine cables without any such problem, simply because the connections were so noisy all the time. In a local environment, however, where a caller in a quiet room phones a machine room, the contrast between silence and sound can be quite unpleasant to listen to, as well as having an effect on intelligibility.) The addition of a voice compression server and a wide area gateway to the ring is shown in Fig. 10.3.

10.5 Control and reliability

So far we have not considered the control of the PABX or how the very many features that are expected from a modern PABX, such as call forwarding, are to be implemented. This is deliberate, for in designing a PABX it is beneficial to split the system into two parts. The low level provides an environment in which any phone can be connected to any other phone or to any resource in as general way as possible. This is what was attempted in the previous section. By making the connection mechanism as general as possible we facilitate the design of the high-level part of the PABX. This part, which one may term a 'feature processor', is responsible for knowing the current state of the PABX. For example, certain phones are in use. Others have

INTEGRATED SERVICES ON LOCAL NETWORKS 345

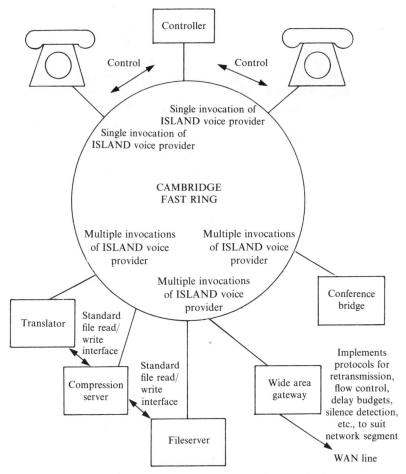

Figure 10.3 Voice compression server and wide area gateway

been 'forwarded' or transferred to another extension. Other phones cannot receive incoming calls or may only be able to make outgoing calls within a local area. Particular extension numbers do not actually correspond to a single phone, but a 'hunt group' of extensions corresponding, for example, to all staff able to handle sales enquiries. A moderate-sized PABX has a long list of such features and therefore a substantial amount of configuration information. (Although the average PABX user cannot remember how to use more than one or two of such features, PABX buyers will usually not consider a system without them!) When an internal or external caller wants to make a call, much of this configuration information needs to be interrogated. Eventually, out of the feature processor drops a command to connect two entities together, that is a command to the low-level portion of the PABX system.

We will not discuss the way in which the software in the feature processor is to be developed here. It is important to realize that the provision of standard connections, as discussed in Sec. 10.4.4 above, makes the task easier but does not solve all the problems. Although the reader will appreciate that the feature processor is a complex entity in itself, it is similar to that in any other PABX. What is relevant here is the reliability of a PABX system built using ISLAND techniques.

The question of reliability is an important one. It was noted above that much importance is attached to the maintenance of a basic level of service in a PABX. When an office catches fire and the local power fails it is important that a sufficient telephone service is available at least in order to call the fire brigade. Traditionally the same requirement for a fallback service has not been found in computing systems: users accept that the system 'goes down' from time to time and that they can always call the computer room to ask for a reboot. Clearly, if a PABX is built out of distributed computing components then the question of reliability needs to be addressed.

A basic level of telephony is usually provided in PABXs using 'drop-out', which means simply that in the event of a power failure all of the incoming lines from the town exchange are connected to selected internal telephones. This is accomplished using relays which upon losing power drop out and connect the stand-by lines.

It is clear that no such trick can be used in an ISLAND environment, for the ISLAND telephone and an exchange line are very different in both protocol and electrical interface. This may therefore seem to be a problem which we have introduced for ourselves in designing an ISLAND-style system. In fact, however, it is a wider problem than this. The new generation of town exchanges currently being installed by PTTs will interface to large PABXs via digital trunk links, carrying 30 calls on a 2 Mbit/s line (at least within Europe). The telephones attached to the PABX may be analogue or they may be digital, for example according to the ISDN 2B+D reference point. In either case, we can no longer expect drop-out to be provided by simple relays since the phones are not at all compatible with the trunk lines. If a basic level of service is still to be provided it will be achieved by keeping a set of basic components available even in the event of power failure or an individual component failure.

This can be achieved by two means: firstly by careful design of core components and secondly by replicating the feature processor.

The ISLAND telephones were designed to be as simple as possible. This was partly to reduce the likelihood of their failing. The simpler the hardware, the less likely it is to fail; the smaller the amount of software, the less likely it is that there are bugs in it.

The implementation of ISLAND uses common hardware modules for as many of the PABX components as possible. (Conference server, translator, voice compressor, etc., can all run in identical processors

INTEGRATED SERVICES ON LOCAL NETWORKS 347

attached to the LAN.) Enough hardware can be provided to run all the components and have a spare processor or two ready in case of failure. In the event of a failure, the module is reloaded into a stand-by machine and all the relevant traffic is redirected to it.

This requires some means to detect failed servers. In the ISLAND design this is the responsibility of the feature processor, although individual machines may in certain circumstance be able to notify the feature processor of their or another machine's malfunction.

Replication of the feature processor is a topic in itself. The ISLAND implementation of the reliable PABX controller is described in some detail in Want (1988). A few notes will suffice here.

Clearly the feature processor needs to be run in replicated hardware if high integrity is to be achieved. This can be done by:

1 Simple redundancy, with each processor computing the same functions and a voting mechanism to handle disagreements.

2 An active processor and 'hot stand-by' machines. The active processor may be a fixed machine until a failure or it may be allocated from the pool on a rotating basis.

3 Load sharing and state sharing between processors, which handle different call connection requests in parallel.

The weaknesses in the second option are twofold. Firstly, a stand-by machine needs to pick up the state of the system upon failure of the active processor. This may imply delays if, for example, the data is backed up on a fileserver. For a PABX, an outage of facilities upon failure of the active processor diminishes the quality of service significantly. Secondly, a 'hot stand-by' is not being exercised fully while it is standing by: the time to discover that the stand-by machine has itself failed is preferably not the time when we wish it to take over from a failed machine!

Either simple redundancy or load sharing will overcome these two weaknesses. Each processor is active all the time and hence failures are not likely to lurk unnoticed. Also, each processor holds the current state of the system and therefore there is no service outage when one processor fails.

The simple redundancy is more straightforward to implement, but a load-sharing approach makes better use of the available computing power. In order to maintain a smooth service in the event of a failure, the processors in the load-sharing scheme must pass current state to one another regularly. In the ISLAND implementation the processors also pass around a 'dead man's handle', so that failed machines can quickly be detected and their duties be reallocated. Calls that were in the process of being set up may be lost, but otherwise the users of the system should not notice any effect of a failure.

The ISLAND experimental work concentrated on failure of

individual components; an equally important thing for the PABX designer is the failure of power. Typically, upon loss of power a PABX should be able to provide just basic internal, incoming and outgoing calls. No other PABX features need necessarily be provided (although many PABXs provide 'hot line' facilities and it is desirable that these also be available.) The other highly desirable feature is that calls in progress are not disconnected.

Stand-by power is clearly needed to maintain basic service in an ISLAND system (just as it is for a PABX interfacing to a digital trunk from the town exchange). The aim is to minimize the number of components that need to be powered. These will be the LAN, the phones, certain gateways and a primitive version of the feature processor.

Note that even complete loss of the normal feature processor functions will never lead to a call in progress being dropped, for the phones merely continue talking to the same party until otherwise instructed. The LAN provides the switching required during a call and calls in progress do not need any active control. A typical way to handle loss of power is to provide one 'knocked-down' feature processor which contains a small amount of non-volatile storage (such as EEPROM). In this is stored the data needed to provide basic service—the mappings of telephone numbers onto ring addresses and a few other things. Upon a power failure, the knocked-down feature processor reads this data and then polls each phone to find out what it is currently connected to. This gives enough information to continue to run the basic phone service.

10.6 High-level integrated applications

As stated at the beginning of this chapter, a principal motivation for investigating integrated services on a LAN was to make available new applications that take advantage of the integrated environment. This chapter therefore concludes by discussing some applications that have been implemented using ISLAND and various other systems.

10.6.1 A better PABX interface

The weakest aspect of most PABXs is the user interface. The user is limited to 12 or so buttons for input and a few progress tones for response from the system, in order to access some very sophisticated facilities. The average user does not read the large instruction book which explains how to harness the power of the PABX—if users did they would not remember the sequence of obscure commands required anyhow. Most users can remember how to use less than half a dozen features, with forwarding a call and short-code dialling the most usually memorized.

With the gradual appearance of a personal computer per desk, a much more obvious interface is now becoming available. A surprisingly high proportion of office workers are as capable of using a PC keyboard as a telephone keypad. (The user interface for Prestel was designed on the assumption that the average user could not handle a keyboard; for many users this is no longer true and that interface now seems extremely clumsy.) It is therefore possible to replace short-code dialing with a 'phone<name>' command. There is no need for the user to remember telephone numbers: these can be held in local directories—both PABX-wide, to replace the standard short-code dialling facility, and personal to the individual user.

Similarly, the process of forwarding a phone can be made very much easier, by using commands in plain English, typed at a personal computer. 'Visit Fred' might be typed at one's own machine before wandering down to Fred's office, or when one gets to Fred's room one can alternatively type 'Stephen is here' or words to that effect. This process can even be automated in the case of users who log on to a machine at a particular place, since the system then knows where to direct calls for that user. Another possibility is some type of infrared or other locator by which individuals indicate their position to the PABX, so that calls can be routed to an appropriate extension.

Various researchers have produced systems with some of these features, a notable example being the Etherphone system at Xerox PARC. A nice touch implemented at Xerox PARC is the replacement of a buzzer or bell in the phones by a 'personal ringing tune'. This has advantages both when several people are in the same room, since the appropriate person can answer the phone, and where a person is a little way from his phone—it is clear from the corridor whose phone is ringing!

Other features that can be provided by the personal computer include visual notification of incoming calls rather than a ring tune/tone—very useful when making a voice recording for example (see Sec. 10.6.2). For the true masochist, the cost of a call can be displayed while it is in progress. The names of callers can be displayed before a call is answered and an automatic filter can even be set up, for example to divert all calls to a secretary except those from certain individuals.

An important design issue that must be tackled when combining the telephone and the personal computer is the issue of reliability. Features such as filters require that the computer be offered a chance to process the call before the phone rings. Such mechanisms to intercept calls need to be planned carefully, so that if the computer crashes, or is powered off, the phone still works correctly. Various ways of doing this have been developed. They are all based around a common requirement: the phone must be effectively separate from the computer. Whether the phone is a board in the computer that is designed to function regardless of the state of the computer or whether it is physically detached and only associated in the feature processor's database does not matter. What is important is that the phone continues to function in the absence of or in the event of a crash of the computer. There typically needs to be some timeout associated with notifications of incoming calls to the computer, for if the computer is to intercept incoming calls they must not be notified to the phone until the computer has decided whether to handle them, yet if the computer crashes during a 'do you want to handle this one' request, this request must not be allowed to get lost and disable basic operation of the phone.

10.6.2 Voice in documents

The ability to add voice annotations to a document is a powerful feature in many work situations. Often a document is passed around for reading and comment. The reader generally scribbles shorthand notes in the margin and then takes it back to the author—it takes too long to write the comments out in longhand. If the reader could annotate the document vocally with his comments, the process would be much easier.

Various voice-in-document systems have been built (e.g. Forsdick *et al.*, 1984; Schmandt, 1981; Thomas *et al.*, 1985; Maxemchuk, 1980; Nicholson, 1983, 1985; Ruiz, 1985; Poggio *et al.*, 1985). Many have been less than satisfactory, because the designers have tried to use a common computer–human interface for handling voice and text for example, whereas in fact the two often need to be handled very differently to be effective. A deliberate decision to treat voice in a special way was made in the case of the Xerox PARC 'Tiogavoice' editor, of which a complete description and discussion can be found in Ades and Swinehart (1986) and Ades (1987). This section will compare and contrast these different approaches.

To illustrate the problem of using a single common interface, editors that allow text and graphics to be merged (such as 'Star' (Smith *et al.*, 1982), 'Diamond' (Forsdick *et al.*, 1984; Thomas *et al.*, 1985) or 'MacDraw') generally require the user to create regions in which each will be put and then to insert the required material into each region. Several researchers have tried to apply a similar approach to inserting voice. The feature rarely gets used if implemented in this way, because it is slow and cumbersome. Voice annotations are the medium for off-the-cuff remarks, and so adding a vocal annotation to a textual passage needs to be a lightweight, fast and spontaneous action. Besides, voice has no inherent visible structure and it is undesirable that space should be set aside for it; it should be possible to add a voice annotation saying 'I do not like the shape of this part of the page' without altering the shape in the process!

In contrast to many other voice editors, the basic premise of Tiogavoice was therefore to handle voice and visual media using a common interface in cases where the user views the operations required as being essentially the same, but to provide special operations on voice for cases where the user wishes to treat it differently from other media. For example, if the user wishes to copy some voice from one place to another, he or she should use the same command interface as for copying text from place to place. However, adding voice into a text block is very different from inserting graphics into a text block.

A few brief examples from the Tiogavoice interface are useful to illustrate the approach further; they also illustrate how the designers felt that users—as opposed to scientists—wish to manipulate voice at a workstation. In the Tiogavoice system, annotations are added by selecting an area of the document using a mouse, selecting a menu command and then speaking into a microphone. When the operation is complete a small icon appears, overlaying the selected area. (The icon used for Tiogavoice is a 'talks bubble', as found in comic strips.) The annotation can then be heard back by selecting the icon and then selecting another menu command.

The icon indicates the presence of voice, but nothing else. The user does not enter a caption for the voice, as is done in other systems; this seems to detract from the spontaneity that is desired. (The user who wished to append explanatory text around an annotation could do so using the normal text editing functions. For most purposes, an 'I'm here: play me' marker is sufficient.)

This type of interface is sufficient for most short annotations. However, for composition and playback of longer speech passages, a formal voice editor becomes desirable, since the chances of composing a long spoken passage correctly on the first attempt are fairly small. A problem that immediately confronts the designer of any screen-based voice editor is one of representation. Unlike the case of editing visual media, what you see is not actually what you are editing—it is only a *representation* of what you are editing. There is no wholly satisfactory visual representation of voice, yet some visual cues are needed to tell the user where in a voice passage he or she is.

Many of the voice editing systems that have been built give the user some type of energy profile or waveform to act as a visual 'clue' (Forsdick *et al.*, 1984; Thomas *et al.*, 1985; Schmandt, 1981). In practice, however, it appears that this is very little clue at all, for even experts cannot reliably interpret such profiles as sounds. Such profiles also cause the user to 'think small'—to edit at the phoneme level. For anything but a speech laboratory system this is unsatisfactory. When typing or correcting text at a word processor, it is natural to replace mistakes letter by letter. However, when using a dictaphone it is more natural to replace whole phrases. Single-word replacement rarely gives a natural flow to the speech. Certainly replacement of only parts of words, on the basis of a visual profile, will not give satisfactory results. (For a description of a voice editor using a phone keypad rather than a workstation as an interface, and for users' reactions to it, see Calnan, 1989.)

Tiogavoice and a few other systems (e.g. Nicholson, 1983) have therefore used a simple 'capillary' display, with periods of sound and silence coloured black and white respectively. This focuses the user's attention on to whole phrases. In Tiogavoice, the user can replace, copy, move or delete portions of voice. A replacement portion can consist either of voice already on 'display' or of new sounds recorded via the microphone.

With such a simple display and for a long message, the user may, however, lose track of which utterance is where; therefore with the capillary approach a voice editor needs to provide a variety of other means by which the user can mark and locate portions of voice. The most important is probably a 'playback cue', a marker that moves along the capillary when a portion of voice is selected and played. The Tiogavoice user interface allows a portion of voice to be selected during playback, so that it can be located as it is heard. Once a portion of voice has been located, it can be labelled for future reference.

There are a variety of ways of providing such labels, for both short-term and long-term recall. The rest of this section will give a few examples. The first is to attach a textual label to a region of sound. This is most relevant for long-term recall. (In the future it will be possible for these labels to be generated from the text itself, by voice recognition. There have already been demonstrations of this (Schmandt, 1981) using recognizers of limited vocabulary.) For shorter-term recall, it is useful to be able to add temporary graphic markers to a portion of sound. This is useful where one portion of voice is to be replaced by another. The user first locates that portion and then marks it while locating the portion with which to replace it.

Tiogavoice also makes novel use of colour to help the user locate passages of interest. The use of colour exploits the concept of 'fresh' and 'stale' voice. This concept says that voice which the user has just input is rather different from voice which has been in the system for some time. The difference lies in the ease with which the user can regenerate that voice. Regeneration is comparatively easy when a sentence has just been spoken and is fresh in the mind. In practice, most of the editing operations performed on voice in a multimedia editor concern fresh voice.

When a Tiogavoice editing window is first opened, the preexisting voice appears in a deep colour. When a passage of voice is copied or input, it appears in a light colour. As successive operations are performed, the newest passage of voice appears in the light colour and each recently added portion becomes a shade darker, eventually assuming the deepest colour, in which the unedited portions all appear. The colour shades, five in the prototype Tiogavoice system, were chosen to evoke an 'aging' process. In early use of the prototype it was found that the provision of colour markers often gave the user sufficient clues to locate voice fragments of interest: it noticeably reduced the tendency of users to set their own markers explicitly.

This appears to be because users spend most time operating on fairly fresh voice. In fact much of the time a voice editor will be used rather like a dictation machine. For this reason, the user interface should be designed to make it very easy to perform the functions that characterize an idealized dictation machine, with commands to play from a selected point to the end of the freshest region of voice; erase from that point to the end of the freshest region and record from there; and to record without erasing, starting from the end of the freshest region. These correspond to the user of a dictation machine recording; making an error; listening to it back and then either recording over it or recording from after the passage if it in fact sounded satisfactory on playback.

10.6.3 Managing multimedia documents

The previous section described possible approaches to implementing a voice editing interface in a workstation environment (see also Calnan, 1989). Inclusion of voice raises some interesting systems issues, in terms of how the voice is stored and accessed. These will be discussed briefly here, although they are more a matter of infrastructure than applications.

Consider a data fileserver. It will have a mechanism designed to reduce the risk of its contents becoming inconsistent. There will naturally be a performance penalty associated with this mechanism, but it is crucial to ensure consistency of the contents—a fileserver that suddenly 'lost' 4 kbytes of text within a document or program would not be widely accepted! In contrast, if one is dictating a voice annotation, 4 kbytes may represent as little as a half a second of input: it could easily be reinput if required. The user of a voice editing system who is recording and playing back large amounts of voice samples is very concerned about fast response to a playback request and much less about occasional consistency problems.

Providing a special fileserver access mechanism for voice is another part of the purpose of the ISLAND translator. In the Cambridge distributed system, under which ISLAND is implemented, the fileserver does not in fact enforce locks, atomicity or other means of ensuring consistency. It provides the relevant mechanisms, but it is up to the fileserver's clients to decide whether to use them.

Applications do not communicate with the fileserver directly, but through one of its 'front ends'. The first, the filing machine (Needham and Herbert, 1982), is a front end to the fileserver which is suited to data transactions and provides directory caching and locking, atomic update committal and so on. The other front end is the translator, which not only converts between 2 ms packets and long blocks of samples but also acts as a lightweight front end to the fileserver. These two front ends offer an excellent filing mechanism for documents containing both voice and data.

The other systems issue to be raised here is the structure used to represent a document in the fileserver. For text files, the most common way to represent a document is a linear array of characters. Editing a document consists of copying the array to computer memory, altering the contents of the array and then copying the modified file back to the fileserver.

If the number of changes made per edit is small, this is wasteful of space. It would be better to append new text to the end of the original file and then build up a system of pointers into the file to indicate which versions of the document contain and omit which characters. This technique, where each version of a document is just a pointer array pointing into a single character array, has been used in version control systems (Tichy, 1982; Leblang, Chase and McLean, 1985). These systems are used to hold a series of different versions of the same software, each version usually differing only in minor ways from the previous one. Each additional version typically requires only 1 to 2 per cent of the space required to hold the latest version as a simple array of characters.

Structured file representations of this kind become more attractive, the larger the amount of data moved around in a single editing operation. For a voice editor such as Tiogavoice, the amount moved rarely falls below 1 second, which means 8 kbytes of uncompressed voice. Even in a compressed form it is still unlikely to be useful to copy the actual voice when editing it—much better to manipulate pointers to it.

Structured files using pointers raise a variety of systems issues, including garbage collection. These are discussed further in Ades (1987) and Terry and Swinehart (1987). We will conclude this section by noting that document definitions proposed by a variety of researchers and standard committees are tending to become structured, with the concept of 'structure files' pointing to 'content files' (Forsdick *et al.*, 1984; Thomas *et al.*, 1985; Reynolds *et al.*, 1985; Horak, 1985). These ideas are very much compatible with the above ideas on organization of voice within documents.

10.6.4 Call distribution and following

This section has so far discussed integrated services applications developed by researchers; it is appropriate to mention here two that are widely available commercially. 'Automatic call distribution' is offered as an add-on package for many medium or large digital PABXs. When a call comes in to, for example, a customer support department, it is referred to either a single operator or to one of a pool of telephonists. The operator asks for the caller's name or support contract number. The nature of the problem is then described and entered by the operator at a computer keyboard, on the basis of which a suitable person is selected by the system to handle the call. As the call is put through to the selected person, the system also transfers the computer screenful, showing the call details so far: the name of the caller and the outline description of the problem. If the recipient wishes to transfer the enquiry again this can be done; naturally the computer screen details are also transferred.

There are also similar types of package offered with various PABXs to support telephone-based selling. A pool of sales personnel may be calling existing customers, cold calling or whatever. Numbers are dialled automatically once a target customer or prospect is selected. Where the person has been contacted before, relevant details are displayed on the computer screen and can be updated during the call. If this is a cold call, the sales person may enter details on the prospect during the call, for future use.

These are two very specific examples of integrated services applications. They are mentioned because they are two important examples of an integrated feature for which there is a very large commercial market, now that the technology has made the features possible.

10.6.5 'Phone slave'

To conclude this section, we will try to paint a picture of what a completely integrated office support package might look like. We began the chapter with a glimpse of the picture, and all of the components of the picture have been discussed; this section is merely intended to pull them together. Much of the scenario to be outlined in the following paragraphs has been demonstrated in the MIT 'phone slave' project (Schmandt and Arons, 1984). There it was done using expensive prototype hardware, because sufficiently cheap and versatile technology was not then available to make it all a reality—but it will be soon!

Each office worker has an electronic diary, by means of which appointments can be entered and into which other people can 'pencil in' meetings, etc. There is also an automatic mechanism by which a person trying to schedule a meeting can find out when everybody is free.

On entering the premises, each employee is logged in by means of a card-key or locator. As employees move around a building, they carry locators which tell the PABX where they are, so that their calls can reach them automatically.

When somebody calls in for an absent member of staff, a synthesizer tells the caller that the person is out, or when he or she is due back, or even where he or she is, according to the wishes of the absent person. The person may have left a voice message for callers. The synthesizer then asks if the caller wishes to be put through to a secretary or to leave a voice message. A voice recognizer enables the system to respond to the caller's wish. This recognizer may also recognize particular voices and play back special messages or reveal various details as to whereabouts, depending on who is calling in.

When voice recognizers have improved to a sufficient robustness and reliability, we will see systems where the message left by a caller finds its way into the person's electronic mailbox in textual form.

10.7 Where next?

The applications that are discussed above rely on two things: advances in technology for certain key components (e.g. speech recognizers) and the spread of office workstations. The latter is happening now: as stated above, the annual growth in personal computers installed in offices and connected to some type of local area network is currently 30 per cent per year. LANs suited to voice and data integration are still not widely and cheaply available however.

At the same time, ISDN development has reached an interesting stage. Widespread availability of ISDN access is still some way in the future and it has not yet been established that there is a commercial case for wide area ISDN. However, very many IC vendors now offer the basic terminal interface ICs. A handful of these comparatively cheap ICs makes up an ISDN phone or data interface; the silicon manufacturers have worked very hard to make sure that their devices have the right functionality and interfaces to enable systems designers to construct very low chip count ISDN terminal equipment.

A few manufacturers offer all the ICs that are required to make an ISDN PABX. These ICs make voice and data switching at 64 kbit/s a reality at very low cost indeed. Very many digital PABXs are now claimed to have some ISDN compatibility.

Cheap circuit-switched data connections at 64 kbit/s, the ISDN office desk interface, are hardly what the 'vision' of this chapter has been looking forward to, although for many companies using predominantly asynchronous or synchronous serial line terminal networks they will seem a great leap forward. For those companies now accustomed to the performance of IEEE 802 LANs, integrated services in the local area will call for something more than plain ISDN B channels.

No LAN-based integrated services product has yet been brought to market. Californian-based Ztel (Kay, 1983) seems to have come the closest but collapsed before bringing the product out. (This appears to have been due to a poor business plan, rather than to problems with the technology.)

Products using LAN switching for data and non-blocking switching for voice, with all terminal equipment access via ISDN B channels, are an interesting new possibility, given the new wave of ISDN ICs. The Philips SOPHO/S PABXs claim to offer this configuration as an option.

Given a suitable high-level system design, combinations of the two types of switching can still form the basis for a coherent integrated services product. Products aimed at proving this assertion are already in development by various telecommunications companies.

This is a very exciting time for commercial office automation products. For integrated services, it is hard to predict whether LANs, already with a substantial installed base and 30 per cent year-on-year growth, or the ISDN office interface, offering voice and data over the existing analogue telephone wires, will have the stronger commercial foothold in a few years' time. They may even co-exist and perform complementary functions within the same system, at a significant number of installations. Whichever way round it is, with a growing commercial awareness and appreciation of the power of Information Technology, the vision with which this chapter started is not so very far from reality.

References

Adams, C. J. and S. Ades (1985) 'Voice experiments in the UNIVERSE network', Proceedings of IEEE ICC 85 Conference, Chicago.
Ades, S. (1987) 'An architecture for integrated services on the local area network', Technical Report no. 114, University of Cambridge Computer Laboratory, September.
Ades, S. and D. C. Swinehart (1986) 'Voice annotation and editing in a workstation environment', Technical Report CSL-86-3, Xerox Palo Alto Research Center, Palo Alto, Calif.
Anderson, H. (1986) 'How to build an intelligent network', *Telecommunications Products and Technology*, August.
Bullington, K. and J. M. Fraser (1959) 'Engineering aspects of TASI', *Bell Systems Technical Journal*, March, vol. 38, no. 2, 353–364.
Calnan, R. S. (1989) 'The integration of voice within a digital network', Ph.D. thesis, University of Cambridge Computer Laboratory, Cambridge.
CCIR (1976) 'Effects of bit errors on transmission of speech through PCM systems', Document 4/75-E, 17 February.
CCITT (1974) 'Planning of digital systems', Special Study Group D, Contribution 103, June.
Dallas, I. N. (1980) 'A Cambridge ring local area network realisation of a transport service', Proceedings of IFIP WG6.4—Workshop on Local Area Networks, Zurich.
Forsdick, H. C., R. H. Thomas, G. G. Robertson and V. M. Travers (1984) 'Initial experience with multimedia documents in Diamond', Proceedings of IFIP 6.5 Working Conference, May.

Gruber, J. G. and L. Strawczynski (1983) 'Judging speech in dynamically managed voice systems', Telesis 1983 Two (Bell Northern Research), 30–34.

Hopper, A. and R. M. Needham (1986) 'The Cambridge fast ring networking system', Technical Report no. 90, University of Cambridge Computer Laboratory.

Horak, W. (1985) 'Office document architecture and office document interchange formats: current status of international standardization', *IEEE Computer*, vol. 18, no. 10, October.

Kay, P. M. (1983) 'A new distributed PBX for voice/data integration' in *Proceedings of Localnet 83*, 485–493, Online Publications, New York.

Leblang, D. B., R. P. Chase and G. D. McLean (1985) 'The DOMAIN software engineering environment for large scale software development efforts', IEEE 1st International Conference on Computer Workstations, San Jose, Calif., November.

Maxemchuk, N. F. (1980) 'An experimental speech storage and editing facility', *Bell System Technical Journal*, vol. 59, no. 8, October.

Metcalfe, R. M. and D. R. Boggs (1976) 'Ethernet: distributed packet switching for local computer networks, *Communications of ACM*, vol. 19, no. 7, July.

Needham, R. M. and A. J. Herbert (1982) *The Cambridge Distributed Computing System*, Addison-Wesley, London.

Nicholson, R. T. (1983) 'Integrating voice in the office world', *BYTE Magazine*, vol. 8, no. 12, 117–184, December.

Nicholson, R. T. (1985) 'Usage patterns in an integrated voice and data communications system', *ACM Transactions on Office Information Systems*, vol. 3, no. 3, July.

Poggio, A., J. J. Garcia Luna Aceves, E. J. Craighill, D. Moran, L. Aguilar, D. Worthington and J. Hight (1985) 'CCWS: a computer-based, multi-media information system', *IEEE Computer*, vol. 18, no. 10, October.

Reynolds, J. K., J. B. Postel, A. R. Katz, G. G. Finn and A. L. DeSchon (1985) 'The DARPA experimental multimedia mail system', *IEEE Computer*, vol. 18, no. 10, October.

Ruiz, A. (1985) 'Voice and telephone applications for the office workstation', Proceedings of 1st International Conference on Computer Workstations, San Jose, Calif., November.

Schmandt, C. (1981) 'The intelligent ear: a graphical interface to digital audio', Proceedings of IEEE Conference on Cybernetics and Society, October.

Schmandt, W. and B. Arons (1984) 'Phone slave: a graphical telecommunications interface', Proceedings of International Symposium of the Society for Information Display.

Smith, D., R. Kimball, B. Verplank and E. Harslem (1982) 'Designing the star user interface', *Byte Magazine*, vol. 7, no. 4, 242–282, April.

Swinehart, D. C., L. C. Stewart and S. M. Ornstein (1983) 'Adding voice to an office computer network', Proceedings of IEEE GlobeCom Conference.

Temple, S. (1984) 'The design of a ring communication network', Technical Report no. 52, University of Cambridge Computer Laboratory.

Terry, D. B. and D. C. Swinehart (1987) 'Managing stored voice in the Etherphone system', Xerox Palo Alto Research Center, Palo Alto, Calif.

Thomas, R. H., H. C. Forsdick, T. R. Crowley, R. W. Schaaf, R. S. Tomlinson, V. M. Travers and G. G. Robertson (1985) 'Diamond: a multimedia message system built upon a distributed architecture', *IEEE Computer*, vol. 18, no. 12, October.

Tichy, W. F. (1982) 'Design, implementation and evaluation of a reference control system', IEEE Proceedings of the 6th International Conference on Software Engineering, 58–67, September.

UK Office of Telecommunications (1986) 'Code of practice for the design of private telecommunications networks'.

Want, R. (1988) 'Reliable management of voice in a distributed system', Technical Report no. 141, University of Cambridge Computer Laboratory, July.

Wilkes, M. V. and D. J. Wheeler (1979) 'The Cambridge digital communication ring', Local Area Communications Network Symposium, sponsored by Mitre Corporation, Boston, Mass., May.

Acknowledgement

Section 10.6.2 is based on material in Ades and Swinehart (1986) and is published here with the permission of the Xerox Corporation.

11 Human factors

BOB DAMPER and GRAHAM LEEDHAM

11.1 Introduction

The subject of human factors is not a single discipline as it encompasses many areas of expertise. It involves physiologists, psychologists, ergonomists, engineers and practitioners from many other scientific fields. According to Kantowitz and Sorkin (1983), human factors is 'the discipline that tries to optimize the relation between people and technology', that is optimizing the specification, design and construction of computational, electronic and mechanical equipment to make it easily and efficiently operated by its users. This is not a trivial task and requires the detailed consideration of many aspects of the user, the task to be carried out and the capabilities of the available equipment. The users' physiological capabilities in terms of their senses, mobility, reach and verbal communication skills need to be considered as well as their psychological capabilities in terms of reasoning power, memory, conceptual view of the task to be performed and potential willingness or otherwise to accept the technology. As well as analysing the users and their requirements to perform the intended task, the task itself must be analysed into subtasks and each subtask allocated to the technology (predominantly computers) or the human user so as to optimize the overall performance and, most importantly, maintain the interest and acceptance of the technology by the users.

These days, computers are many people's main contact with technology so that human–computer interaction (HCI) is an increasingly important branch of human factors (see Baecker and Buxton, 1987). One way that the desired optimization can be achieved is by careful selection of the *medium* of communication between human and computer. In this chapter, we explore the question of what speech interaction with computers offers over the use of alternative, more conventional communication media.

The dream of being able to communicate with machines by speech has long fascinated engineers and scientists. Over recent years, a number of national and multinational research programmes have been aimed at realizing major advances in the technology of speech communication with computers. These programmes have been motivated by the challenging nature of the problem, together with the belief that speech input and output (I/O) offers the key to dramatic improvements in the effectiveness of the user interface. For instance, Lea (1980) writes:

> ... you will want to use speech whenever possible because it is the human's most natural communication modality

and Lee (1989) states:

> Voice input to computers offers a number of advantages. It provides a natural, fast, hands free, eyes free, location free input medium.

Given such powerful theoretical advantages, it is reasonable to ask why speech is not more widely used in practice. Undoubtedly, the above-quoted authors are basing their statements on the nature of human-to-human communication, but current technical capabilities in speech I/O offer a far lower level of interaction. Thus, one obvious answer is that the technology has not yet advanced to the stage where adequate performance for mass applications is available at a competitive cost. Another possibility, however, is that even present capabilities in speech technology are not being appropriately exploited because of lack of understanding of the requirements of a successful speech interface.

The basic thesis advanced in this chapter is that study of the human factors of speech I/O is important in two ways. Firstly, by analysis and by task-related experimentation (case studies), we can tell how best to utilize current technology. This is likely to involve an understanding of natural (human-to-human) communication and an appreciation of how speech communication with machines differs. Secondly, by simulation of *future* generations of technical device, we can determine useful features and so identify priorities for development.

The chapter is structured as follows. We first introduce the methodology of human factors. Since speech is just one of many possible interface media, we treat the topic in a general way aimed at revealing basic principles. We next examine human capabilities (and, indeed, limitations) as they impact on user interface design. Then, current technical capabilities in synthesis and recognition are outlined, since these determine the spectrum of possibilities available to the interface designer. We deal next with the issues that must be addressed in designing any interface featuring speech I/O. Finally, the interesting research topic of the simulation of future speech systems is described, before drawing conclusions.

11.2 Human factors methodology

The branch of human factors concerned with human–computer interaction has as its aim 'to develop or improve the safety, utility, effectiveness, efficiency and usability of systems that include computers'

(Diaper, 1989). In the design and development of a specific interface, however, the implementor needs some means of quantifying criteria such as efficiency and usability. Shneiderman (1987) has suggested the following measurable human factors goals as suitable for this purpose:

- Speed of performance
- Rate of errors
- Subjective satisfaction
- Time to learn
- Retention over time

These, then, are the yardsticks against which any interface, including one based on speech, should be evaluated.

11.2.1 Input requirements and abstract devices

It seems that many researchers proceed on the almost implicit assumption that speech offers overriding advantages, relative to alternative media, on all of the above dimensions. (Indeed, this is the import of the quotations from Lea (1980) and Lee (1989) quoted in Sec. 11.1 of this chapter.) We believe, however, that speech is most unlikely to be a panacea for all user interface problems; rather, there are particular applications for which speech is appropriate and others for which it is less good (as assessed by the measurable goals listed above). This being so, we need some analytical framework in which to predict the likely success of a specific application.

One very useful classification of *input* tasks has been put forward by Foley, Wallace and Chan (1984). Foley and his colleagues were primarily concerned with graphics input, but their scheme can be usefully applied to other domains. They identify the following six types of interaction:

- Select
- String
- Quantify
- Orient
- Position
- Path

The *select* function is illustrated, for example, by the common requirement to pick an item from a menu. The *string* function involves composing a sequence of characters selected from some set as in text composition; hence, it can be viewed as a sequence of selection operations. *Quantify* calls for the specification of some scalar (unidimensional) quantity denoting, for example, the point in some file where editing is to take place. *Orient* specifies an angular quantity, such as the orientation of a line segment. *Position* identifies a point in (usually) two-dimensional space. Finally, *path* describes an arbitrary curve in the applications space and can be seen as a sequence of either position or of orient.

We can think of these input requirements as representing ideal, 'abstract' devices. Real, physical devices map onto the abstract devices in vastly differing ways. Thus, whereas the conventional QWERTY keyboard is essentially a string composition device, it can be used to simulate the other abstract devices with greater or lesser facility. For instance, augmented with cursor control keys, it can perform the position function. However, a pointing device such as a mouse is a far better realization of the abstract position device. In turn, a data tablet used in conjunction with a stylus is a generally better realization of the path abstract device than is a mouse, even though path can be viewed as a sequence of position, because a stylus has far better dynamics than a mouse.

Examination of the way that automatic speech recognition (ASR) maps onto the abstract devices gives a powerful means of evaluating its strength and weaknesses. Superficially, a speech recognizer would appear to be a very promising string device for use in automatic dictation. For a direct implementation, however, this application really requires connected-word, large-vocabulary capabilities beyond what is currently feasible (see Sec. 11.4.2). However, ASR is potentially an excellent means of one-out-of-N selection, provided N is reasonably small, since this is very close to the recognizer's real mode of operation; in this particular case, isolated-word and small-vocabulary restrictions are relatively unimportant. It should be readily apparent that ASR is a poor match to the requirements of, for instance, quantify and position. The former calls for an essentially analogue, continuous device, but ASR produces a discrete output, whereas position calls for spatial, pointing abilities which ASR does not possess.

11.2.2 The role of case studies

We must be careful of trivializing the multidimensional problems of interface design. While the sort of analysis of input requirements described above can be extremely helpful, human–computer interaction is a complex subject; it cannot realistically be reduced to the simple procedure of mapping abstract devices onto real devices. Typical of the many other dimensions that need to be considered are the user's physical situation and skills, the cognitive load imposed by the task and safety criticality. Because we do not fully understand all the factors impacting on performance, it remains mandatory to carry out task-related 'case' studies. The basic methodology that has evolved is to design a prototype interface which is then evaluated against the measurable goals, listed above, in a case study (see Downton, 1991).

11.2.3 The role of simulation

While methodologies based on case studies can, in principle, use the most up-to-date interface technologies, there will be many occasions when this is not possible. Consequently, much human factors work could be criticized as focusing on the shortcomings of currently available equipment rather than assisting the development of the next generation of interface technologies. For this reason, there is considerable interest in studying simulations of future devices and technologies. We discuss this work in Sec. 11.7.

Thus far, we have rather concentrated on the input side of human–computer interaction. Analysing the computer's requirements for input information in an abstract, task-independent fashion is a far simpler proposition than determining the human operative's requirements for feedback. Consequently, while we can usefully decompose input functions into the abstract categories of selection, quantity, etc., it is much less clear how we might do this for output. In the absence of any analytical framework, we are forced to work with guidelines, that is a catalogue of 'good' and 'bad' applications as assessed by case study and experience. As far as speech output is concerned, therefore, we content ourselves with a review of positive and negative indications for the use of speech output (see Sec. 11.5.2).

11.3 Human capabilities

11.3.1 Modelling the human

In engineering, it is common practice to model systems and devices in order to study their behaviour when subjected to different forms of stimuli. The study of human factors is no exception.

In the literature a number of different models have been proposed to describe the human processing system (see Bailey, 1982, and Card, Moran and Newell, 1983). None of these models are entirely accurate but provide sufficient detail to explain the main features and implications of the human system. One such model (see Bailey, 1982) is shown in Fig. 11.1. In this model, the human–computer interface is shown as occurring at the input and output devices of the computer system. With speech systems we are concerned with speech synthesizers as the computer output device and speech recognizers as the computer input device.

In the model shown in Fig. 11.1, output from the computer is detected by the human sensors (ears for speech communication) and this information is passed to the brain where it is perceived. This channel is a low-capacity information channel because only a small proportion of the signals impinging on our senses are actually processed in our brains. The signals, which are accepted by the brain,

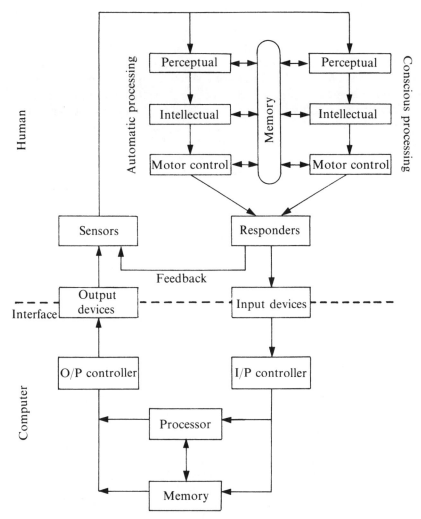

Figure 11.1 Model of the human processing system (after Bailey, 1982)

are processed with reference to human memory to determine a suitable response to the received signal. After analysing the perceived input signals the intellectual processing produces signals to the motor control parts of the brain, which in turn cause suitable responses to be generated by the human responders. This can take the form of several responses, one of which is speech.

The processing of information in the brain can take place at intermediate stages between the two levels of conscious processing and automatic processing. Conscious processing involves considered, reasoned responses and more time is spent in the intellectual processing stage. This form of processing is usually associated with new or infrequent actions which need to be analysed and considered before a suitable response can be initiated. Automatic processing is usually associated with frequently required responses such as reflex actions which involve little consideration of a suitable response and therefore little time is spent on intellectual processing. Automatic responses are generally associated with skill and can take place at the same time as conscious actions. For example, when an experienced driver is driving a car he or she is able to steer, react to other traffic and operate clutch, gears, accelerator, brake and other controls quite automatically while holding a conversation with passengers. The driver is able to carry out observations and driving actions at a low level of conscious processing while concentrating conscious processing on the conversation.

Depending how frequently the same situation is analysed and acted upon, the processing can be accomplished at many intermediate levels between purely conscious processing and purely automatic processing. Conscious processing is generally slower than automatic processing but is more flexible to change and not particularly susceptible to error. Conversely, automatic processing is generally fast but relatively inflexible and is more error prone than conscious processing, especially when an unusual occurrence that requires a switch to conscious processing immediately follows a series of common occurrences that have been processed automatically. For example, consider the case of the driver who is inadvisedly holding a conversation while driving in heavy traffic. If anything unusual or unexpected happens such as another driver carrying out a dangerous manoeuvre then, to avoid a collision, a considered rapid evasion manoeuvre is required. In this situation, the chances of an accident happening are higher than if the driver were concentrating on observing the surrounding traffic conditions rather than on his conversation. Bailey (1982, Chapter 6) gives a fuller description of automatic and conscious processing.

11.3.2 Voice

Earlier chapters of this book have described how speech is a form of communication between humans which has evolved over many hundreds of generations and is used by the vast majority of the world's population as their natural form of communication. The typical rate of normally spoken speech is 180 words per minute and can exceed 400 words per minute for short periods of time. There are also approximately 3000 different spoken languages of which 130 are used by more than one million people (Bailey, 1982, Chapter 13).

The speech signal can be analysed and shown to be a continuously changing acoustic signal composed of combinations of various frequencies, mainly between 80 Hz and 8 kHz, caused by the periodic or aperiodic excitation of resonances in the vocal tract. The information passed from one person to another during a conversion is extremely difficult to decode from this acoustic signal because the speech rate is *variable*, the sound is often *continuous* over word boundaries, speech can be *ambiguous* (e.g. 'wreck a nice beach' is phonetically similar to 'recognize speech') and the signal may be *contaminated* by other voices or noises including the redundant utterances such as coughs and 'umms' and 'ahs' frequently generated by the speaker. In addition much information is conveyed in variations of pitch and tone and also in other gestures such as facial movements and hand and arm movements.

11.3.3 Hearing

The human ear is a complex organ (see Chapter 1) which translates the acoustic signal into a set of signals that are transmitted as electrical impulses through nerve fibres to the brain where they are analysed by the human processing system. The ear is sensitive to acoustic signals in the range approximately 20 Hz to 20 kHz, but this will vary from person to person and from the same person from time to time depending on health, age and temporary degradation of hearing caused by exposure to loud noises.

It is important to remember that hearing defects in the form of various levels of deafness are a common problem among the population and these defects are not immediately obvious to other people. Indeed, many people with a hearing impairment are more sensitive about their disability than they would be if they had some visual impairment. To a large extent, this is due to the general attitude that the wearing of spectacles to counteract visual impairment is more socially and fashionably acceptable than the wearing of a hearing aid. Designers of systems should be aware of and sensitive to this fact when considering speech or other audible output from a system, as some degradation in human–computer performance must occur even with partially deaf users of the system if the audible output cannot be clearly heard. As well as producing degraded performance, the users who experience difficulty hearing the sound output of a system will also show a marked reluctance to accept the system.

11.3.4 Memory

There are three forms of human memory: short-term memory, long-term memory and sensory memory. The main ones are short-term memory (STM) and long-term memory (LTM). STM is a temporary store of a small amount of information. This small amount of information is limited to about seven units, or 'chunks', of information and the duration of this storage is limited to a few seconds (Miller, 1956). The term 'units' or 'chunks' of information is used because the user will encode incoming information into meaningful pieces of information. For example, the number sequence 4730216 will, to most people, have no immediately discernible pattern and occupy approximately seven chunks of information while the pattern 1234567 has an immediately apparent pattern and therefore represents only one or two chunks of information in STM.

LTM is, as the name suggests, a long-duration memory. Experiments have been unable to ascertain any upper limit on the capacity of LTM and it is, therefore, considered to be a virtually unlimited store. LTM is considered by some psychologists to be accessed by context links and therefore access times are closely related to the time since the last access to that particular piece of information and the number of contextual links that reference it. Items are not forgotten but instead the context links are broken through lack of use or confusion with other closely related links. The transference of information from STM to LTM is by repeated experience or rehearsal which gradually transfers information frequently loaded into STM into LTM. The more frequently this rehearsal is repeated the stronger are the links to that memory.

Some memory is also present in the senses themselves and is termed sensory memory. This is associated with persistence of the sensory stimulus after the stimulus has stopped. For hearing, this persistence is usually less than 250 ms in duration (Bailey, 1982).

The above is a simplistic explanation of human memory. There are more complex theories describing the detailed nature of human memory which are supported by experimental data but the full explanation of how memory operates has still not been discovered. Research is still being carried out to increase our understanding. More detailed discussion of human memory can be found in Baddeley (1983) and Rumelhart and McClelland (1986).

11.3.5 Perception and reasoning

Perception and reasoning involve the initial collection of data entering the brain from our senses and the subsequent processing to understand the incoming data and determine a suitable response.

The process of perception is strongly influenced by memory; that is our perception, cognition and reasoning ability are highly dependent upon our past remembered experiences of similar situations. For example, our ability to understand the words in spoken language is based upon our previous knowledge and memories of earlier hearings of the same words. Compare, for example, your own ability to understand your first language with your ability to understand a less familiar language in which you are only able to pick out occasional words and your inability to understand a totally unfamiliar language where no words are familiar.

11.4 Technical capabilities in synthesis and recognition

11.4.1 Synthesizers

Earlier in this book (Chapter 6) the production of synthetic speech by several techniques has been described. These techniques can be as simple as playing back parts of recorded messages using a tape recorder or as complex as generating any combination of spoken words from a textual representation of the word or sentence by the application of rules. These rules convert the text into basic speech segment sounds and concatenate these basic speech segments together to generate spoken words.

Although a description of speech in terms of its spectral components or basic speech segments is a valid exercise to obtain some understanding of human speech, it does not illustrate the nuances of speech which combine to convey a large amount of information in the speech signal. Speech is not a blank monotone; all voices vary in quality and tone due to health, physical features, the emphasis and stress of a word (including silent pauses) and many other causes. This variation is termed prosody and conveys meaning and emotion. For example, the word *extract* can mean the extract from a book or magazine (noun) or the verb 'to remove'. Different stresses can convey a difference between the two words which is usually sufficient for a listener to determine the intended usage without the aid of its context within a sentence. Prosodic information provides the naturalness of speech and improves the intelligibility. If we take a sentence and change the place of maximum stress it is possible to show that this changes the meaning of the sentence quite considerably. For example, consider the question 'Where shall we go?'. If the major stress is placed on the word 'where' or on the word 'we' the sentence has two subtly different meanings.

We are often in danger of forgetting that speech is our natural interaction medium and do not realize the difficulties that exist in creating synthetic speech by computer. We are lulled into a false sense of technology by science fiction films which portray computers or robots that can communicate using speech as correctly and naturally as if speaking to an articulate human. This level of speech synthesis and speech understanding is far from the current reality. The reality is that current speech synthesizers are not fully articulate and few make any attempt to add prosodic information. Speech synthesis by means of copy synthesis can achieve very high quality speech which is indistinguishable from real speech when prerecorded words or phrases are spoken, but is too inflexible for use with large vocabularies and general sentence construction. The result of concatenating isolated words together produces understandable but unnatural sounding speech and cannot be used effectively to create unlimited vocabulary speech.

The statement 'It's not what you say but how you say it' is clearly exemplified when listening to the output of currently available speech synthesizers. As already mentioned, prosodic information can change the meaning of the same set of words quite considerably. It is possible to control rhythm, pitch and stress to some extent in isolated words or short phrases in limited vocabulary systems, but to incorporate these features fully machine understanding of the speech is required in unlimited vocabulary systems if the prosodic information is to sound natural. It is possible to apply some generalized rules that work reasonably well in most situations but there is further research work needed before the voice is indistinguishable from a human speaker.

A feature of human acceptance of synthetic speech is that the more you listen to a particular speech synthesizer the more accustomed you become to its voice. This is a natural process which takes place in our everyday encounters with people. We may talk to people who speak English with a strong regional or foreign accent which at first we find difficult to understand. After a period of time, however, we usually become attuned to the accent and find little difficulty in understanding the speech. It is often the case that a designer who incorporates speech output into a system has listened to the synthesized speech for so long during the development stage that they have no difficulty in understanding it and are surprised to find that on demonstrating their system most other people are unable to understand the speech output. It is important for designers to be aware of this fact and assess reaction and acceptability of the synthesizer using naive listeners before expending a great deal of effort incorporating it into a system.

11.4.2 Recognizers

The majority of currently available speech recognizers are speaker-dependent, isolated-word devices with highly restricted vocabularies. The vocabularies of these systems can be several thousand words but unless the number of 'active' words at any time is restricted to less than 100, say, their performance is poor. Speaker-dependent connected-word recognizers are also available but have a greater restriction on allowable vocabulary (around 200 words), and unless the syntax is also fully defined and restricted their performance is also poor. The only speaker-independent recognizers are isolated-word devices and typically cope with about 10 words. All of these systems are reported to achieve 95 per cent or better recognition performance (Wallich, 1987) but these figures are dependent upon the vocabulary chosen, the experimental details and on the amount of co-operation the user is prepared to give to the recognizer. The figures should, therefore, be treated with a certain amount of caution and are usually upper limits on the performance that can be expected.

More ambitious recognizers which attempt continuous speech recognition on very large vocabularies are starting to become available (see *Speech Technology*, 1989). Due to the highly restricted recognition performance and relative high cost of speech recognizers, they can only be used in restricted applications where an error rate of around 10 per cent is acceptable. With carefully selected words and highly restricted vocabularies, the error rate can be reduced to 1 to 2 per cent, but the cost of such a system is usually high in comparison to other restricted selection devices (e.g. keyboards) and they must prove themselves cost effective in any application (see Sec. 11.5.2).

11.5 The role of speech in the user interface

11.5.1 Speech communication

Before deciding to use speech input or speech output in an interactive environment it is necessary to consider the implications of speech as a communication medium. Chapanis *et al.* (1977) carried out a number of experiments which concluded that two people working on a problem were able to reach a solution to that problem faster using speech communication than using manual means. This, however, assumed co-operative speakers and listeners who were adept at communicating in the language spoken and was restricted to the solution of a problem by two people. Difficulties start to be encountered when one of the 'speakers' or 'listeners' is a machine. Based on the performance of current technology, a co-operative speaker (and listener) is required and a highly restricted vocabulary must be used. The error rate is likely to be high.

One of the main disadvantages of speech output in some interactive systems is the obtrusive nature of speech. It is not possible to avoid the sound. While you can ignore an image or display you do not wish to see or part of a display screen that you find distracting, it is very difficult, if not impossible, to ignore an obtrusive sound. This can produce an unacceptable working environment when a number of workstations that use speech input/output are sited together. The only solutions in such a situation are for each user to wear headphones and a head-mounted directional microphone or to place each workstation and its user in a separate sound-insulated enclosure. Both of these solutions impose environmental constraints which are unacceptable to users in many working situations.

Speech output can give a computer personality and generate

attitudes and expectations on the part of the user (Monk, 1985). Newell (1986) has suggested that the computer's apparent personality is the cause of most complaints from users of systems employing speech output. However, if the system is silent and information is presented to the user on a display screen, this personality is not so apparent.

Synthetic speech will always be in direct competition with people. Because speech is peoples' natural medium of interaction, they are very aware of all the nuances that make human speech such a complex and effective communication channel. Therefore people will always be aware of all the minor mistakes made in synthetic speech and will be very intolerant of them.

Speech output from a system in the form of synthetic speech provides an information transfer rate which is less than that for vision. The ability to browse, which we use frequently in visual analysis, is greatly reduced in speech because the speech is a serial presentation of information and it is difficult to use it to search a document rapidly for the required information. Vision, on the other hand, allows the reader to scan a document rapidly and assimilate information about the size, layout, relative length and positions of sections of text and pictures, the location and content of the main headings. These important scraps of information are then used to decide which part of the document to examine in detail. Obtaining these navigational aids is vital in the understanding of a document. It is difficult to provide this ability quickly, to browse and explore a document and to obtain a mental image of the whole document's layout using speech output. The problem can perhaps be compared to trying to read a document visually by looking through a hole in a piece of paper placed over the document which is little larger than individual characters within the document.

The serial and transient nature of speech causes an additional problem. If you miss hearing a word or phrase it is gone and it may be a time-consuming task to cause the speech to be repeated. Because short-term memory has a limited capacity, detailed information contained in spoken text is quickly forgotten and needs to be refreshed several times. With vision this refreshing is a rapid process involving frequent scanning of the same parts of a document to extract the required information. This can only be achieved effectively when the document's layout is understood and even then speech output is considerably slower than visual reading.

11.5.2 Potential application areas

Despite all the limitations of synthetic speech output and automatic speech recognition devices there are a number of application areas where their use is effective. In general, speech input/output is useful when:

- The user's hand and eyes are fully occupied with other tasks,
- The user needs to be mobile and other input/output devices are ineffective, or
- The user is remote from the system and can only use voice communication over conventional telephone or radio links.

Deatherage (1972) investigated the use of non-speech audio cues and suggested the guidelines shown in Fig. 11.2 when determining whether to use auditory or visual presentation of information. Most of these guidelines can also be applied directly to speech output.

Speech input/output systems are particularly suitable as aids for the disabled (Damper, 1990). For the blind and persons with severe speech defects speech synthesizers in such forms as talking typewriters or reading machines (e.g. the Kurzweil reading machine) and keyboard-entered text-to-speech systems as a replacement voice provide useful effective devices. Speech recognition systems can also be used as voice-operated control devices for the severely physically disabled. This can take the form of, for example, simple environmental controllers to operate lights, televisions, telephones, etc. While the quality of synthetic speech is barely tolerable to the majority of the population, many disabled persons are more tolerant of the imperfections in current speech synthesizers because they are highly motivated and are

Use auditory presentation if:
1. The message is simple.
2. The message is short.
3. The message will not be referred to later.
4. The message deals with events in time.
5. The message calls for immediate action.
6. The visual system of the person is overburdened.
7. The receiving location is too bright or dark-adaptation integrity is necessary.
8. The person's job requires continual mobility.

Use visual presentation if:
1. The message is complex.
2. The message is long.
3. The message will be referred to later.
4. The message deals with location in space.
5. The message does not call for immediate action.
6. The auditory system of the person is overburdened.
7. The receiving location is too noisy.
8. The person's job does not require continual mobility.

Figure 11.2 Guidelines for using visual or audio presentation of information (based on guidelines produced by Deatherage, 1972)

prepared to accept the limitations of current technology. Even a poor-quality speech channel provides them with access to information, communication and control channels which were previously inaccessible to them.

Finding suitable applications for speech output is very difficult since application areas are not easily defined. Unfortunately a number of designers have added speech output into systems without considering the implications of doing so and have produced very gimmicky results which have received a lot of criticism from users and have made the product unpopular. Generally, speech output may be useful in areas where there is high visual concentration and the speech output provides an additional channel for warnings or short information bursts or in areas where the visual channel is not available either through blindness or in conventional (audio-only) telephone communication.

When considering the use of synthetic speech output from a system, the designer must always consider whether it is more sensible to provide a speech output, giving what is probably a verbal warning or information burst in a rather poor quality voice, when a simple flashing light or non-speech sound output may be adequate. Therefore, it is recommended that when considering speech as an output for a system it should be used with a great deal of caution and the designer should consider carefully whether the speech is absolutely necessary. Figure 11.3 summarizes some possible application areas for speech input and output.

Hands and eyes busy
Inventory checking
Inspection
Sorting (parcels, baggage)
Aviation
Cartography
Control (e.g. eye surgical camera)

Remote from the system
Surveys
Database enquiry
Telephone doctor

Other applications
Aids for the disabled
Security/validation
Language translation
Warning messages
Teaching aid

Figure 11.3 Potential application areas of speech input/output

In many applications, speech input offers no real advantage over many other standard input devices such as keyboards and pointing devices. Figure 11.4 provides a comparative overview of speech input with other alternative interaction devices for the abstract operations of string, orient, position, etc., as proposed by Foley, Wallace and Chan (1984) and discussed in Sec. 11.2.1. The potential benefits of speech are that they reduce the mental and physical load and can, in the right circumstances, produce a faster interaction and can be more efficient in hand- and eyes-busy environments.

When considering speech for input applications it is necessary for the designer to consider what tasks are to be performed and whether speech will help. The users of the system must also be considered and the question that must be asked is 'Will they use it?'. In addition, the environment must be considered: 'Is the background noise too high or is the user wearing protective clothing which will make it difficult to use a microphone or to operate a keyboard?' It is also necessary to consider the capabilities of the recognition technology. It is of little use attempting to employ a recognizer for a particular application when the error rate will be unacceptably high or the vocabulary too small.

Speech recognition is a very expensive technology and needs to prove its cost effectiveness and acceptability in any given application before it is used. Applications have been found in parcel sorting in postal offices and in a number of avionic applications. There have also been a number of useful applications in aids for the disabled, but the high price of recognizers has limited this.

INPUT

		'Abstract operation'				
	String	Quantify	Selection	Position	Orient	Path
Speech	Poor	Poor	Good	Poor	Poor	Poor
Mouse	Poor	Reasonable	Good	Good	Good	Reasonable
Tracker ball	Poor	Reasonable	Reasonable	Good	Reasonable	Good
Lightpen	Poor	Good	Good	Good	Good	Good
Touch screen	Poor	Poor	Good	Good	Reasonable	Good
Keyboard	Good	Reasonable	Reasonable	Poor	Poor	Poor
Joystick	Poor	Reasonable	Reasonable	Reasonable	Reasonable	Reasonable

OUTPUT

	Textual	Pictorial	Numerical
Speech	Reasonable	Impossible	Reasonable
Display	Good	Good	Good

Figure 11.4 Comparative overview of speech and alternative input/output devices

11.6 Design issues

In Sec. 11.2 above, we outlined measurable human factors goals such as speed, error performance, user satisfaction and skill retention. The design of any user interface must attempt to balance and optimize these goals by the selection of suitable media of communication between human and machine, by providing appropriate dialogue structures and so on. In this section, we first compare speech with the most obvious alternative medium—key pressing. Thereafter, we outline some studies of the use of speech in a high workload environment: one focusing on recognition and the other on synthesis. These studies seem to be of particular import for the problem of deciding how best to use speech. Finally, we consider the design of human–computer dialogue featuring spoken interaction.

11.6.1 Speech versus keyboard entry

A number of investigations have attempted to assess the relative merits of speech and keyboard input. However, according to Simpson *et al.* (1985):

> Research comparing speed and accuracy of voice versus manual keyboard input has produced conflicting results, depending upon the unit of input (alphanumerics or function) and other task-specific variables.

In our view, however, the analytical framework provided by Foley, Wallace and Chan (1984) provides an explanation for much of this conflict without ascribing it to imponderable, 'task-specific' factors.

Consider first the issue of one-out-of-N selection. Certainly, the entry of numeric data by keypad in a simple, primary task (i.e. without concurrent, secondary tasking) is much faster and less error prone than entry by speech (e.g. Welch, 1977). This is only to be expected as a comparison of speed of keypressing with the time to utter a word makes clear. Hershman and Hillix (1965) found that targeting the finger and depressing a key on a QWERTY keyboard took some 200 ms for a practised typist; the corresponding figure for unskilled typing is some 1000 ms when typing meaningful text (Devoe, 1967). By contrast, it takes some 400 to 800 ms to utter a typical spoken command (Damper, 1988) with present-day recognizers imposing a further processing delay. In the case of discrete-word input, an additional overhead is introduced by the need for a distinct pause between inputs. Given these figures, it seems unlikely that spoken entry of primary data of low cardinality (i.e. small N) using isolated words could ever be competitive with keying by even a moderately practised typist. Speech, then, is only a candidate input medium for selection tasks when keyboard entry cannot be used for some reason or secondary tasks need to be performed—particularly if these are of a hands- and eyes-busy nature.

One investigation which apparently contradicts this view was performed by Poock (1980). In his study, subjects entered commands (from a set of 90) typical of a naval application, such as 'go to echo' and 'forward message', either by speech or keyboard. The stated findings were that speech input was some 17 per cent faster than typing while typing had 18.2 per cent more errors than speech command entry. Crucially, however, while the spoken commands were entered as *single tokens* (speaking, for example, 'go to echo' as a connected phrase), the keyed commands had to be typed character by character (so that 'go to echo' was composed of 10 or 11 separate input acts).

According to one possible view, this failure to equalize the number of input tokens to effect a given command represents a gross methodological error. Arguably, Poock could be criticized for confounding two different tasks—that of one-out-of-N selection using a speech recognizer and string composition using a QWERTY keyboard. (It would be interesting to see how a 90-key pad allowing commands to be selected by a single keypress—a much closer physical realization to the underlying abstract device than either of the alternatives studied—would have performed.) Since speech and keypress devices are not being compared under equivalent conditions, and also because the task domain is relatively impoverished, the usefulness of any general conclusions that can be drawn from the study is strictly limited. A counter view, however, is that Poock is merely exploiting the inherent advantages that accrue from replacing string by selection, and speech input increases the scope for such replacement.

Certainly, *unrestricted* speech recognition could offer significant potential advantages over keyboard entry of text strings. While arbitrary text must necessarily be entered character by character on the keyboard, an unrestricted speech recognizer could operate essentially as a high-cardinality selection device, choosing words from an effectively unlimited vocabulary. This, of course, is the long-term goal of speech recognition research but such a degree of performance is currently unattainable. In the short to medium term, it is likely that a restricted but large-vocabulary (several thousand words, say) recognizer would be an extremely worthwhile device indeed, since the provision of several thousand keys on a keypress device, to allow direct selection of words, is not an attractive proposition. The objection to Poock's work, therefore, is not so much the principle of replacing string by selection— which is well founded—as the apparent belief that string composition on a QWERTY keyboard is a sensible way to implement small-cardinality selection.

Recently, it has become possible to perform preliminary, experimental comparisons of speech and keyboard entry of text strings using large-vocabulary recognizers. Brandetti *et al.* (1988) used the

'Tangora' recognizer developed by Jelinek and his colleagues at IBM (Jelinek, 1985) in such a study. According to Brandetti *et al.* (1988), the prototype for the Italian language 'recognizes in real time natural language sentences built from a 20,000 word vocabulary'. Subjects entered a 553-word text on two occasions: 'once they used the keyboard only'. (Note the implication, confirmed by a personal communication, that the 'speech' condition is actually speech *plus* keyboard.) For eight non-typist subjects, it required less time to enter the raw text using speech input than using the keyboard; the relative average figures are 15.5 minutes for speech and 22.4 minutes for keyboard. Errors were significantly higher for speech; an average of 7.75 per cent of words in error for speech as against 1.75 per cent for keyboard. The total time to produce corrected text was comparable for the two conditions: 29.7 minutes for speech and 28.6 minutes for keyboard. These figures, averaged for all non-typist subjects, are not given explicitly in Brandetti *et al.*'s paper but have been computed from their published data. For two professional typist subjects, total times were 40.0 minutes for speech as against 22.0 minutes for keyboard, with errors of 8.8 and 0.5 per cent respectively. Subjectively, it is said that 'users found more pleasure and satisfaction in the usage of voice rather than keyboard'. Finally, the authors believe that use of a 'voice activated text editor indicated that large-vocabulary speech recognition can offer a very competitive alternative to traditional text entry'.

It is not clear how they arrive at this conclusion given that speech input is apparently inherently faster than keying, but this advantage is entirely negated by far higher error rates. Figure 11.5 shows a comparison of methods for text input to computer with respect to *maximum* speed (unconstrained by technological limitations) and the number of potential users. It is apparent that speech (along with both cursive and printed handwriting) are by far the most popular forms of textual or verbal communications. The QWERTY keyboard is currently the most popular means of entering textual information into a computer system and there are very few serious contenders to change this situation. Handwriting is approximately half the speed of the QWERTY keyboard while the speed of normal spoken English is approximately twice the speed of the QWERTY keyboard. At a first glance this would appear to suggest that the QWERTY keyboard is inferior to speech but superior to handwriting and that all problems would be solved if speech input were possible. The slow speed of handwriting would indicate that it is too slow to be of any advantage over keyboard entry. This is not the case, as there are good reasons why all three text input methods should be considered. The major advantage and reason for studying the automatic recognition of handwriting and speech is that they are both natural forms of communication as opposed to forms based on the requirement of some mechanical tool (excluding pens and pencils) and therefore have

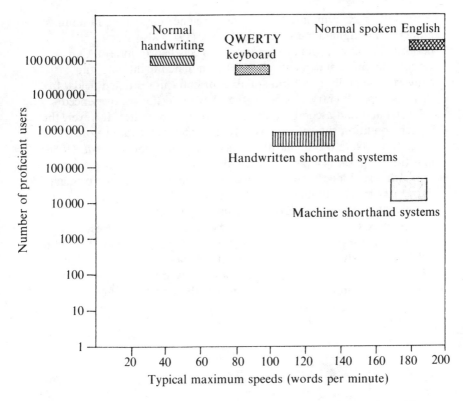

Figure 11.5 Comparison of methods for text input to computer with respect to speed and number of potential users

advantages in office or industrial environments where the need to learn typing skills is unpopular or unlikely to be cost effective. The preceding discussion of the relative merits of speech and keyboards for one-out-of-N selection of command and general text entry has only partially reinforced the advantage of using speech for input purposes.

11.6.2 Speech I/O in a high workload environment

The broad conclusion to be drawn from Sec. 11.6.1 is that, with the current state of technology, speech input is not yet competitive with keyed input for *primary* data entry. However, advantages of speech I/O may accrue when hands and eyes are busy. Thus it seems that speech interaction might come into its own in cases of high workload or concurrent tasking when the use of an additional sensory or motor channel becomes important. In this section we explore this issue further.

As well as examining data-entry performance for simple, random, numeric and alphanumeric strings, Welch (1977) also studied performance in a 'complex scenario'. Subjects had to interpret an English language statement relating to simulated flight data and convert it mentally to a form suitable for entry in restricted fields. In this case, speech entry was faster than keyboard (for inexperienced subjects) and had a comparable 'operational' error rate. However, the speech condition showed a substantially higher error rate before correction. (Note the similarity to the results of Brandetti *et al.* (1988) in a transcription task as described above.)

Welch also added a button-pressing secondary task to the primary data-entry task. Although the secondary task did not impact significantly on speed or error rates for the simple data-entry scenario, the complex scenario revealed that the input speed using speech was degraded less severely than with keyboard or lightpen entry (Fig. 11.6).

Welch also studied the relative merits of speech prompting of the data to be entered versus reading it, and showed that speech prompting reduced errors in the complex scenario both before and after the

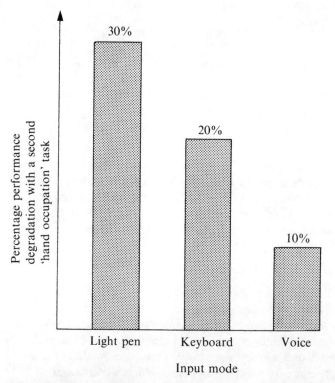

Figure 11.6 Percentage input speed degradation for lightpen, keyboard and voice when a secondary task is added to the primary data entry task (after Welch, 1977)

correction phase. This is interpreted as a direct consequence of freeing the eyes from the reading burden during data entry. By contrast, feedback of the entered data by speech for purposes of verification was found to be too slow and to increase the error rate in most conditions. Broadly similar results from Schurich, Williges and Maynard (1985) confirm that speech feedback should be reserved for situations where a visual display is, for some reason, not feasible. Note that these findings are counter to the predictions of the so-called stimulus central-processing response (S-C-R) model of Wickens and Vidulich (1982) which holds that visual feedback should follow manual data entry whereas auditory feedback should be used in conjunction with speech input. Rather, the conclusion is that the *separation* of instructions and data is helpful, and that instructions should be heard and data seen as previously argued by Hammerton (1975).

Mountfield and North (1980) studied a dual task in which pilots had to keep their (simulated) aircraft on course using a joystick while, at the same time, having to select radio channels either by speech or keypress. They found that tracking performance using the joystick was degraded very little when radio channels were selected by speech; however, errors were much increased with keyboard selection. Also, radio channel selection errors were higher for the keyboard condition. These results are summarized in Fig. 11.7.

Broadly similar results have been obtained by Coler *et al.* (1977) and by Simpson, Coler and Huff (1982). Unfortunately, as for the work of Poock (1980) reviewed above, interpretation of these results is again complicated by use of a full QWERTY keyboard as the keypress device rather than a dedicated function keypad.

It is tempting to explain the speech-input advantage in the dual-tasking studies described above on the basis that the speech channel is additional and does not interfere with the motor channel employed for keypressing in one or other of the tasks. Indeed, we have implicitly taken this to be true in justifying the utility of speech in the hands- and eyes-busy situation. However, on the basis of experimentation, Berman (1984) cautions 'the assumption that . . . freeing a particular channel for inputs or responses must necessarily increase the number of tasks or items that can be attended to is not always true'. One relevant finding is the common observation that speech recognition performance under single-task (e.g. list reading) conditions in the laboratory is always significantly higher than in real or simulated multitask conditions (e.g. Biermann *et al.*, 1985; Damper, Lambourne and Guy, 1985). The implication is that task stress competes for information processing resources which would otherwise be allocated to speech production, even when the concurrent task is manual in nature.

Berman (1984) considers a range of models of human information processing and their implications for dual-task interference when speech I/O is used in conjunction with other modalities. His favoured model is

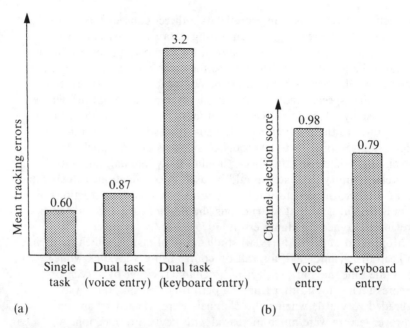

Figure 11.7 (a) Tracking performance in a primary flight-simulation task was degraded much less by speech selection of radio channels in a secondary task than by keyboard selection. (b) Channel selection errors were also lower when using speech selection than using keyboard selection (after Mountfield and North, 1980)

that of McLeod (1977), which extends Kahneman's (1973) model by replacing the 'undifferentiated' single reservoir of processing resources by multiple reservoirs or resources. Berman states:

> The implications for speech recognition are that, if it permits the use of a previously less utilised pool of resources, then the total processing resources involved will have effectively increased. However, it also raises the possibility that the change in response modality and, potentially, in encoding structures, may act to overload a previously less burdened pool of resources, and hence cause an ... increase in task interference.

In a study of the use of speech recognition in television subtitling, Damper, Lambourne and Guy (1985) obtained results that can be interpreted in exactly this way. Speech was compared with keypad for the entry of 'style' parameters (colour, on-screen position, etc.), with a keyboard being used for subtitle text entry in both conditions. It was found that speech input of style increased total preparation time by some 9 per cent in spite of the fact that the time spent transferring between text and style entry was reduced. Significantly, and counter to our initial hypothesis of diminished task interference, there was a more than offsetting increase in the 'time for apparently unrelated activities as evidenced by longer text-entry, "between subtitle" and idle times'. This lends weight to Berman's belief that 'One should not be misled into anticipating workload reductions by adherence to an inappropriate information processing model' (see also Wickens, Sandry and Vidulich, 1983) and reinforces the view that task-specific case studies remain essential given the current state of our knowledge in this area.

We have already seen from the work of Welch (1977) that speech prompting can be beneficial and that the advantage is presumed to result from freeing the eyes from reading during data entry. However, this and subsequent studies have tended to use real, rather than synthetic, speech. A question which naturally arises is the extent to which synthetic speech, and rule based speech in particular, competes for information processing resources with concurrent tasks. This problem has been studied extensively by Pisoni and his co-workers using the MITalk text-to-speech system; the results are summarized in Pisoni, Nusbaum and Greene (1985).

Synthetic speech can be evaluated along several dimensions, including the intelligibility of phonemes, isolated words or words in sentence context, or one can estimate listening comprehension or cognitive load (Klatt, 1987). Pisoni's point of view is that: 'without some model [of human information processing] . . . to serve as a guide to the perceptual process of listening, one can never investigate all the valid differences . . . that affect word recognition and comprehension' (Pisoni, 1982). His experimental results on naming latencies for words and pseudo-words demonstrate 'the existence of both perceptual and cognitive difficulties in perceiving synthetic speech signals' relative to natural speech. A subsequent attempt was made to localize the source of this difficulty within the human information processing model. Indications were that 'synthetic speech in some manner disrupts the subject's ability to maintain information in short term memory . . . under conditions of memory stress. The obvious interpretation . . . is that subjects are "borrowing" from the limited STM capacity needed for maintenance rehearsal . . . in order to encode the synthetic word lists.' Overall, Pisoni and his colleagues feel that these processing and cognitive demands may place important constraints on the usability of synthetic speech in high information-load situations.

11.6.3 Dialogue design

Within any dialogue style, the desirable features according to Kidd (1982) are: initiative, flexibility, complexity, power and information load. *Initiative* can be on the part of the user or the computer, depending on the type of user and the frequency with which the user operates the system. *Flexibility* involves allowing a user to achieve the desired objective by a number of different methods. This should be based on a detailed study of the user's view of the tasks to be carried out, which are not necessarily the same views as the designer. Dialogues which are unnecessarily *complex* should be avoided. It is wise to group commands into logical groups. Provision should be made in the dialogue for *powerful* commands which short-circuit several simple commands. Many users (particularly experienced frequent users of a system) become frustrated when they have to repeat several simple operations to achieve the desired objective. *Information load* is concerned with the amount of cognitive processing and memory load required to use a system. If the load is too high this leads to dislike of the system and an increased error rate while too low a load also results in dislike of the system because it is seen to be unnecessarily pedantic and limiting the user's ability.

In most cases, the justification for the use of speech I/O will almost certainly be economic: that speech improves system throughput and, thereby, productivity. Thus, in the words of Simpson *et al.* (1985): 'it is essential to design a speech system dialogue that facilitates rapid information transfer between human and machine'. Clearly, the basic performance of the recognition algorithms on which the system relies plays a key part in this. However, related factors that impact on the efficacy of spoken dialogue are now reasonably well understood and include:

- Choice of vocabulary
- Syntax constraints
- Prompting and feedback
- Error control
- Response latency

In small-vocabulary, speech command applications, the choice of commands is important for two reasons. Firstly, recognition accuracy is highly dependent upon the acoustic distinctiveness of the words used. Secondly, the vocabulary 'shapes' the dialogue, which must be as natural as possible to promote acceptability, speed up the information entry and reduce the incidence of 'out of vocabulary' utterances. Thus the words chosen must be easy to say, come readily to mind, conform with feedback and prompting messages and be consistent in meaning with the responses they elicit. Not infrequently, these requirements are in conflict. For instance, considerations of naturalness would lead us to prefer use of the familiar letter names to select alphabetic characters. However, reasons of acoustic distinctiveness effectively rule this out; practical systems tend to use the so-called pilot's alphabet. Recent attempts to produce large-vocabulary systems—such as Tangora (Jelinek, 1985) and SPHINX (Lee, 1989)—have unrestricted speech-to-text conversion as their long-term goal. Accordingly, the system designer is severely limited in exercising any choice of vocabulary, which tends to be firmly based on usage in natural language.

For command and control applications, recognition performance is commonly improved (and throughput enhanced) by 'structuring' the vocabulary so that only a subset of commands can follow any given command. Such structuring corresponds to imposing a primitive task syntax—or language model—and, because it constrains the possible utterances in a rigid way, may or may not be appropriate to a particular task. (See Damper, Lambourne and Guy (1985) for an application where structuring was deliberately avoided.) In the large-vocabulary systems being developed and aimed at speech-to-text applications, a more flexible language model is obviously required. This is usually a probabilistic, finite-state grammar based on a Markovian assumption.

If the task to be accomplished is in any way complicated, prompting can be used to good effect to guide the user. This prompting can be as simple as a visual indicator showing when a discrete-utterance recognizer is ready to accept a new command. In discussion of the work of Welch (1977) above, we have already seen that task guidance in the form of speech prompts can be very useful. If we assume that some recognition errors are inevitable, user feedback becomes an important aspect of the dialogue design (Schurich, Williges and Maynard, 1985) because it promotes error detection. Feedback can be either *primary* or *secondary*. With primary feedback, the commands recognized cause some observable system action from which the user can infer the correctness of the recognition. Secondary feedback echoes the recognized command explicitly to the user. Note that the importance and type of feedback depends in large measure on the error rate; if this is very low, explicit (secondary) feedback may not be worth while since it will command the attention of the operator more than is warranted.

The purpose of error-correction procedures is to integrate an error-prone input device (the speech recognizer) into an error-free speech-operated system. In its simplest form, error correction consists of merely cancelling the previously recognized utterance—for example by saying 'delete'—and reentering the desired command. There are a couple of problems with this simple scheme. Firstly, misrecognition, of the 'delete' command, will result in disruption of the dialogue and probable puzzlement on the part of an inexperienced user. Secondly, the requirement to reenter the very utterance which caused the initial misrecognition can easily elicit an unnaturally stressed command from the speaker. Further misrecognitions or rejections follow which, in turn, lead to increasingly anxious but unsuccessful attempts to get the recognizer to accept the intended command. For reasons such as these, a number of authors have advocated various kinds of automatic error correction. For instance, the 'delete' input can be followed by automatic selection of the second choice for the previous utterance (e.g. Damper, 1984). The use of error-correcting parsers has also been suggested (e.g. Spine, Maynard and Williges, 1983).

Finally, we turn to consideration of response latencies. At the most basic level, latencies should be minimized to speed throughput. As mentioned above, this factor by itself tells against the use of speech response; faster visual response is preferred. Consistency of response is also important, since any non-determinancy predisposes users to slow delivery so as to accommodate the worst case. This is a potential problem for connected-word recognizers and may explain, in part, the observation that some users default to a discrete utterance style of input with such devices.

11.7 The simulation of future speech systems

When input/output is to be based on emerging, developing technologies (as in the case of speech) the usefulness of case studies is limited by the performance of available devices. These limitations can be overcome by using humans to simulate future, less-restricted speech I/O devices (Life, Long and Lee, 1988). Data collected from such studies can be invaluable in determining the maximally useful properties of future speech systems. To date, work in this area has concentrated on speech input perhaps because technical capabilities in synthesis are greater or because it is felt to be rather easy to simulate powerful speech output systems—merely by employing a human speaker.

11.7.1 The simulated listening typewriter

According to Simpson et al. (1985): 'By simulating speech recognition hardware, various levels of speech-recognition capability can be controlled and evaluated experimentally'. The work of Chapanis et al. (1977) detailed earlier is, of course, an implicit instance of the use of humans to simulate speech I/O devices. In that case, however, the 'simulated technology' was of such power and sophistication that 'relevance to current speech recognition capabilities is limited' (Simpson et al., 1985). By contrast, Gould, Conti and Hovanyecz (1983) performed an important study in which the capabilities of a simulated speech-input device were constrained in various ways.

Gould and his colleagues asked the question: 'Would an imperfect listening typewriter be useful for composing letters?' A human typist entering text on a conventional (QWERTY) keyboard was the basis of the simulated listening typewriter (SLT). The particular imperfections studied were limitations of vocabulary size and the need for artificial pauses between utterances. Thus, the various different versions of SLT used had:

- Either 1000-word, 5000-word or unlimited vocabulary;
- Either isolated-word or connected-word capability.

The vocabulary restriction was simulated by matching the typist's keyboard entries to words stored in a fixed-size dictionary. Words not in the dictionary could be entered in a spell mode. Feedback to the subjects used a visual display unit.

Subjects were professionals used to office work, grouped into those with and without dictation experience. They were required to compose letters by various means: namely, speech input using the SLT, writing or dictation onto audio tape for later transcription. Only those with dictation experience produced letters by dictation. Subjects without dictation experience were asked to produce a corrected (FIRST FINAL) letter by writing but used the SLT to produce both DRAFT and FIRST FINAL letters. Dictators were allowed to choose their own strategy—possibly because they did not have sight of the dictation letters. Results were assessed in terms of speed, subjective preference and the 'quality' of the letters produced as judged by a panel of vetters.

Use of the SLT was claimed to be faster than writing but slower than dictation, although the latter does not produce hard copy output in 'real time'. As might be expected, the isolated-word SLT was slower than the connected-word input and speed generally increased with vocabulary size. Figure 11.8 shows mean composition times for the subjects without dictation experience under the FINAL strategy. The subjects mostly preferred the speech condition, presumably because of the combination of reasonable speed and observable results in 'real time'. Interestingly, large-vocabulary size was judged more important than connected-word capability by the subjects. Finally, and not entirely surprisingly, letter quality was more a function of composition time than of means of input. This latter finding argues that, for creative writing at least, speed of input is not a vital issue since the slow process of composition is the limiting factor.

11.7.2 Methodological objections

Gould, Conti and Hovanyecz's work was influential in establishing simulation as a legitimate methodology in the study of the human factors of speech input. It has, however, a number of fundamental shortcomings as recognized by the researchers themselves. Firstly, the use of a QWERTY typist leads to slow response times set by the speed at which this keyboard can be operated. Secondly, the SLT does not simulate error patterns (substitutions, rejections, deletions and insertions) at all realistically. In fact, all errors were either rejections (typographical errors on the part of the typist or out-of-vocabulary utterances), appearing as XXXXs on the screen, or simulated homophones (the highest frequency token of a homophone pair was always selected initially). Additionally, Simpson et al. (1985) implicate 'inconsistent restriction of discrete data entry when the spelling mode was used . . .' as a shortcoming.

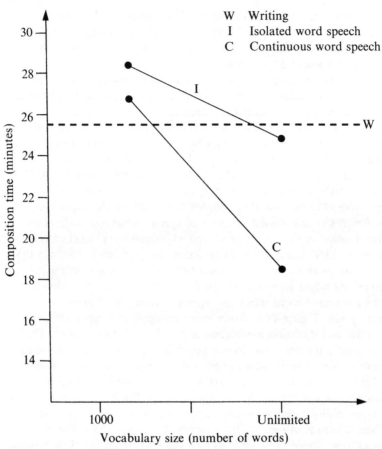

Figure 11.8 Mean composition times for letters produced by subjects without dictation experience (after Gould, Conti and Hovanyecz, 1983)

The apparent superiority of connected-word over isolated-word input also warrants further discussion. For instance, Biermann *et al.* (1985) state: 'users seem to be able to learn to speak machine recognizable discrete speech more easily than they are able to speak machine recognizable connected speech'. This is borne out by some (unpublished) work in our laboratory in which we observed that practised speakers chose to enter spoken commands to a recognizer with connected-word capability (Marconi SR-128) in isolated fashion because this reduced the error rate. In criticism of Gould, Conti and Hovanyecz's work in this connection, Biermann *et al.* state: 'no ultimate conclusions can be drawn for long-term interactions in the applications environment', presumably on the grounds that the simulation of connected-speech capability is unrealistically powerful for the foreseeable future. It must be remembered, however, that the SLT work indicates that vocabulary size is more important than speech mode; to this extent, there is perhaps less conflict between Gould et al. and Biermann et al. than might appear.

11.7.3 Recent work

Recent work by Newell and his co-workers (e.g. Newell, Arnott and Dye, 1987; Carter, Newell and Arnott, 1988) has overcome several of these objections in a simulation of speech-driven word processor (SDWP). In particular, these researchers use a Palantype (machine shorthand) keyboard in place of the QWERTY keyboard employed by Gould, Conti and Hovanyecz, thereby much reducing the response time of the simulated system. Input is effectively connected speech with a very large vocabulary of some 13 000 words. Using university students as subjects, composition rates for the SDWP (7.9 words/minute) were not as high as those obtained by Gould and his colleagues (11.5 words/minute) for the unlimited vocabulary, connected-word SLT using inexperienced dictators (Carter, Newell and Arnott, 1988). Also, subjects were less impressed with the SDWP than were Gould's subjects with the SLT, all ranking speech as worse than writing even though the simulation was for full-speed, connected speech with a very large vocabulary. Interestingly, those subjects who thought they were using a real SDWP (the 'covert' group) were less impressed than those who were told the SDWP was a simulation (the 'overt' group).

Overall, simulation is a powerful way of assessing what would be useful features of future speech systems and, therefore, what ought to be the priorities for development. For instance, the finding of Gould, Conti and Hovanyecz (1983) that a listening typewriter restricted to isolated-word input might prove useful, provided it had a large enough vocabulary, would be difficult to obtain by other means. There is, however, another potentially important application for such simulations. At present, machine shorthand (such as Palantype) is the only feasible, real-time means of large-vocabulary text-to-speech conversion in existence. As such, it offers a vehicle for conducting research at the 'higher' (linguistic) levels of speech understanding without first having to solve the lower-level problem of acoustic-to-symbolic transformation. At present, this potential appears to be unrealized.

11.8 Conclusions

In this chapter we have considered the human factors of speech communication with machine and have shown that a detailed study of human factors is crucial to the design of a good speech-driven interface. Speech input/output has the potential advantages of naturalness, improved speed and efficiency in a number of applications when compared to other input/output devices. However, interaction with machine has a number of problems associated with it which must be considered before choosing to use speech in a particular application.

There is infinite variety between people both psychologically and physically. The physical factors of speech and hearing can vary considerably from one person to another and even with the same person from time to time. The psychological features of humans in terms of their intellectual processing power, memory and perception can also vary considerably. Indeed, the mechanisms of human processing of information are far from understood and several conflicting models of the processes involved have been proposed. In addition, the capabilities of current speech technology are far inferior to the speech and hearing capabilities of humans.

Thus we have the job of matching a poorly understood model of the user to a technology of limited capability which is attempting to emulate the natural communication processes of the user in order for the user to carry out some useful task. To achieve this the task to be accomplished must be analysed into subtasks and allocated to person or machine to optimize the overall performance and acceptability of the whole system. There is no simple process by which this can be achieved. In this chapter we have shown that while there are a number of basic guidelines that can be applied to the use of speech, and indeed any interactive device, the use of detailed case studies and simulations of the tasks is an important aspect in the design of a speech-based interactive system. In carrying out these case studies and tasks the basic measurable goals that can be utilized are speed of performance, rate of errors, subjective satisfaction, time to learn and retention over time.

We have discussed the concept of 'abstract' computer interactive devices and their potential use for performing the types of 'abstract' operations required in any interactive environment: selection, string, quantify, orient, position and path. While speech is a natural communication medium it is poor at describing, for example, the position of an object, but is good for selecting one-out-of-N categories. However, our discussion of the relevant literature would suggest that even in the one-out-of-N selection case speech is inferior to keyboard entry, except in environments where a keyboard cannot be used or hands-busy, eyes-busy secondary tasks need to be performed.

In general, *current* speech technology is only competitive in applications where the user's hands and eyes are busy, when the user needs to be remote from the system and speech communication is the only available channel, when the user needs to be mobile and cannot operate other input/output devices, or when other input/output devices are ineffective. The potential use of *future* speech technology featuring high-quality speech synthesis and recognition performance is high and can be assessed by simulation studies. However, the decision when to use speech and when not to will still remain. In general, when employing speech input or speech output from a system the designer should first ensure that the use of speech is indeed the most effective solution to the problem and that the technology is capable of eliciting the levels of human performance and user acceptability required of the application.

References

Baddeley, A. D. (1983) *Your Memory: A User's Guide*, Penguin, Harmondsworth.
Baecker, R. M. and W. A. S. Buxton (1987) *Readings in Human-Computer Interaction: A Multidisciplinary Approach*, Morgan Kaufmann, Los Altos, California.
Bailey, R. W. (1982) *Human Performance Engineering: a Guide for System Designers*, Prentice-Hall, Englewood Cliffs, N.J.
Berman, J. V. F. (1984) 'Speech technology in a high workload environment' in *Proceedings of 1st International Conference on Speech Technology*, J. N. Holmes (ed.), 69–76, IFS and Elsevier (North Holland).
Biermann, A. W., R. D. Rodman, D. C. Rubin and J. F. Heidlage (1985) 'Natural language with discrete speech as a mode for human-to-machine communication', *Communications of the ACM*, vol. 28, 628–636.
Brandetti, M., P. D'Orta, M. Feretti and S. Scarci (1988) 'Experiments on the usage of a voice activated text editor', Proceedings of Speech '88, 7th FASE Symposium, Edinburgh, 1305–1310.
Card, S. K., T. P. Moran and A. A. Newell (1983) *The Psychology of Human Computer Interaction*, Lawrence Erlbaum Associates, Hillsdale, N.J.
Carter, K. E. P., A. F. Newell and J. L. Arnott (1988) 'Studies with a simulated listening typewriter', Proceedings of Speech '88, 7th FASE Symposium, Edinburgh, 1289–1296.
Chapanis, A., R. N. Parrish, R. B. Ochsman and G. D. Weeks (1977) 'Studies in interactive communication: II. The effects of four communications modes on the linguistic performance of teams during cooperative problem solving', *Human Factors*, vol. 19, no. 2.
Coler, C. R., R. P. Plummer, E. M. Huff and M. Hitchcock (1977) 'Automatic speech recognition research at NASA–Ames Research Center', Proceedings of the Voice-Interactive Real-Time Command/Control Systems Applications Conference, NASA–Ames Research Center, Moffett Field, Calif., 143–163.
Damper, R. I. (1984) 'Voice input aids for the physically disabled', *International Journal of Man–Machine Studies*, vol. 21, 541–553.
Damper, R. I. (1988) 'Practical experiences with speech data entry' in *Contemporary Ergonomics*, E. D. Megaw (ed.), Taylor and Francis, London.
Damper, R. I. (1990) 'Speech aids for the handicapped' in *Advances in Speech, Hearing and Language Processing*, W. A. Ainsworth (ed.), JAI Press, London (in press).
Damper, R. I., A. D. Lambourne and D. P. Guy (1985) 'Speech input as an adjunct to keyboard entry in television subtitling' in *Human–Computer Interaction—INTERACT '84*, B. Shackel (ed.), 203–208, Elsevier (North Holland), Amsterdam.
Deatherage, B. H. (1972) 'Auditory and other sensory forms of information

presentation' in *Human Engineering Guide to Equipment Design*, H. P. Van Cott and R. G. Kinkade (eds), revised edn, US Government Printing Office, Washington DC.

Devoe, D. B. (1967) 'Alternatives to handprinting in the manual entry of data', *IEEE Transactions on Human Factors in Electronics*, vol. HFE-8, 21–31.

Diaper, D. (1989) 'The discipline of HCI' (editorial), *Interacting with Computers*, vol. 1, 3–5.

Downton, A. C. (ed.) (1991) *Engineering the Human Computer Interface*, McGraw-Hill, London.

Foley, J. D., V. L. Wallace and P. Chan (1984) 'The human factors of graphics interaction techniques', *IEEE Computer Graphics and Applications*, vol. 4, 13–48.

Gould, J. D., J. Conti and T. Hovanyecz (1983) 'Composing letters with a simulated listening typewriter', *Communications of the ACM*, vol. 26, 295–308.

Hammerton, M. (1975) 'The use of same or different sensory modalities for information and instructions', *Ergonomics*, vol. 18, 683–686.

Hershman, R. L. and W. A. Hillix (1965) 'Data processing in typing: typing rate as a function of kind of material and amount exposed', *Human Factors*, vol. 7, 483–492.

Jelinek, F. (1985) 'The development of an experimental discrete dictation recogniser', *Proceedings of the IEEE*, vol. 73, 1616–1624.

Kahneman, D. (1973). *Attention for Effort*, Prentice-Hall, Englewood Cliffs, NJ.

Kantowitz, B. H. and R. D. Sorkin (1983) *Human Factors: Understanding People–System Relationships*, Wiley, New York.

Kidd, A. L. (1982) 'Man–machine dialogue design', Research Study 1, no. 1, Martlesham Consultancy Services, BTRL.

Klatt, D. H. (1987) 'Review of text-to-speech conversion for English', *Journal of the Acoustical Society of America*, vol. 82, 737–793.

Lea, W. A. (1980) 'The value of speech recognition systems' in *Trends in Speech Recognition*, W. A. Lea (ed.), 5–18, Prentice-Hall, Englewood Cliffs, N.J.

Lee, K-F. (1989) *Automatic Speech Recognition: The Development of the SPHINX System*, Kluwer, Boston.

Life, M. A., J. B. Long and B. P. Lee (1988) 'Human simulation of speech technology: an illustration of the ergonomic approach' in *Contemporary Ergonomics*, E. D. McGaw (ed.), Taylor and Francis, London.

McLeod, P. (1977) 'A dual task response modality effect: support for multi-processor models of attention', *Quarterly Journal of Experimental Psychology*, vol. 29, 651–667.

Miller, G. A. (1956) 'The magical number seven plus or minus two: some limits on our capacity for processing information', *Psychological Review*, vol. 63, 81–97.

Monk, A. (ed.) (1985) *Fundamentals of Human–Computer Interaction*, Academic Press, London.

Mountfield, S. J. and R. A. North (1980) 'Voice entry for reducing pilot workload', Proceedings of the Human Factors Society, 185–189.

Newell, A. F. (1986) 'Speech communication technology—lessons from the disabled', *IEE Electronics and Power*, vol. 32, no. 9, 661–664, September.

Newell, A. F., J. L. Arnott and R. Dye (1987) 'A full speed speech simulation of speech recognition machines', Proceedings of European Conference on Speech Technology, Edinburgh.

Pisoni, D. B. (1982) 'Perception of speech: the human listener as a cognitive interface', *Speech Technology*, vol. 1, 10–23.

Pisoni, D. B., H. C. Nusbaum and B. G. Greene (1985) 'Perception of synthetic speech generated by rule', *Proceedings of the IEEE*, vol. 73, 1665–1676.
Poock, G. K. (1980) 'Experiments with voice input for command and control', Naval Postgraduate School Report, NPS55-80-016, Monterey, Calif.
Rumelhart, D. E. and J. L. McClelland (1986) *Parallel Distributed Processing*, vol. 1, *Foundations*, MIT Press, Cambridge, Mass.
Schurich, J. M., B. H. Williges and J. F. Maynard (1985) 'User feedback requirements with automatic speech recognition', *Ergonomics*, vol. 28, 1543–1555.
Shneiderman, B. (1987) *Designing the User Interface: Strategies for Effective Human–Computer Interaction*, Addison-Wesley, Reading, Mass.
Simpson, C. A., M. E. McCauley, E. F. Roland, J. C. Ruth and B. H. Williges (1985) 'System design considerations for speech recognition and generation', *Human Factors*, vol. 27, 115–141.
Speech Technology (1989) Special Issue on Large Vocabulary Recognisers, vol. 4, no. 4, April/May.
Spine, T. M., J. F. Maynard and B. H. Williges (1983) 'Error correction strategies for voice recognition', Proceedings of the Voice Data Entry Systems Applications Conference, Chicago, Ill.
Wallich, P. (1987) 'Putting speech recognisers to work', *IEEE Spectrum*, vol. 24, no. 4, 55–57, April.
Welch, J. R. (1977) 'Automated data entry analysis', Rome Air Development Center Report RADC TR-77-306, Griffiss Air Force Base, N.Y.
Wickens, C. D. and M. Vidulich (1982) 'S-C-R compatibility and dual task performance in two complex information processing tasks: threat evaluation and fault diagnosis', Technical Report no. EPL-82-3, Department of Psychology, University of Illinois, Champaign, Ill.
Wickens, C. D., D. L. Sandry and M. Vidulich (1983) 'Compatibility and resource competition between modalities of input, central processing and output', *Human Factors*, vol. 25, 227–248.

Index

Accent mark, 193, 207
Accuracy of recognition, 249
Acoustic cue, 32
Acoustic feature extractor, 230
Acoustic knowledge, 232
Acoustic model, 237
Acoustic realization (phoneme), 247
Adaptation, 20
Adaptation control, 94, 95
Adaptation rate, 95
Adaptive differential pulse code modulation (see Modulation, ADPCM)
Adaptive predictor, 94
Adaptive quantizer, 94
ADC (see Converter, analogue-to-digital)
ADPCM (see Modulation, ADPCM)
Affricate sound, 19
  phoneme symbols, 6
A-law, 87, 89, 90, 95, 341
Algorithm, for ASR, 235
Aliasing, 40, 81, 179
Aliasing distortion, 82
Allophone, 200, 204
Allophonic segment, 188
Alveolar ridge, 15, 37, 200
Alveolar tap, 201
Amplifier distortion, 115
Amplitude density function, 41, 83
Amplitude distribution, Gaussian, 90
Amplitude quantization, 75, 82
Amplitude resolution (of ADC), 98, 101, 130
Analogue-to-digital converter (see Converter, analogue-to-digital)
Analysis filter (see Filter, LPC analysis)
Analysis, linear predictive, 162
Analysis of sentence, 71
Application level, 35
Application of ASR, 252
ARMADA, 248
Articulation, 20
Articulator, 5, 214

Articulatory apparatus, 1
Articulatory model, 214
ASR (see Automatic speech recognition)
Assessment, subjective, 79
Assimilation, 20
Asynchronous transfer mode, 303–304, 317
ATM (see Asynchronous transfer mode)
Auditory feedback, 22
Auditory nerve, 26, 27
Auto-correlation, 56, 57
Auto-correlation method (of LPC), 165
Auto-covariance, 56
Auto-covariance function, 57–59
Auto-covariance matrix, 63
Automatic pattern recognition (see Pattern recognition)
Automatic speech recognition, 223, 226, 257
  early approaches, 227
Average energy, 43
Average magnitude, 39, 41

B-channel (ISDN), 302–303
Bandwidth:
  channel, 91
  telephone, 77, 82
  transmission, 89, 93–95
Basilar membrane, 25–26, 31
Bass voice, 4
Baum–Welch algorithm, 242, 245
BBC computer, 205
Beam search, 276–277, 282
Bearer services, 302
Bit error rate (see Error rate)
Bit rate, variable, 288–289
Bitstream (DAC), 115, 137, 150
Block processing, 47
Boundary, synchronous/asynchronous, 310
Branching factor, 249
Breath group, 208

394

INDEX

Breathing, 2
Breathing cycle, 3
British English, 184
Broadband ISDN, 303, 317, 319
Buffer, circular, 340
BYBLOS, 248

Cable, undersea, 344
Call distribution, automatic, 355
Call forwarding, 349
Cambridge fast ring (*see* Network, Cambridge fast ring)
Capability profile, 249
CCITT I series, 299
CCITT I.462, 303
CCITT G.711, 87
CCITT G.712, 78, 82
CCITT G.721, 94
CCITT G.722, 96, 181
CCITT X.25, 338
CCITT X.31, 303
CELP (*see* Coding, code excited linear predictive)
Cepstral analysis, 64
Cepstral coefficients, 66–67
Cepstral function, 65
Cepstral smoothing, 66
Cepstrum, 65
Cerebral cortex, 32
Channel, digital transmission, 75, 91
Checksum, 288
Cholesky decomposition, 164
Circuit switching, 289, 332
Classifier, 259, 261
    Bayesian, 235, 260
    maximum likelihood, 63, 236, 242
    minimum distance, 260
    nearest neighbour, 63, 260
    Viterbi, 243
Clipping, subjective effect, 307
Clipping distortion (*see* Distortion)
Clipping level, mean square, 84
Cliping point, 78, 82–84, 90
Closure, vocal tract, 15
Co-articulation, 21, 203, 214
Cochlea, 23–25
Cochlear nerve, 24
Codebook:
    binary search, 172
    full search, 172
    VQ coder, 171–172
    VQ decoder, 172
Codebook generation, 173
Coder:
    ADPCM, 94

noise shaping, 115, 118
recursive model, 118
sub-band, 178
Coding:
    code excited linear predictive, 177
    comparison of LPC methods, 168
    differential, 91
    embedded, 309
    LPC, 159, 162, 233, 268
    parametric, 60, 158
    residual excited linear predictive, 175
    waveform coding system, 75
Collision detection, 293
Command system, voice operated, 266, 278
Compander, 85, 87, 90
Comparator, 228, 257
Compressor, 85–86, 88–90
Computer, input task, 362
Computer input:
    joystick, 375, 381
    keyboard, 363, 375–6, 378, 380–381, 386–387
    lightpen, 375, 380
    mouse, 375
    speech, 363, 375, 377–378, 380–381, 386–387
    touch-screen, 375
    tracker ball, 375
Computer output, speech, 383
Computer, speech I/O, 360–361
Conference bridge (ISLAND), 342, 345
Conference call, 288, 340
Conference server, 340, 343, 346
Conferencing, audio-visual, 77, 322
Congestion control, 309
Consonant, voiced, 5
Consonant cluster, 203
Constriction, 1, 18, 20
Content word, 207
Continuous process, 46
Contralto voice, 4
Control (PABX), 344
Control:
    centralized, 310
    distributed, 310
CSMA (*see* Network, CSMA/CD)
Conversational speech, 29
Conversion speed (of ADC), 98
Converter:
    analogue-to-digital, 81–82, 85, 89, 91, 93, 97, 129, 135, 152

digital-to-analogue, 75, 89, 91, 97,
    129, 131, 135, 150, 152
  flash, 137
  pcm to 1 bit, 143
Convolution, 39
Copy synthesis, 184
Correlation, 57
Covariance method (of LPC), 163
Critical band, 30–31

D-channel (ISDN), 302–303
DAC (*see* Converter,
    digital-to-analogue)
Data characteristics, 328
Data structure, for text-to-speech, 186
Data, voiceband data signal, 95
Database, speech, 43, 261
d.c. offset, 46
Decimation, 179
  three-stage system, 107
Declarative sentence, 209
De-emphasis filter, 56
Delay element, 61
Delay:
  in data connection, 330
  mean transfer, 299
  in packet network, 304, 338
  variable, 305–307, 309
  in voice connection, 328–329
Delay tolerance, 288, 331
Density function:
  amplitude, 41, 83
  Laplace, 43
DFT (*see* Discrete Fourier transform)
Dialling, short code, 349
Dialogue structure, 251–252
Dictionary, 184
  exceptions, 188, 139–194
Dictionary search, 194
Differential coding, 91
Digital telephony, 78–79
Digital to analogue converter (*see*
    Converter, digital-to-analogue)
Digital transmission, 74–75
Digital transmission channel, 91
Diphone, 202, 206
Diphthong, 14
Diphthong sound, phoneme symbols, 6
Direct Memory Access, 287
Discomfort, threshold, 29
Discrete Fourier transform, 46, 48, 64
Discriminant function, 260
Distance, cumulative, 273
Distance measure, 62, 171, 226, 268,
    272

city-block, 269
Euclidean, 63, 229, 269
Itakura–Saito, 64, 173
Mahalanobis, 63
squared Euclidean, 63
Distortion, 78
  aliasing, 81–82
  decorrelation of, 101, 152
  peak clipping, 82, 90
  quantizing, 75, 78, 82–83, 93, 118
  slope overload, 93
Distortion measure (*see* Distance
    measure)
Distortion reduction factor, 123
Distributed blackboard, 322
Dither, 130, 152
  digital, 133
  high-level, 142
  intelligent, 98, 123
  self, 98, 141
Document retrieval, 288
D-traffic (Magnet), 314
Document, multi-media, 353
Document processing, 322
Down-sampling, 40, 46, 175, 178–179
DRAGON-DICTATE, 248
Drop-out (PABX), 327, 346
Durbin's recursive procedure, 166
Dynamic element matching, 134
Dynamic programming, 233, 237, 241,
    270–271, 273, 278, 281
  global constraint, 275
  local constraint, 275
  penalty function, 274–275
Dynamic range, 77, 91, 94, 101
Dynamic time warping, 234, 241, 270

Ear, 24, 78
  inner, 24–25
  middle, 23–24
  outer, 23–24
Eardrum, 24
Electronic mail, 322, 356
Encoding, for model, 235
End effects, 49
End point location, 267
Endolymph, 25
Energy, short time, 44
Environment, affecting ASR, 250, 267
Epiglottis, 2
Error:
  mean square, 90, 162
  transmission, 306
Error detection, 339
Error rate, 291, 328–330, 332, 338

INDEX

Error recovery, 291
Error tolerance, 288, 331
Esophagus, 2
Estimation, in model, 236
Ethernet, 293
Etherphone (Xerox PARC), 349
Euclidean distance (*see* Distance measure)
Eustachian tube, 24
Exceptions dictionary, 188, 193–194
Exchange, 302
   private (PABX), 312, 322–329, 333, 336–337, 340, 343, 347, 355–357
Excitation, speech, 11, 158, 215
Excitation function, 66
Excitation sequence, 64
Expander, 86, 89
Expert system, 184
Expiration, 2
External canal, 24

Facsimile, 322, 324
Fast Fourier transform, 47
Fatigue, 28
Feature, 63
Feature extractor, 259
Feature processor (PABX), 344–348
Feature space, 259
Feature vector, 259
Feedback, 22
   encoder feedback loop, 91, 94
FFT (*see* Fast Fourier transform)
File transfer, 322
Fileserver, 341–342, 353
Filter, analogue, 98
   anti-aliasing, 98–99, 131
   band-limiting, 75, 82, 91
   decimation, 100–101, 107
   de-emphasis, 56
   digital, 152, 219, 220
   FIR, 61, 110
   FIR direct structure, 104
   FIR transposed structure, 104
   half-band, 105
   IIR, 109
   interpolation, 100–101, 107
   LPC analysis, 159–162, 168
   LPC synthesis, 162, 169
   moving average, 40, 46
   multi-stage, 105
   noise shaping, 119
   prediction, 47
   pre-emphasis, 54
   quadrature mirror, 179
   reconstruction, 75, 80, 91, 98–99, 150
   time-variant, 11
   transversal, 61
   vocal tract, 64, 213, 215
Filter bank, 268, 277
Filter design:
   FIR equiripple method, 112
   FIR half-band design trick, 113
   FIR Remez exchange method, 112
   FIR window method, 110
Flow control, 339
Formality of language, 191
Formant frequency, 7–8, 10, 12, 30, 54, 60, 66, 215
Formant position, 13
Formant resonator, 219
Forward–backward algorithm, 242
Fourier transform, 41, 64, 66, 80, 245, 268
Frame relaying, 303
Frame switching, 303
Frequency domain, 39, 49, 80
Frication, 7, 37
Fricative, voiced, 158
Fricative sound, 18, 36, 60, 215
   phoneme symbols, 6
FSM (*see* Machine, finite state)
Function word, 207
Fundamental frequency, 4, 7–8, 29, 208
   synthesis of, 211
Fundamental pitch, 54, 57

Gateway, wide area, 344–345
Guassian amplitude distribution, 57, 90
Generation of sentence, 69
Glide sound, 15
Glottal pulse, 4, 37, 60, 62
Glottal pulse frequency, 158
   (*see also* Pitch frequency)
Glottis, 3
Grammar, 67, 72, 230, 233, 263
   context-free, 67, 68
   context-sensitive, 68
   finite state, 67, 72, 228–229
   formal, 238
   pattern, 261–262
   stochastic regular, 238
   toy grammar, 68
Grammar network generator, 233

Hair cells, 26
Half-band filter (*see* Filter, half-band)
Hamming window (*see* Window function, Hamming)

Hangover, speech detector, 306
H-channel (ISDN), 302
Hann window (*see* Window function, Hann)
Hard palate, 2
Harmonics, 5, 7, 54
HARPY, 231, 281
Hash function, 195
HCI (*see* Interaction, human–computer)
Head syllable, 209
Hearing loss, 28
Hearing, threshold of, 28
Helicotrema, 25
Hidden Markov model, 223, 238, 245
 sub-word, 246–247
 whole word, 245, 247
 sequence of, 244, 247
HMM (*see* Hidden Markov model)
Human capability, hearing, 367
 memory, 367
 model of, 364
 perception, 368
 physiological, 360
 psychological, 360
 reasoning, 368
 voice, 366
Human factors, 360
 methodology, 361
Human memory:
 long-term, 367–368
 short-term, 367–368, 372
Human–computer interface, 252

IEEE 802, 293–297, 312, 317, 356
Imperative sentence, 209
Incus, 24
Infrastructure, (LAN), 325
Integrated services, 287, 299, 303
ISDN, 288, 299
 packet mode access, 303
Integrated services, 323, 325, 327, 333, 355–357
Integration, voice and data, 356
Intelligibility, 77, 80
Interaction, human–computer, 360–361, 364
Interface, computer–human, 350
Intermodulation products, 101
International Phonetic Alphabet, 6
Interpolation, 179
Inter-symbol interference, 105
Intonation, 21, 72
 (*see also* Prosody)
Inverse discrete Fourier transform, 49, 66

Inverse filter, 159, 161, 169
IPA (*see* International Phonetic Alphabet)
ISDN, 323, 332, 346, 356
ISDN private exchange, 356
ISLAND project, 325–327, 334–338, 340–343, 346–347, 353
Itakura–Saito distance (*see* Distance measure)

Juncture, 22

Kempelen, Wolfgang von, 10
Keyword spotting, 278
Knowledge, 235, 263
 lexical, 281
 syntactic, 281
Knowledge base, 184
Knowledge engineering, 227, 231, 264
Knowledge network, 281
Knowledge source, for understanding speech, 231, 234, 236

LAN (*see* Network, local area)
Language, 67
 natural, 68, 71–72, 249, 251
 regular, 67
 spoken, 191
 stress-timed, 211
 syllable-timed, 211
 written, 191
Language model, 237
Laplace density function, 43
Larynx, 3, 19
Latency, 291
Lattice, HMM data structure, 239, 241, 244
Lattice method (of LPC), 167
Layer, ISO model, 298
Least mean square (1ms) error, 162
Letter-to-phoneme, conversion by rule, 196–199
Levinson's method, 166
Lexical expander, 233
Lexicon, 71, 230
Lexicon rules, 68–69
Line spectrum, 49
Linear discriminant function, 260
Linear prediction, 60
Linear prediction coefficients, 67
Linear predictive analysis, 47
Linear predictor, 159
Lips, 2
Liquid sound, 15
Listen before talk, 293
Listen while talking, 293

Local line, 302
Local transmission path, 77
Logarithmic law, 88–89
Loudness, 20, 28, 207
  equal loudness contours, 28
LPC: *see* Coding, LPC
Lung, 2, 208

μ-law, 87–89, 95
MAN (*see* Network, metropolitan area)
Machine, finite state, 237
Mahalanobis distance (*see* Distance measure)
Main lobe, 41, 47, 50, 54
Mains ripple, 46
Major phrase, 208
Malleus, 24
Markov chain, 238
Masking, 30–31, 78, 85
  backwards, 31
  forwards, 31
  suppression, 31
  swamping, 31
Maximum likelihood classifier (*see* Classifier)
Mean opinion score, 79–80, 95, 305–307
Meaningful sentence, 69
Measurement, in model, 235
Media, for packet network, 292, 294
Metric, 63, 226
Microphone, 43, 266
Mini-packet, 296–297
Minor phrase, 208
MITalk, 198, 210, 212
Mobile phone, 329
Model, 239, 245
  acoustic, 237
  generative, 237–238
  hidden Markov (*see* Hidden Markov model)
  language, 237
  source-filter, 9, 64
  speech pattern, 236
Modulation, ADPCM, 74, 94–95, 178, 181
  PAM, 75
  PCM, 74, 95
Modulator:
  delta, 97, 122
  delta sigma, 97, 115, 135, 141, 143, 150–151
Monopulse coding (LPC), 174
Morph, 198, 203, 210, 263
Morphological decomposition, 199–200
Motor theory, 32
Multipulse coding (LPC), 174

Nasal cavity, 2, 24
Nasal sound, 15
  phoneme symbols, 6
Nasal tract, 5
Natural language (*see* Language, natural)
Navy rules, 196–197
Nearest neighbour classifier (*see* Classifier)
Negative feedback, 115
Nerve fibre, 26–27, 31
  firing, 27, 31
Network:
  Cambridge fast ring, 296–297, 313, 334–335, 338, 342, 345
  circuit switched, 323, 343
  CSMA/CD, 293–294, 298
  Fasnet, 312, 333
  FDDI-2, 312, 333–334
  FXNet, 312, 333
  local area, 289, 293, 311, 325, 329, 333–334, 344
  Magnet, 312–313, 333
  metropolitan area, 291, 317, 343
  Orwell, 312, 315, 333–334
  packet switched, 288
  Philan, 312, 333
  Prelude, 312, 333–334
  QPSX/DQDB, 312, 317–318, 333–334
  SCPS, 312, 333–334
  slotted ring, 296, 335
  telephone, 288
  telex, 288
  token passing, 294–295, 299
  wide area, 291, 343
Network bandwidth, 328, 330–331
Network characteristics, 331
Network design, 328, 333
Network interface, 326
Network performance, 298, 299
Network toplogy, 291
  broadcast, 292
  bus, 291
  fully connected, 292
  loop, 291
  mesh, 292
  star, 291
  treee, 292
Neurobiology, 32
Neuron, 27
  firing, 27

Neutral vowel, 9, 14
Noise:
  bandlimited, 84
  idle channel, 150
  quantization, 130
  random, 62, 78
Noise component of speech signals, 158
Noise generator, 10
Noise shaping, 97, 114–115, 133, 150
Noise shaping advantage, 133
Non-terminal symbols, 68
Nuclear syllable, 209
Nyquist bandwidth, 170
Nyquist sampling rate, 99
Nyquist sampling theorem, 40

Observation sequence (HMM), 239–244
Octet, 294, 319
Office environment, 322–325, 332, 349, 355
Ohm's acoustical law, 30
Open systems interconnection, 297
Operating system, 326, 340
Opinion, user opinion, 78
Optical fibre, impact on network, 304
OSI (see Open systems interconnection)
Optimality, principle, 273
Oral tract, 2, 5, 12
Organ of Corti, 26
Orthography, 184
Ossicles, 24
Oval window, 24–25
Overlapping window function, 48
Oversampling, 97, 99, 115, 133, 150
  high, 135
  mild, 129
Oversampling ratio, 99, 103, 123, 139

PABX (see Exchange, private)
Packet collision, 294
Packet destination, 291
Packet interleaving, 288
Packet loss, 308, 339
Packet network, 287, 289
  integrated, 310–311
Packet radio, 292
Packet satellite, 292
Packet switching, 287
Packet voice terminal, 308
Packetization, 308
Packetized speech, 304
Pain, threshold of, 29
Palantype keyboard, 389

PAM (see Modulation, PAM)
Parameter re-estimation (HMM), 237, 242
parametric coding (see Coding, parametric)
PARCOR (see Partial correlation coefficients)
Parser, 71–72
  chart, 71
  halted, 71–72
  phrase level, 199
  syntax, 186, 230
Parsing program, 71
Part of speech, 72, 210
Partial correlation coefficients, 167
Path pruning, 275, 282
Pattern:
  input to recognizer, 257–259, 269
  reference, 228
Pattern class, 258
  boundaries, 259
Pattern classifier, 259
Pattern matching, 227–229
  whole word, 229, 234
Pattern primitive, 261–263
Pattern recognition, 256
  decision theoretic method, 258
  statistical method, 257, 258
  structural method, 257, 261, 263
  syntactic method, 261
  template matching, 257
Pattern variation, inter-class, 259
PCM (see Modulation, PCM)
PCM system, 75, 77–78, 85
  differential, 91
Peak clipping, 78
Perception, 23, 31–32
  categorical, 32
Performance, 80
Performance bound, CCITT, 78
Performance characterization (ASR), 248
Performance scores, 95
Perilymph, 25
Periodic segment, 60
Periodic waveform, 48
Periodicity pitch, 30
Pharyngeal cavity, 2
Phon, 28
Phonation, 4
Phoneme, 5, 20, 206, 263
  definition of symbols, 6
  key word examples, 6
  in speech model, 236
Phoneme substitution, 189
Phonetic decoder, 230

Phonetic knowledge, 232
Phonetic patterning, 247
Phonetic representation, 6
Phonetic transcription, 192–193, 199, 247
  morph based, 198
Phonological recoding, 200
Phonology, 184
Phrase boundary, 72
Pinna, 23
Pitch, 9, 20, 29, 37, 54, 57, 60
Pitch movement, 207–208
Pitch pattern, 209
Place theory, 29
Plosive sound, 7, 15, 215
  phoneme symbols, 6
Positive definiteness, 62
Power spectral density, 57
Pragmatic knowledge 232
Prediction, 91
Prediction filter, 47, 159
Predictor:
  adaptive, 94
  backward, 167, 169
  forward, 167, 169
  long-term, 177
  short-term, 177
Predictor coefficients, 162
Pre-emphasis filter, 54
Prefix morph, 198
Preparation of text for synthesized speech, 192
Pre-processor, 54, 228, 230, 267
Prioritization, 309
Processor bank, 326
Production, 68
Pronounceable text, 189
Prosodic quality, 194, 203, 219
Prosody, 21, 185, 206–208, 369
Protocol:
  dedicated access, 292
  distributed queuing, 318
  logical link control, 297
  medium access control, 297, 308, 313, 318
  multiple access, 292
  network voice, 308
  packet voice, 337
  polling, 293
  random access, 293
  reservation access, 293
  token passing, 293
Protocol efficiency, 298
Pulse amplitude modulation (*see* Modulation, PAM)
Pulse code modulation, 74

Pulse generator, 10
Pulse train, 11, 62

Quality:
  speech, 79, 93
  segmental, 185, 203, 219
  subjective, 74, 80
  suprasegmental, 185
Quantization, 82
  amplitude, 75, 82, 100
  logarithmic, 84, 86–89
  LPC, 162
  scalar (of LPC parameters), 171
  time, 100
  uniform, 84, 89
  vector (of LPC parameters), 171
Quantization noise (sub-band coder), 179
Quantizer, 75, 82, 118, 122, 142
  adaptive, 94
  inverse adaptive, 94–95
  logarithmic, 89
  non-saturating, 121
  uniform, 84, 89
Quantizing distortion, 75, 82–84
Quantizing error, 82, 84
  mean square, 83
Quantizing interval, 83, 88–89, 93
Quantizing precision, 93
Quefrency, 65
Question:
  tag, 209
  Wh, 209, 211
  yes/no, 209, 211

Random noise, 11, 62
Ratio:
  bit error ratio, 78
  peak-to-mean amplitude, 84
  signal-to-distortion, 78
  signal-to-noise, 98, 123, 128
  signal-to-quantization noise, 179
  signal-to-quantizing distortion, 84, 87, 89
Receiver synchronization, 309
Recognizer:
  application, 323, 356
  technical capability, 370
Recognition, 72
  connected word, 234, 265, 279, 388
  connection with synthesis, 234
  continuous speech, 249
  isolated word, 229, 246, 249, 265, 268, 388
Recognition accuracy, 249

Rectangular window (*see* Window function, rectangular)
Recursive rules, 67
Recursively enumerable set, 68
Redundancy in speech signal, 158
Regular language, 67
Reissner's membrane, 25–26
Reliability, 327, 346
RELP (*see* Coding, residual excited linear predictive)
Repeater, 296
Re-transmission, faulty packet, 288
Representation, in model, 235
Residual signal, 61–62, 160, 168, 174
Residue pitch, 30
Resonance, 8
Resonant frequency, 60
Resonator, 215–216
Retransmission, faulty packet, 338–339
Retransmission tolerance, 331
Rhythm, 72
  (*see also* Prosody)
R.M.S. value, 43
Root morph, 198
Round window, 24–25
Rule base, 231
  phonetic, 230
Rules:
  lexicon, 68–69
  recursive, 67
  syntax, 68–69, 262, 263

Sample and hold circuit, 75, 81
Sample by sample, 39
Sample stretcher, 75
Sample-and-hold, 130
Sampler, 75, 80, 91
Sampling frequency, 80
Sampling rate, 36, 82, 94, 97, 99
Sampling theorem, Nyquist, 40, 75, 80, 82
Satellite link, 329, 344
Scala media, 25–26
Scala tympani, 25–26
Scala vestibuli, 25–26
Schwa, 7, 36
Search:
  binary, 195
  hash function, 195
Segment:
  duration of, 21, 207, 211, 213, 245
  phonetic, 233
  speech, 45–46, 49
Segment level, 35, 46, 66

Segmental quality, 185, 203, 219
Segmental rate, 62, 159
Segmentation, optimal boundaries, 244
Segmentor, 228, 230, 233
Semantic content, 231
Semantic knowledge, 232
Semi-vowel sound, 15
  phoneme symbols, 6
Sentence, 186
  analysis of, 71
  generation of, 69
  meaningful, 69
  well-formed, 69
Sentence type, 72, 209, 211
Sequencing, 339
Short-time characteristics, 39
Short-time energy, 44
Side lobe, 54
Sideband, 82
Signal level, 35
Silence detection, 289, 329, 344
Silence within speech, 7, 45, 158
Simulation, of noise shaper by computer, 122
Slope overload, 93
SNR (*see* Ratio, signal-to-noise, 98
Soft palate, 2
Soprano voice, 4
Sound pressure level, 27
Source-filter model, 9–10, 64, 161
Speaker indentification, 265–266
Speaker variation, 22
Speaking clock, 204
Speaking style, affecting ASR, 250
Spectral analysis, 48
Spectral components, speech, 77
Spectral envelope, 66, 158
Spectral flattening, 56, 160
Spectral folding, 175
Spectral whitening, 160
Spectrogram, speech, 16, 46, 56, 224–225, 245, 276
Spectrum:
  of aliasing distortioin, 81
  of band-limited speech, 81
  distance measure based on, 64
  of friction noise, 11
  of glottal pulse train, 11
  of sampled speech, 81
  of sampling pulse train, 81
  of vocal tract filter, 11
  of voiced speech, 54
Speech detector, 305
Speech:

ambiguous nature, 226, 264
bandlimited, 80
complex nature, 226
contaminated, 226
continuous nature, 224
levels of description, 235
obtrusive nature, 371
packetized, 287
telephone quality, 77
variable nature, 224, 245, 270
Speech analysis, 35
Speech database, 43
Speech detector, 45
Speech event, 47
Speech file, 46, 43
Speech I/O, 379–381
application, 372–374
dialogue design, 383–385
simulation, 386
Speech pattern modelling, 223
Speech recognition, 72
Speech sample sequence, 60
Speech segment, 45–46, 49
Speech signal, 61, 84
redundancy in, 158
noise component, 158
Speech sound, 19
spectral range, 4
Speech understanding system, 231
SPHINX, 248
SPL (*see* Sound pressure level)
Spoken language, 191
Standard:
ISDN, 182, 323
international 75
(*see also* CCITT *and* IEEE)
Stapes, 24–25
Starting symbol, 68
State output probability (HMM), 240
State sequence (HMM), 239–241, 244
State transition probability (HMM), 240
Statistical model, structured, 235
Stop, 7, 15, 37
Stored template, 268
Stream processing, 39
Stress, 21, 194, 199, 207
primary, 192
secondary, 192
String comparison, 194–195
Style, difference between speaking and writing, 191
Sub-band coder, 178
Subjective quality (*see* Quality)
Subsidiary lobe, 41, 47

Suffix:
derivational, 199
inflectional, 199
Suffix morph, 198
Suppression:
cause of making, 31
two-tone, 27
Suprasegmental quality, 185
Swamping, cause of masking, 31
Syllable, 203, 205
emphasis, 208
in speech model, 236
nuclear, 209
Symmetry, 62
Syntactic knowledge, 232
Syntactical structure, 72
Syntax rules, 68–69, 262–263
Synthesis:
algorithmic approach, 188
concatenation of diphones, 206
concatenation of phonemes, 206
concatenation of syllables, 205
concatenation of words, 204
connection with recognition, 234
table-look-up approach, 188
text-to-speech, 72, 184, 186, 219
Synthesis filter (*see* Filter, LPC synthesis)
Synthesis of LPC speech, 169
Synthesized speech, 72
application, 323, 356
user intolerance, 372
technical capability, 369
System integration, of ASR, 252

Talking machine (von Kemepelen's), 10
TANGORA, 248
TASI advantage, 329, 344
Teeth, 2
Telephone, 74, 342
ISLAND, 336
Telephone handset, 348
Telephone numbers, pronouncing, 190
Telephone terminal, 74
Telephony, 322, 324
Teleservices, 302
Telex, 322
Template matching, 257, 268, 271
Template pattern, 257–258, 269
Temporal theory, 29
Tenor voice, 4
Terminal symbols, 68
Testing, subjective, 79
Text-to-speech (*see* Synthesis, text-to-speech)

Textual input for synthesis, 189
Threshold of hearing, 28
Threshold zone, 29
Throughput, information, 299
Time alignment, linear, 270–271
 non-linear, 270, 272
Time alignment path, 241
Time domain, 39, 49
Tiogavoice editor, 350–353
Toeplitz matrix, 165
Tone group, 208
Tone unit, 208
Tongue, 2
Toy grammar, 68
Trachea, 2
Traffic, data, 324–326, 328, 330–332, 335
Traffic video, 325–326, 331, 336
Traffic, voice, 324–326, 328, 331–332, 335, 338
Training data, 236, 261
Transformation, signal-to-symbol, 235
Transitional sounds, 37–38
Translator (ISLAND), 341–343, 345–346, 353–354
Transmission, digital, 74–75
 local transmission path, 77
Transmission rate, 80
Triangle inequality, 62
Triphone, 247
Trunk link, 287, 329, 346
Tympanic membrane, 23

Unrestricted text, 189, 210
Unvoiced speech, 3, 7, 37, 45–46, 56–57, 59, 158
Up-sampling, 178–179
User access, 287, 289, 325
User channel, 302
User interface (PABX), 348
User opinion, 78–79
User population, adaptation by ASR, 251
Utterance level, 35, 67, 72

Validation, of model, 236
Variance, 43, 94
Vector quantization (VQ), 171, 239
Velum, 2, 5, 15
Vestibular apparatus, 24
Vestibular nerve, 24
Video compression, 332
'Visible speech', 12, 47
Vital capacity, 3
Viterbi classifier, 233, 241, 243, 244

Vocabulary:
 confusable, 258
 influence on ASR, 250
Vocabulary independence, 247
Vocal cords, 1–3, 208
Vocal tract, 1, 18–19, 37, 60
Vocal tract closure, 15
Vocal tract constriction, 208
Vocal tract filter, 10–11, 64
Vocal tract model:
 cascade resonator, 215
 control parameters, 218
 electronic, 213
 Klatt model, 217–218
 parallel resonator, 216
Vocal tract resonance, 37, 54, 60, 158
Vocal tract shape, 12, 18
Vocal tract transfer function, 66
Voice messaging, 288
V-traffic (Magnet), 314
Voice, characteristics, 77, 328
Voice compression server, 343–346
Voice editor, 351–352, 354
Voice in document, 350
Voice processing, 322
Voice record/playback, 341
Voiceband data signal, 95
Voiced fricative, 215
Voiced speech, 4, 6, 37, 45, 56–58, 158, 208
Voiced/unvoiced switch, 10
Voicing, 3
Vowel cluster, 203
Vowel sound, 12–13, 30
 phoneme symbols, 6
Vowel triangle, 13

WAN (see Network, wide area)
Waveform, 37
 periodic, 48
 repetition of, 158
 speech, 36, 48
Waveform asymmetry, 43
Waveform coding, 74–75, 80, 203
Waveform quantization, 82
Waveform sampling, 80
Waveform tracking, 93
Weighting, 47, 61
Well-formed sentence, 69
Wideband audio, 77, 82, 96, 181
Wiener–Kintchine theorem, 60
Window function, 39, 41, 43–44, 67, 111, 165
 Hamming, 44, 46–47, 50, 54
 Hann, 47, 112

Kaiser, 111
  overlapping, 48
  raised cosine, 47
  rectangular, 39, 41, 54, 111
Woodward's notation, 80
Word boundary detection, 278–279
Word processor, speech-driven, 389
Word substitution, 189
Word-to-phoneme:
  conversion by dictionary, 194
  conversion by rule, 193
Workstation, 322, 324–325, 327, 351–352, 356
Written language, 191

Zero, transmission, 170
Zero-crossing rate, 46
Zero-padded sequence, 50, 54
Zero-padding, 50, 54, 67